Industrial Training Handbook

THE TRAINING FRAMEWORK

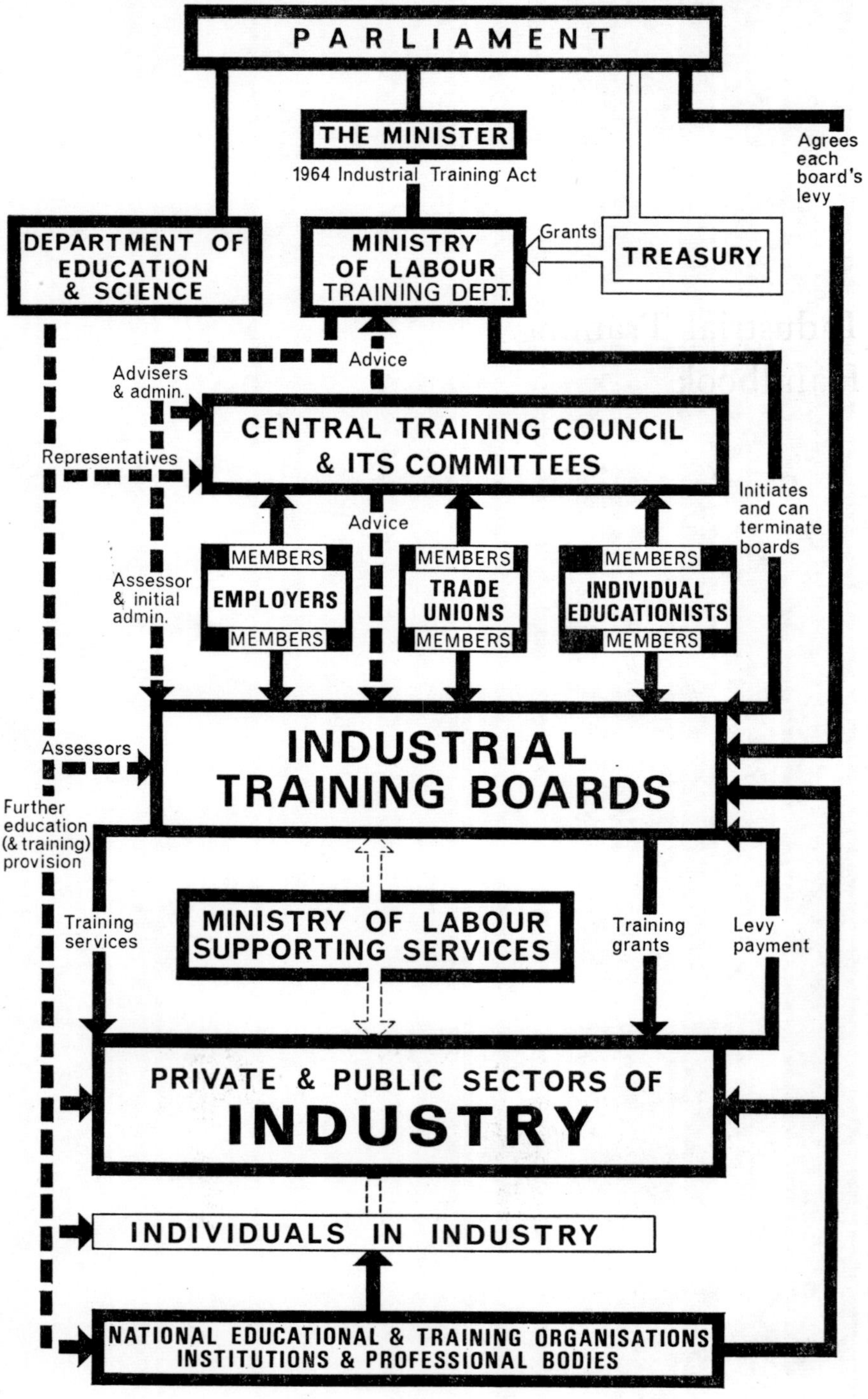

Industrial Training Handbook

Edited by John W. Barber, M.A. (Oxon.)
Senior Lecturer in Management Studies,
Portsmouth College of Technology

London ILIFFE Books Ltd.

ILIFFE BOOKS LTD
42 RUSSELL SQUARE
LONDON W.C.1.

First published in 1968

592 03510 7

Printed in England by
The Chapel River Press
Andover, Hants

Advertisements through T. G. Scott & Son Limited,
1, Clement's Inn, Strand, London, W.C.2.

Contents

Preface

This book should be regarded as an introduction to the enormously wide and fascinating subject of industrial training. It has been written mainly by training practitioners each of whom could no doubt have written a book rather than a chapter on their chosen aspect.

The original idea of the book derives from sets of reference notes on industrial training. These notes were written as a course project by members of an eight weeks' full-time course for industrial training officers. Since the beginning of 1966, the Department of Management Studies at Portsmouth College of Technology has run at least one of these courses each term. The reference notes were made on the sessions taken by guest speakers and college lecturers.

For the purpose of the book, however, the editor invited these lecturers to write chapters on their lecture topics and, in addition, commissioned several more chapters from other experts on this subject. The book is, therefore, closely linked to the Portsmouth programme. The editor is greatly indebted to the College for permission to publish on this basis and to Mr. H. A. Nobes, Head of the Department of Management Studies, for his encouragement and support; and to Colonel A. L. Work, a member of the Department, for his help and advice.

Thanks are also owed to the many members of the Training Department of the Ministry of Labour, who have so readily, and with such patience, given help and advice. This applies in particular to Mr. J. G. Stewart (now retired), formerly Permanent Under-Secretary in charge of Training, and to Dr. A. O. Martin, the Department's Principal Psychologist.

The editor also wishes to thank all others who have helped to bring the book into being including, of course, all of its authors for their patience and co-operation.

For selection and arrangement of the material and for any errors or omissions, the editor holds himself entirely responsible.

Southampton. J.W.B.

Foreword

JAMES G. STEWART, C.B., C.B.E., M.A.(Edin.), Hon.F.I.T.O.

Formerly Permanent Under-Secretary in charge of Training at the Ministry of Labour.

This book grew out of a series of lectures given at the Portsmouth College of Technology as part of one of the introductory courses for training officers recognised for grant by the Ministry of Labour and by industrial training boards.

It is perhaps hardly necessary to emphasise the importance of training officers at the present time. Great Britain is engaged in the establishment of a new structure of industrial training which could make a vital contribution to the improvement of its industrial efficiency. But the structure will not accomplish the objective without an adequate body of training officers in industry, that is of members of the management team concerned with planning and organising the training of people at all levels. (For the smaller firms, it is no doubt unnecessary and impossible to have their own full-time training officers, but like larger firms these undertakings will need the advice of a full-time expert on training obtained either through joining in a group training scheme, or calling in some kind of consultancy service.) A great deal of the actual work of training must fall on line management, but someone is necessary in each undertaking who understands the functions and techniques of training and can show the other members of the management team why systematic training is necessary and how it should be done.

Unfortunately in the past, comparatively few British firms have employed full-time staff to carry out this training function. Therefore, there is a great shortage of training officers and the efforts of the industrial training boards to improve training in their industries is impeded by lack of the necessary experts. The measures required for supplying this need have been considered by the Committee on the Training of Training Officers of the Central Training Council, but to meet the immediate needs, the Committee made recommendations in its Report on the Training of Training Officers entitled *Introductory Courses*, published by H.M.S.O. (1966). The main source of potential supply for this short-term need is to be found among people who already have a good background knowledge of industry, together with the intelligence, education and interest in the subject necessary to carry out the duties of a training officer. People who meet these qualifications can, if given a relatively intensive training on the principles and practice of industrial training, at least begin to shape up to the responsibilities of training officer work. The Education Departments and the Ministry of Labour have therefore, with the support of the Central Training Council, and in accordance with the recommendations of the Committee on the Training of Training Officers, been encouraging technical colleges to provide special introductory courses for this purpose. One of these is the course at Portsmouth on which this book has been based.

Courses of this kind, which normally consist of periods of supervised project work as well as lectures and seminars, are devised to show the scope and responsibilities of a training officer's job, the skills required for the discharge of these responsibilities and some of the means by which a training officer can obtain guidance in his job. The courses are essentially of an introductory nature. They are intended to be followed up by further courses in particular skills and preferably by a period of work as an assistant in a training department rather than by the assumption of full responsibility.

In its second report, *A Pattern for the Future*, H.M.S.O. (1967), the Training of

Training Officers Committee lays down guide lines for longer term developments.

For the purpose of this book, the subjects dealt with in the original course have been developed and expanded under John Barber's editorship. Even in this expanded form they should no doubt be regarded still as an introduction to, rather than a manual on, industrial training. But in this form they should be of value not only to the potential training officer, but also to higher management generally and, in particular, to those concerned with any kind of personnel function.

Under the guidance of the industrial training boards, existing or still to be appointed, a new and more systematic approach to training as a means of improving the efficiency of industry is being developed in Great Britain—and already there are signs that this development is being watched with interest and perhaps envy by other countries. But if the hoped for results are to be achieved, it requires a widespread appreciation in industry and in the world of technical education of the objectives and techniques of industrial training. This book is devised to meet this need.

James G. Stewart

PUBLISHER'S NOTE

After the book had been printed, the name of the Ministry of Labour was changed to the Department of Employment and Productivity. Readers are asked to note this change throughout the book.

PART 1

THE TRAINING FRAMEWORK

I

The Need for Effective Use of Manpower

A. W. BROWN, M.S.

Principal, Manpower Research Unit, Ministry of Labour

THE MANPOWER GAP

The outstanding characteristic of the labour market situation since the war has been the low margin of reserve manpower on which the economy has been run. In part, this has been a result of the Government's acceptance of full employment as a corner stone of its economic policy, and in part, a result of the general economic expansion which has made this possible. Operating on small margins, however, has led to recurrent acute shortages of particular types of labour—especially skilled labour—which have been, and remain, a serious obstacle to economic growth.

Practically every industry in recent years has experienced shortages of skilled and highly qualified manpower, but none more so than Engineering, Construction and Electronics, on which much of the rest of the economy depends. In July 1966, for example, in the engineering group of crafts, there were five times as many vacancies as unemployed; and in particular occupations within the group the proportion was nearly ten to one. In some areas, especially the Midlands and the South-East, the shortages have been worse than in others, but even areas with relatively high levels of unemployment have not always had the skilled people they needed.

Recent attempts to forecast future developments give little cause to hope that the situation will resolve itself. Although demand will level off slightly over the next few years, supply will fall off still more because of the slowing down in the rate of growth of the working population. In the last decade, the working population of the U.K. has been increasing almost as fast as the population as a whole. In the period ahead, however, although the total population will continue to increase at a similar rate, the working population will increase much more slowly. This will be due mainly to demographic factors—the trend towards earlier marriages, the fact that fewer women are remaining single and the ending of the 'bulge' of young people who started work in the early 1960s. Other contributory factors are the increased numbers of young people in full-time education—to which the raising of the school-leaving age to 16 in 1973 will contribute appreciably—and the trend towards earlier retirement. Moreover, the general trend towards greater leisure, especially in the form of longer holidays and a shorter working week, is likely to contribute to a reduction in the average number of hours worked and therefore to the pressure on manpower resources. Meanwhile as new technologies emerge and as techniques and methods of work change in existing technologies, the need for systematic training continues to grow apace.

Clearly, if the predicted manpower shortage is to be made good, the available labour force must be used more efficiently: by introducing labour-saving equipment, by deploying labour more economically and by improving the skills of individual workers. To achieve these objectives a range of complementary measures is essential. Industry must be ready to invest in technologically advanced equipment; and the trade unions to permit the new machines to be fully utilised. Movement of labour to expanding and successful industries and firms has to be facilitated; and every effort made to help employees to advance to more skilled and responsible work. Overmanning and underemployment should be discouraged, so as to release valuable skills for more productive and demanding work, and training must be extended and improved.

The problem of maximising the effectiveness of the labour force is, in a sense, ultimately one of training. The introduction of new plant may require the retraining of the existing workers. Redeployment may depend partly on the ability of those redeployed to master another job; and this ability will almost certainly be greater if early education and training have been broad-based and thorough. The manpower 'gap', therefore, reflects, to an important extent, a training—and retraining—'gap'. It is a gap which will have increasingly serious consequences for the country's economic growth if it is not closed. That is one reason why the Government has been so concerned in recent years about the development of training in industry.

APPRENTICE TRAINING

Discussion of industrial training in the period following the Second World War inevitably focuses mainly on the apprenticeship system, partly because this is the most significant form of training for skill, partly because apprenticeship has failed to provide the number of craftsmen and technicians the country needs and partly because of its deficiencies as a method of training. As concern about the supply of skilled people mounted during the latter half of the 1950s, so also did criticism of the system under which most craftsmen and technicians were trained. For some, apprenticeship was a medieval institution dressed up in twentieth century clothing, concerned primarily with the maintenance of the status and security of 'guildsmen'—a system regulating entry to trades. For others, it was a means of providing cheap labour. Most critics believed apprenticeship to be an inefficient and wasteful system of training, fettered with a variety of restrictions which made it far too inflexible to meet the needs of a modern economy.

Nevertheless, the most commonly heard plea in the years before the Industrial Training Act was passed was not for the abolition of apprenticeship, but for more apprenticeship opportunities—for a widening of the scope of apprenticeship, for more employers willing to take on apprentices, for some relaxation in the regulations governing the age of entry to apprenticeship and for a reduction in the length of the period of apprenticeship. These pleas reflected the concern quite generally felt about the recurrent shortages of skilled workers, and about the lack of opportunities for progressive and satisfying employment for young people.

Two main attempts were made to increase these opportunities:

1. The Ministry of Labour did a great deal to encourage the development of national apprenticeship agreements in a variety of industries, particularly in the period immediately after the Second World War. Moreover, the Ministry used its powers to defer call-up for National Service to ensure that apprentices trained under these agreements were given a reasonable training with opportunities for further education (normally, for example, deferment was granted only to apprentices who were released one day a week to attend an appropriate course of further education). This pressure undoubtedly contributed to expanding the field of apprenticeship (e.g., to commerce and catering) and in general increased the numbers of school-leavers entering apprenticeship.

2. *The Carr Committee and the Industrial Training Council*—In 1956 a sub-committee of the Ministry of Labour's National Joint Advisory Council was established under the chairmanship of Mr Robert Carr (then Parliamentary Secretary to the Ministry) to look into the question of industrial training and, in particular, to examine the adequacy of existing institutions to cope with the challenge presented by the so-called 'bulge' of school-leavers expected to come on to the labour market in the early 1960s. The Carr Committee Report was published in 1958. It clearly recognised the need for more skilled workers, and the need for training to take account of changing occupational requirements. The Committee accepted, in whole or in part, some of the criticisms that had been made of the apprenticeship system—the inadequacy of the numbers being trained, for example, and the rigidity of the restrictions on the age of apprenticeship. It considered, however, that 'one of the best traditional features of apprenticeship is its capacity for change'. It recommended, therefore, that industrial training should be developed on existing foundations; and that responsibility for training should remain firmly with industry, while the Government should concentrate on improving and expanding facilities for technical education.

The Committee's most important recommendation was that a national apprenticeship council should be formed to follow up its recommendations and to collect and disseminate information on training matters. This resulted in the establishment, jointly by the British Employers' Confederation, the Trades Union Congress and the nationalised industries (with financial support from the Ministry of Labour), of the Industrial Training Council. The Council's function was largely educational. It had little executive or policy making power. But it did establish a small team of training consultants, the Industrial Training Council Service (I.T.C.S.) (now known as the Industrial Training Service—I.T.S.), which came to play a valuable role not only as adviser to firms on training problems but also as expert consultants to the Council itself. Moreover, the work of the I.T.C.S. was supplemented by the appointment by some industries, with the support of the Council, of training development officers who were responsible for examining and advising on the need for training in these industries.

Although the activities of the Council were on a limited scale, there can be no doubt that it assisted materially in informing industry about the urgency of tackling training seriously. What is more, it focused attention not only on the need for more apprenticeships but also on other forms of training—for example, operative training and the training of girls. Finally, the Council gave a stimulus to the establishment of a number of group training ventures through which smaller firms were able to make a better contribution to industry's training effort.

It may be fairly said of the Carr Committee and of the Industrial Training Council that they were prepared to acknowledge many of the shortcomings in existing training arrangements, and that they helped to prepare the ground for more radical changes. The fact that these changes —when they came—proved broadly acceptable to industry was due largely to the work of these two bodies. Nor should it be overlooked that, judged by the increase in the numbers of school-leavers obtaining apprenticeships, the Carr Committee and the Industrial Training Council were largely successful in their major objectives. The number of boys undertaking apprenticeships rose to record level, and did more than keep pace with the number of boys leaving school; the 'bulge' was not left high and dry.

GOVERNMENT INTERVENTION

The criticisms of allegedly out-of-date training arrangements were not, however, abated by the setting up of the Industrial Training Council. If anything, the reverse was true. Studies of apprenticeship in the mid 1950s by Gertrude Williams and Kate Liepmann had commented upon the inefficiency of a training system that had no means of quality control, and that consisted largely of 'sitting by Nellie'. These criticisms were echoed and emphasised by more and more people as the years went by. For many apprentices, the only formal instruction received was through attendance at college classes on 'day release'; and no standards of competence were insisted upon before an apprentice was regarded as qualified to practise as a skilled craftsman. Supervision of training was often lax where it was not just incompetent; and the range of training and experience was frequently much too narrow. Whilst all commentators acknowledged the high standards set by the more progressive and far-sighted employers, it was generally pointed out that many were content to 'poach' skilled labour rather than make the effort to train.

Moreover, it was noted that, whilst rather less then 40% of boy school-leavers entered apprenticeship, a similar proportion received no formal training at all. The position among girls was even more unsatisfactory. Training opportunities—and further education—outside the apprenticeship field were very limited. Both management training and supervisory training were largely neglected, except by a small number of large firms; and some fruitful experiments in operative and adult training were followed up by a mere handful of employers. Training for commercial and office work was largely confined to technical colleges and was pursued in the employee's spare time. Though the Association of British Chambers of Commerce had established a Commercial Apprenticeship Scheme for potential recruits to the commercial professions, the number of apprentices registered under this scheme increased only very slowly.

If the picture painted by some critics was unduly gloomy, the force of many of their arguments became more evident in the early 1960s when the Government established the National Economic Development Council and became committed to an economic growth target of 4% p.a. The aim of securing a higher rate of economic growth, without risk of serious inflation, clearly depended in part on having adequate reserves of skilled manpower. Moreover, the approach to the Common Market, the increasing competition in world markets and the redeployment of coal-miners and railwaymen pointed to the need for a more active labour market policy on the part of the Government, in which, of course, training would have a vital part to play.

During the period from 1960 to 1962 a number of possible initiatives in the training field were examined by the Government. One innovation was the introduction in 1960, on a relatively small scale, of first year apprentice training courses at Government Training Centres, which had previously concentrated largely on training the disabled, the ex-servicemen and the unemployed. These apprenticeship courses, which were mainly in the engineering and electrical trades, were intended to demonstrate to industry the value of full-time training in a special training centre or school. It was hoped, in addition, that they would be of assistance to the smaller firms with relatively slender training resources. In addition to these courses, the Ministry of Labour provided training in its Instructor Training College at Letchworth and, later, at Hillington, Glasgow, for instructors in industry who were responsible for training apprentices and others.

Of perhaps greater long-term significance were the steps being taken by the Education Departments in the later 1950s and early 1960s to expand and reorganise the further education system. A big investment programme in college facilities was heralded by the publication of a Government White Paper in 1956; and in 1961 a further White Paper—*Better Opportunities in Technical Education*—outlined proposals to revise the system of professional, technical and business education. Whilst these measures could not in themselves secure a corresponding increase in training opportunities, they did provide an important stimulus to the development of day release as part of a soundly based training programme.

But the central problems remained: how to encourage more employers to establish systematic training, and how to improve the quality of training in industry. The idea of a training levy raised by the Government on a uniform basis across industry, on the model of that in France, was much discussed. At length, however, the Government agreed on a variation of the training levy concept: a levy which would not be the same for all employers but would be determined industry by industry according to the decisions of the industry training authorities.

A White Paper setting out the case for action, and embodying the Government's proposals, was published late in 1962. Although responsibility for industrial training was still to rest with industry, the White Paper gave stronger recognition to the Government's interest in this field. The change in the climate of opinion between the publication of the Carr Committee Report in 1958 and of the White Paper in 1962 is best indicated by the fact that the radical proposals in the latter were welcomed by the main employers' organisations and trade unions no less than by the educational world. Indeed, some critics claimed that the proposals were not sufficiently far reaching. After extensive discussion with interested bodies during the early part of 1963, the Government was sufficiently confirmed in its opinion of the workability and acceptability of its schemes to present a Bill in Parliament embodying the White Paper proposals. The Bill became law in March 1964 as the Industrial Training Act.

BIBLIOGRAPHY

WILLIAMS, G., *Recruitment to Skilled Trades*, Routledge and Kegan Paul (1957)

LIEPMANN, K., *Apprenticeship: an enquiry into its adequacy under modern conditions*, Routledge and Kegan Paul (1960)

MINISTRY OF LABOUR, *Industrial Training: Government Proposals* (*Cmnd. 1892*), H.M.S.O. (1962)

WELLENS, J., *The Training Revolution*, Evans Bros. (1963)

BRITISH ASSOCIATION FOR COMMERCIAL AND INDUSTRIAL EDUCATION, *Cmnd. 1892: the next step*, B.A.C.I.E. (1963)

MINISTRY OF LABOUR, *The Pattern of the Future: Manpower Studies No. 1*, H.M.S.O. (1964)

FEDERATION OF BRITISH INDUSTRIES, *Education in Transition. The implication for industry of the Government Reports on Education and Training: 1959-64*, F.B.I. (1965)

2

The Industrial Training Act

A. W. BROWN, M.S.

Principal, Manpower Research Unit, Ministry of Labour

It was noted in Chapter 1 that the White Paper of December 1962 outlined the Government's proposals for securing an improvement in the amount and quality of training, and a more equitable distribution of the cost of training. These proposals were generally well received in industry, the educational world and in Parliament; and they were translated into law by the Industrial Training Act in March 1964.

The Act is what is technically known as an enabling measure. It gives the Minister of Labour powers to establish industrial training boards* for such activities of industry and commerce as he thinks appropriate; and it sets out the duties and functions which these boards are to discharge. It preserves a discreet silence, however, on such matters as the content and length of training, and the standards to be achieved. These are to be determined by each training board in the light of the particular needs and circumstances of its industry, subject to the approval and general oversight of the Minister. In a word, the Act establishes machinery and gives powers; it does not determine training policy.

It may be as well to summarise at the outset the basic assumptions underlying the Government's policy in the field of training, to which the Industrial Training Act gives effect.

Firstly, the Act does not require an employer to provide training; or, if he does train, to meet any minimum standards. What it does is to establish a system of financial inducements by which the employer is presented, in effect, with a choice between training and contributing towards the costs of other employers who do train. If he chooses to train it provides him with an incentive to do it well.

Secondly, the major responsibility for expanding and improving training continues to rest with industry and not with the Government. The training boards are established by the Minister, but the majority of their members are appointed from employers and trade unionists in the industry concerned; and their operations are financed largely out of the levy paid by employers. Notwithstanding this, the effect of the Act is to

*Details of the industrial training boards are given in Table 2.1 on page 23.

create a much closer working partnership between industry, the educational interests and the Government, each having a special contribution to make.

Thirdly, the training boards operate on an industrial, rather than an occupational or regional, basis. That is to say, the boards have responsibility for the training of all categories and types of employee in establishments coming within a defined industrial sector (for example, the production of man-made fibres, or the construction industry, or agriculture). They are not concerned exclusively with particular occupations, though some boards will obviously cover the majority of people in certain trades. The board for road transport, for example, will have a special interest in lorry drivers, though its responsibilities will extend to the managerial, supervisory, clerical and maintenance staff, as well as the drivers, employed in transport undertakings. The organisation of training boards implies, therefore, a focusing of attention on the total training needs of an undertaking, an establishment or a firm, as well as on the occupational divisions within the business.

Fourthly, the provisions of the Act extend to all industrial and commercial undertakings, and to all levels of employment. This is not an apprenticeship act; nor is it restricted to young people. Managers, clerks and sales assistants are the concern of boards just as much as craftsmen and technicians. Moreover, industries as diverse as agriculture, distribution, banking and catering come within the scope of the Act as well as manufacturing industries.

Finally, the Act may be well described as a measure to develop vocational education in its broadest sense as well as industrial training. Formally, responsibility for further and higher education rests with the education authorities; but the Act specifically enjoins training boards to commend to industry the educational courses which should be associated or integrated with industrial training. In this way, the Act underlines the principle that 'further education and training are complementary aspects of a single process'; and that 'the more effective integration of the elements of technical education, systematic instruction in basic skills and practical experience . . . on the job must be one of the main objectives for industry and the education service alike . . .'*

FUNCTIONS OF THE INDUSTRIAL TRAINING BOARDS

As stated earlier, it is the responsibility of the various industrial training boards to give effect to the provisions of the Act. These boards have two main duties. The first is to ensure that sufficient courses and training facilities are provided in their respective industries. The second is to make recommendations about the nature, length and content of training for different occupations. This latter duty involves determining who should

*Central Training Council Memorandum No. 1, *Industrial Training and Further Education*, p. 3.

be trained (criteria for selection), how they should be trained and what they should be taught (content and methods of training), the qualifications of trainers and instructors, the standards to be reached at the conclusion of training and the educational course or courses which should be associated with the training.

The boards' duties are, therefore, quantitative and qualitative. The main instruments by which these duties are carried out are the levy and the grant. All boards are required by the Act to impose a levy on employers in their industries; and it is mainly through the money raised in this way that the boards finance their operations. The most important of these operations is assessment of the quality and quantity of training provided by firms, and the payment to employers concerned of grants in respect of this training. This means, briefly, that an employer will be under an obligation to pay the board a periodic levy; but he will be eligible, if he provides suitable training, for training grants from the board. The amount of grant will depend generally on the extent and quality of the training provided. It is, therefore, possible for an employer whose training costs are considerably more than the average in his industry to get back in grant more than he has paid in levy. On the other hand, those employers whose training effort is relatively small—or non-existent—will pay levy and get little or nothing back. In this way, the training costs borne by different employers are more fairly distributed.

grant + levy

At the same time, the levy and grant system is intended to provide an incentive to employers to train to a reasonable standard. It is hoped that the very fact of having to pay a training levy will stimulate many employers to re-examine critically their existing training provision, and to expand and improve it. Moreover, in general, the boards will only pay training grants to employers where the recommended standards have been achieved. It will not be worthwhile, therefore, for an employer to establish a training scheme unless that scheme follows the training board's recommendations. It is perhaps this aspect of the levy and grant system which is likely in the long term to prove the most significant.

A board is not limited to financing the training activities of employers in its industry. It may pay training grants to other organisations which run courses for employees in its industry. For example, a board may finance the costs of training provided by voluntary organisations, consultants and colleges of further education in order to supplement the training carried on in firms; and a board may set up its own training centres, and take on its own trainees. Thus, while the main training effort will undoubtedly have to come from individual firms, in the event of any deficiency it is possible for a board to enlarge the training resources available either by supporting suitable external courses or by establishing its own.

It is a logical result of the establishment of boards for different industries that the rate of levy, and the conditions for grant, will vary from one industry to another, depending upon the circumstances of the industry and the policy which each board chooses to adopt. The cost of training the skilled people that one industry needs may be very different from the cost of training in another; and one would expect this to be reflected in the rate of levy proposed by the respective boards for the two particular industries. Similarly, the basis of assessment of levy may vary from one board to another—though in practice the majority of training boards have so far decided to assess levy as a proportion of the total wages and salaries bill of a firm.

Although an employer cannot challenge the rate of levy imposed by a board, he has, however, the right to appeal against his assessment. This means that if any employer considers that he is not liable to pay a levy to the board which has demanded it, or if he thinks the amount assessed to be wrong, he can appeal to the Industrial Tribunal (established under the Industrial Training Act) to have the assessment set aside or reduced. The Tribunal's main function is, in effect, to interpret the various industrial training and levy orders made by the Minister, for it is these orders which determine whether or not an employer is liable to pay levy to a particular board, and the rate and basis of assessment of that levy.

One of the most common grounds of appeal, and consequently one of the issues which the Industrial Tribunal will frequently have to consider, is whether or not the activities of a particular employer come within the scope of one training board or another. This is primarily a question of deciding the meaning of the words included in the relevant industrial training order; but in a number of cases a decision turns on the question of what is an 'establishment'. It is the major activity of an establishment which determines whether an employer's activities come within the scope of a particular industry and board. Neither the Act nor any industrial training order defines an 'establishment', since it was felt impossible to achieve a definition which would prove satisfactory for all situations. The Tribunal, therefore, has to interpret this word in the light of the commercial and other circumstances of the industry and firm concerned. Whilst, in general, an establishment is taken to imply the business carried on at a particular address, the Tribunal has not hesitated to consider the way in which firms are organised and it has divided or combined establishments for purposes of liability to levy where this has seemed appropriate.

MINISTER OF LABOUR'S POWERS

In addition to his responsibility to set up training boards and appoint their members, the Minister of Labour is required under the Act to approve boards' proposals for the exercise of their powers; in particular, the Minister's approval is necessary before a board may implement its proposals for raising a levy. This means that the boards must work within a framework of policy which has been agreed by the Minister. The Minister is thus able to bring pressure to bear on a board, if necessary, to ensure that it exercises its powers in the national interest; and in the last resort, he has power under the Act to dismiss the members of a board which, in the face of the Minister's directions, fails to submit satisfactory proposals.

In addition to approving a board's plans, the Minister is able to make loans or grants to boards (within a limit of £50,000,000) to assist them in carrying out their policies. The Act itself does not lay down the kinds of activity which the Government grant might be used to support, but it was made clear during the passage of the Bill through Parliament that the Government's grant would be used selectively to encourage forms of training which were regarded as of special importance. Grants have been made available to boards to assist in the training of training officers and instructors, and of those undergoing industrial training as part of a sandwich course leading to a degree or a Higher National Diploma. The Ministry also has agreed to subsidise part of the cost of providing extra off-the-job training places and to meet half the cost of research projects undertaken or sponsored by the boards.

It is evident from what has been said that, while the training boards are the bodies primarily responsible for improving training in industry, considerable powers are vested also in the Minister. The Act provides, however, that the Minister should carry out his functions with the advice of a Central Training Council. The Central Training Council was not mentioned in the 1962 White Paper which outlined the Government's proposals and its establishment was the most significant extension of those proposals agreed by the Government in discussions with industry before the Industrial Training Bill was introduced into Parliament.

The role of the Central Training Council will be described in the final section of this chapter. It is sufficient at this point to say that the Council is composed of industrial and educational members, supplemented by six 'independents' and a maximum of six chairmen of industrial training boards; and its terms of reference are to advise the Minister on the exercise of his powers under the Act, and on industrial training issues generally. It has no formal executive responsibilities, and is staffed and serviced by officials of the Ministry of Labour.

MAIN TASKS OF THE TRAINING BOARDS

So far, this chapter has been concerned mainly with the legal framework within which the boards operate. It is now necessary to have a closer look in practical terms at the main tasks which each training board has to accomplish if it is to carry out its legal obligations. Any list of these tasks is likely to be selective, but there can be little argument that the following are of central importance:

1. Convincing management of the importance of training.
2. Increasing the number of training officers and instructors and raising the standard of instruction.
3. Forecasting manpower and training requirements.

4. Determining levy and grant strategy.
5. Drawing up training recommendations.
6. Implementing training recommendations, and maintaining a reasonable standard of training.

The implications of these tasks and the ways in which the Central Training Council can assist the boards in discharging them will be considered.

TRAINING AND MANAGEMENT

It has been noted already that one of the purposes of the levy and grant system is to encourage firms to provide more and better training for those they employ. There is a real danger, however, that as a result of the Act firms may be tempted to embark on training not as a means of developing the abilities of their employees but rather as a way of obtaining a satisfactory amount of grant for the least possible effort. The first task of any board should be, therefore, to convince directors and top managers that sound and well organised training is essential to the greater efficiency and profitability of an undertaking. The best results can only be obtained if managers regard training as a valuable economic activity rather than as a social duty; and as an essential function of the job of management, and not as a welfare benefit or grant-earning device.

Furthermore, boards have to show clearly that their objective is to help individual firms to meet their own training needs in the most economical and effective way. A considerable communications exercise—involving courses, conferences and seminars for senior managers—will be necessary to persuade firms both of the importance of sound training and of the relevance of the boards' schemes. Unless this exercise is undertaken—and is successful—the boards will fail in their main purpose. In the last resort, the attitude of senior management determines whether a firm responds positively and constructively, or negatively, to the levy and grant stimulus.

TRAINING SPECIALISTS

Although senior line managers will have to head the drive within each company for an improved standard of training, and must bear the major responsibility for training their staffs, they will need to turn to training specialists for expert assistance and advice in planning and implementing training programmes. In some firms, the appointment of a full-time training adviser will be necessary; in others, a group adviser may be shared with a number of firms, or the training officer may be responsible for several functions in addition to training. However, these appointments will not be of great value unless the training officers concerned have had a thorough training to prepare them for their responsibilities.

At present, the number of fully competent training specialists is quite inadequate to bring about generally the improved standard of training which the boards' recommendations are intended to secure. In all industries, therefore, it will be a matter of urgent priority for the boards to increase the supply and develop the skills of training officers and instructors, both on their own staff and in firms. To do this three things are necessary. Firstly, boards can (and do) make generous grant payments to firms employing training officers who have undergone appropriate training. Secondly, the boards can themselves appoint—and, where necessary, train—training officers who will be available to advise and help firms. Thirdly, there must be a considerable effort to establish and expand suitable training courses.

Several full-time introductory courses for training officers have already been established in technical colleges largely on the initiative of the Government and with the backing of the Central Training Council's Training Officer* Committee. In addition, however, there will be a growing need for a range of more specialised courses both to meet the requirements of particular industries and organisations, and to supplement basic training. All the existing boards are well aware of the need to provide a substantial incentive to firms to release people to attend these courses, and the Ministry of Labour makes a special grant to boards to meet a proportion of their expenditure on the training of training staff. Until these measures bear fruit, and most firms have access to sound advice on training problems, the boards will face an uphill struggle in communicating to firms the meaning of training recommendations.

MANPOWER, FORECASTING AND PLANNING

One of the most complex tasks before the training boards will be that of producing realistic manpower budgets as the basis for estimating the training requirements of their respective industries. Forecasts of future manpower requirements have to take into account the existing supply and shortages of manpower by occupations, the expected growth or contraction of the industry, the degree and character of innovation and technological change, wastage rates and demographic and educational changes. The task of preparing these forecasts will not be made easier by the fact that, in many occupations, there is a substantial flow of labour from one industry to another, and that relatively few firms at present think very far ahead about their future manpower requirements. This means that a sophisticated analysis of future requirements may be based on information which is itself unreliable or incomplete. Thus, it will be of prime importance to take steps to improve the quality of information provided by industry and to achieve close co-operation between different boards, and between the boards and the Government. The Ministry of Labour's Manpower Research Unit was set up in part to assist the boards in tackling the problems of manpower forecasting, and to advise them on the collection and analysis of statistical information. In some industries, moreover, the 'Little Neddies' will be in a position to provide useful information regarding the amount and type of training which may be required.

* Since April 1968 known as The Training of Training Staff Committee (Ed.)

However, it will be up to each board to work out training targets for its industry; and to see that these targets are achieved. In the light of the statistical projections of manpower needs, the boards should be able to determine their training priorities, and the grant incentives which it is necessary to offer industry. The level of grants paid to firms will be an important—indeed an essential—instrument for directing attention to those occupations where shortages of manpower are likely to be most serious. It will not be enough, however, simply to increase the amount of grant paid in respect of particular types of training unless the industry is able to recruit a sufficient number of suitable people for the occupations concerned. It may well be found that it is not possible to rely upon an adequate supply of young people coming into shortage occupations. A board's manpower planning effort must, therefore, be directed not only towards stimulating the recruitment of suitable young trainees, but also towards retraining and promoting people already in the industry for more responsible and more highly skilled work. Accelerated training and adult training will probably have to play an increasingly important part in relieving shortages (particularly short-term and medium-term shortages of people in skilled occupations) as well as in adapting the labour force to undertake new jobs.

LEVY AND GRANT STRATEGY

It has been stated that the levy and grant system is intended to secure three main objectives: to increase the amount of training, to improve its quality and to distribute more equitably between employers the cost of training in an industry. The last of these objectives implies that the amount raised in levy should be broadly equivalent to the total cost of training in an industry. If the levy is equal to the total cost of training, it will be possible for a training board to repay to each employer in grant an amount roughly equal to what he spends in providing training for his employees. If the trained person then leaves to work for another employer, his original employer has the consolation that the industry as a whole has helped to meet the cost of his training; the 'poacher' has, so to speak, paid in full for his 'game'.

This is the theory behind the levy and grant schemes. There are, however, a number of reasons why it may not be possible, initially at least, to operate the system with this degree of precision and justice. In the first place, it is extremely difficult to establish what the training costs of an industry are or should be. Different accounting conventions can make a mockery of a board's good intentions. Secondly, assuming a fair guess at what it costs to train, a board will have to apportion this cost between employers with widely varying training needs. Consider the case of one employer with relatively few craftsmen and another with a high proportion. It would not be just to expect the first to train as many apprentices as the second. The levy paid should, therefore, reflect differences of this kind, and should not put some employers at a disadvantage simply because of the type of work force they need. But if the rate of levy is set at a high level in relation to the cost of training, the risk of injustice to one type of employer or another will be considerable.

The essential problem facing the boards is, therefore, to achieve simultaneously a rate of levy which redistributes a fair part of the cost of training over the industry as a whole but does not bear unreasonably on different types of employer. At the same time, boards may have to be careful, especially in the early stages, not to make substantial grant payments for training which is of no very high standard, or which their staffs are unable to inspect or evaluate. It is for these reasons, among others, that most boards have chosen in the first year or two of operation to impose a relatively low rate of levy on employers.

As each board builds up its staff and produces detailed training recommendations for various occupations, so the scope of its levy and grant schemes may be extended, the amount of levy and grant increased and the standards set become more rigorous. This gradual progress towards a comprehensive grant scheme should also enable firms to adjust more easily to the standards which the boards will require them to reach.

TRAINING RECOMMENDATIONS

The purpose of industrial training is to contribute to the efficiency of individuals in the jobs they have to do. It is a means to meeting a firm's requirements for people who are able to carry out certain functions. It must, therefore, in the last analysis be geared to meeting the specific requirements of individual firms. Nonetheless, every board has a duty to make recommendations which are generally applicable throughout its industry. It has also to see that the particular requirements of individual firms are met so far as possible without restricting unduly the range of training provided, and without inhibiting the ability of trained people to move from one job to another. A balance has to be struck. On the one hand, recommendations have to be clear-cut and precise if they are to provide useful guidance; on the other hand, they must be sufficiently flexible to meet the needs of individual firms. The boards have to avoid the twin dangers of too rigid a pattern of training, and of leaving too much to be determined by the individual employer without reference to the broader requirements of the industry and the individual trainee.

There is no doubt that this is a real dilemma for the boards. One of the ways in which it can be resolved is by defining the essential features of a training programme for a given occupation, and by outlining the basic training required. The firm is then left to draw up a detailed programme reflecting its own situation but embodying the essential features prescribed.

A board which adopted this approach might, for example, require a firm to provide a training programme which included induction training, a written-down syllabus or programme of planned training based on job analysis, competent supervision and instruction, the maintenance of training records and appropriate provision for further education. It might also specify a number of fundamental tasks or skills in which the trainee must be instructed. And it would probably feel bound to give some advice on methods of training, on selection criteria, on appropriate courses of further education and the standard of competence expected in instructors and supervisors. Within these limits, however, the firm would be free to decide for itself.

It is implicit in an approach of this kind that a board concentrates more on *how* the firm goes about its training and on the conditions under which training is given, than on the detailed content of the training. This will not necessarily rule out fuller specification of the content of training courses as the boards gain experience of industries' requirements. But in many cases this will probably be undesirable. The concept of a 'trade' as a clearly defined and more or less unchanging grouping of skills and knowledge is one that is scarcely tenable when industrial conditions and technology are changing as rapidly as they have been in recent years. If detailed training specifications are laid down they will soon get out-of-date as well as acting as a disincentive to firms to innovate.

IMPLEMENTING TRAINING RECOMMENDATIONS

Having published its training recommendations, each board will have to see that they are carried into effect by firms. Implementation of training recommendations will involve, firstly, a means of inspecting and evaluating the training provided and, secondly, the establishment of an advisory or consultancy service to provide practical assistance to managements in planning training programmes. The boards' own staffs will have to bear the major responsibility for both these tasks—inspection and advice. Therefore it will be of considerable importance for each board to give priority to the recruitment and training of a corps of training advisers to serve industry in this capacity. The experience of countries like Germany and the Netherlands points clearly to the conclusion that improvements in training, particularly in the smaller firms, depend no less on the quality of the training advisory service than on nationally recognised training specifications. Given the flexible approach to training recommendations outlined in the previous paragraphs, there is little reason to doubt that this conclusion is very likely to hold good in Great Britain.

In addition to the establishment of a training advisory service, a board may find it desirable to institute some system of objective assessment of the training carried out in firms, through tests and examinations. Boards have power to establish tests, and make it a condition of grant that trainees undergo them. By this means, they can hope to establish both the level of competence reached by trainees and the standard of training in different firms. While there can be no question that such tests could become an alternative to the inspection and advisory service provided by the board's full-time staff, there are powerful arguments in favour of boards establishing objective tests wherever this is practicable. These tests need not be established on a national basis—as happens, for example, in the Netherlands—but possibly for an area or even an individual firm so as to reflect regional and local training needs.

THE ROLE OF THE CENTRAL TRAINING COUNCIL

Although the Central Training Council* is an advisory body without formal executive responsibilities, its central and independent position enables it to exercise considerable influence on the direction of policy and to speak with special authority. In particular, the Council is well placed to give a lead on training for occupations which are common to all industries, on research into training problems and on general policy issues affecting all boards. Much of the Central Training Council's activity in the last two or three years has been concentrated in these three fields.

One of the disadvantages of setting up training boards on an industry basis is that it becomes more difficult to secure similar standards of training for occupations which are common to every industry. Every board has responsibility for training all those employed in its industry; and if each goes its own way, there is a real danger that conflicting standards will be set for these common occupations. The Central Training Council decided, therefore, at an early stage that it must assume some responsibility for advising the boards on the training needs of office staff, training specialists, supervisors and managers. The results of the Council's work in this field are already becoming evident. Among its earlier publications was a booklet on the training of training officers which was drawn up by the Council's Committee on the Training of Training Officers. The recommendations in that booklet have become the basis of the introductory courses now provided in a number of colleges. More recently, that same Committee has published a further report[1] recommending new initiatives including longer and more challenging courses aimed at attracting graduates to the training field. The Council has also sponsored a booklet on supervisory training[2] and has published reports of its Committees on Clerical and Commercial Training[3] and on Management Training and Development;[4] this latter gives guidance to the training boards and to individual firms on ways to promote and improve management education, training and developments. In each case, the advice offered has been commended by the Minister of Labour to the training boards as a starting point for their recommendations. In this way, a similar approach to the common occupations can be achieved throughout industry.

So far as research is concerned, the Council has established a Research Committee whose function is to determine in what directions more research is needed, to support research projects and to assist in the dissemination of information about research findings. The Committee considers projects which are submitted by university departments, training boards and others, and recommends whether they should receive grant support from the Ministry of Labour. The work of the Research Committee, therefore, makes it possible for the Council to expand the frontiers of knowledge whilst at the same time reducing the gap between the research worker and the practitioner.

*The relationship of the Central Training Council with the Training Department of the Ministry of Labour is shown in Fig. 2.1 on page 21, and in the frontispiece.

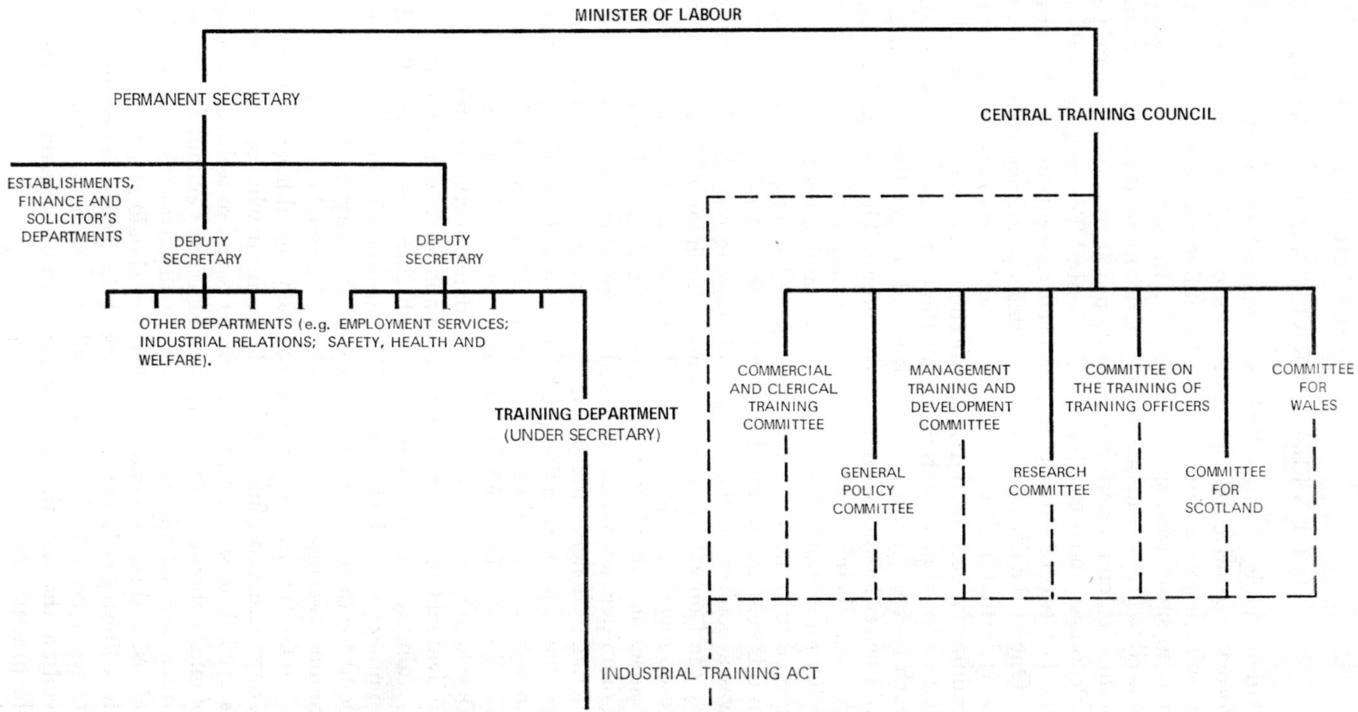
MINISTER OF LABOUR
PERMANENT SECRETARY
CENTRAL TRAINING COUNCIL
ESTABLISHMENTS, FINANCE AND SOLICITOR'S DEPARTMENTS
DEPUTY SECRETARY
DEPUTY SECRETARY
OTHER DEPARTMENTS (e.g. EMPLOYMENT SERVICES; INDUSTRIAL RELATIONS; SAFETY, HEALTH AND WELFARE).
TRAINING DEPARTMENT
(UNDER SECRETARY)
INDUSTRIAL TRAINING ACT
COMMERCIAL AND CLERICAL TRAINING COMMITTEE
GENERAL POLICY COMMITTEE
MANAGEMENT TRAINING AND DEVELOPMENT COMMITTEE
RESEARCH COMMITTEE
COMMITTEE ON THE TRAINING OF TRAINING OFFICERS
COMMITTEE FOR SCOTLAND
COMMITTEE FOR WALES

BRANCH A	BRANCH E	BRANCH B	BRANCH C	BRANCH D
(ASSISTANT SECRETARY)	(CHIEF INSPECTOR)	(ASSISTANT SECRETARY)	(ASSISTANT SECRETARY)	(CHIEF TRAINING ADVISER)
1. ADMINISTRATION OF GOVERNMENT TRAINING CENTRES AND INDUSTRIAL REHABILITATION UNITS. 2. RESPONSIBILITIES CONNECTED WITH TRAINING OF DISABLED PEOPLE. 3. ADMINISTRATION OF GRANTS FOR TRAINING LABOUR IN DEVELOPMENT DISTRICTS.	1. GENERAL AND TECHNICAL SUPERVISION OF GOVERNMENT TRAINING CENTRES AND INDUSTRIAL REHABILITATION UNITS. 2. PROVISION OF TRAINING WITHIN INDUSTRY FOR SUPERVISORS. 3. ADVICE ON COURSES AT INSTRUCTOR TRAINING COLLEGES.	1. POLICY UNDER THE INDUSTRIAL TRAINING ACT ON PRINCIPLES AND METHODS OF TRAINING AT ALL LEVELS, EXCEPT CRAFT TRAINING. 2. SECRETARIAT FOR THE CENTRAL TRAINING COUNCIL AND ITS COMMITTEES (2).	1. ESTABLISHMENT OF, AND RELATIONS WITH, INDUSTRIAL TRAINING BOARDS. 2. POLICY ON GROUP TRAINING SCHEMES, OFF-THE-JOB TRAINING, SANDWICH COURSES AND CRAFT TRAINING. 3. RELATIONS WITH INDUSTRIAL TRAINING SERVICE. 4. TRAINING ARRANGEMENTS FOR YOUNG WORKERS IN INDUSTRIES NOT COVERED BY INDUSTRIAL TRAINING BOARDS.	1. ADVICE ON TECHNICAL AND INDUSTRIAL QUESTIONS ARISING FROM THE IMPLEMENTATION OF THE ACT. 2. ADVICE ON TRAINING RESEARCH. 3. COLLECTION, CLASSIFICATION AND DISSEMINATION OF TRAINING INFORMATION. 4. ADVICE ON THE TRAINING OF TRAINING SPECIALISTS.

(1) THIS TABLE SHOWS THE ORGANISATION OF THE DEPARTMENT, WITH ITS MAIN FUNCTIONS.

(2) THE COMMITTEE FOR WALES AND SCOTLAND HAVE LOCAL SECRETARIATS, BUT MAINTAIN LIAISON WITH THE DEPARTMENTS THROUGH BRANCH B.

Fig. 2.1 The Training Department of the Ministry of Labour showing its relationship with the Central Training Council

In association with the Committee, the Training Department produces an Industrial Training Research Register. It also issues occasional Training Information Papers and operates a Training Abstracts Service.

The third field of activity of the Council may be described as that of 'general policy': the consideration of problems which crop up in every industry. Here, the Council acts, so to speak, as the Minister's 'conscience' to spur the boards to greater efforts. For example, the Council emphasised in its first published statement the importance of the integration of industrial training and further education. In Memorandum No. 5, the Council analysed the main tasks facing any training board, and set down its views about what these tasks involved. Its first report to the Minister underlined, among other things, the need for more adult training. Through statements of this kind, the Council can open controversial issues to informed discussion, and draw attention to matters which call for action on the part of the boards or the Government. This function is of first importance at a time when a radical and forward looking approach to manpower development is the only one likely to be fruitful.

THE INDUSTRIAL TRAINING BOARDS

Details of the industrial training boards are given in Tables 2.1(a) and 2.1(b) on pages 23–31.

REFERENCES

1. CENTRAL TRAINING COUNCIL, *Training of Training Officers: A Pattern for the Future*, H.M.S.O. (1967)
2. CENTRAL TRAINING COUNCIL, *Supervisory Training: A New Approach for Management*, H.M.S.O. (1966)
3. CENTRAL TRAINING COUNCIL, *Training for Commerce and the Office*, H.M.S.O. (1966)
4. CENTRAL TRAINING COUNCIL, *An Approach to the Training and Development of Managers*, H.M.S.O. (1967)

BIBLIOGRAPHY

CENTRAL TRAINING COUNCIL, *Report to the Minister (HC352)*, H.M.S.O. (1965)

CENTRAL TRAINING COUNCIL, *Training of Training Officers: Introductory Courses*, H.M.S.O. (1966)

MANPOWER TRAINING FOR INDUSTRY, *Ninth Report from the Estimates Committee (HC548)*, H.M.S.O. (1967)

Table 2.1(a) INDUSTRIAL TRAINING BOARDS (April 1968)

Boards Listed

1. Wool, Jute and Flax
2. Iron and Steel (Foundry Industry—Joint Committee with Engineering I.T.B.)
3. Construction
4. Engineering
5. Shipbuilding
6. Electricity Supply
7. Gas
8. Water Supply
9. Ceramics, Glass and Mineral Products
10. Furniture and Timber
11. Man-made Fibres Producing
12. Carpet
13. Knitting, Lace and Net
14. Cotton and Allied Textiles
15. Agricultural, Horticultural and Forestry
16. Road Transport
17. Hotel and Catering
18. Civil Air Transport
19. Petroleum
20. Rubber and Plastics Processing
21. Chemical and Allied Products

Being established during 1968:
Printing and Publishing
Paper Making and Paper Products
Distributive
Food, Drink and Tobacco (Processing)
Footwear, Leather and Allied Products

Note: At the time of publication, any additions to this list would be pure speculation, e.g., Clothing; Broadcasting and Film Making; Banking, Insurance and Finance; Shipping; Rail Transport; and many miscellaneous minor industries such as laundry and hairdressing. Meanwhile, a Local Government Board has been established on a voluntary basis (see note at end of Table 2.1 (b)).

Table 2·1 (b)

DETAILS OF BOARDS

Personnel details on all I.T.B.s are conveniently available from the current *Progress Report on I.T.B.s* published annually by B.A.C.I.E. (see Chapter 32).

Serial Number *Initiating Date* *Statutory Instrument Number (S/I)*	*Board Information*	*Order of Magnitude (1968)* *Number of Establishments*	 *Number of Employees*	 *Levy and Yield*
1 Wool 29.6.64 S/I 907 Wool/J/F 27.4.66 S/I 428	WOOL, JUTE and FLAX I.T.B. 55 Well Street, Bradford 1, Yorks. Tel.: 0274 26493 Chairman Chief Executive Secretary Chief Training Adviser	2,074	200,000	1·5 (standard) 0·5–2·0% (differential) £2,090,000
2 3.7.64 S/I 949	IRON and STEEL I.T.B. 4 Little Essex Street, London W.C.2 Tel.: 01–240 2044/8 Chairman Director Assistant Director of Training Assistant Director of Finance and Admin.	669	327,000	£18 p.a. per employee £5,700,000
3 21.7.64 S/I 1079 Amended 13.7.67 S/I (1967) 924	CONSTRUCTION I.T.B. Radnor House, London Road, Norbury, London S.W.16 Tel.: 01–764 5060 Chairman Secretary Chief Executive Chief Training and Development Adviser	56,000 (firms)	1,700,000	1·0% (and 1·0% in respect of payments to labour supplied by labour only contract) £15,000,000

4 23.7.64 S/I 1086 redefined from 9.3.67 S/I 279	ENGINEERING I.T.B. St. Martin's House, 140 Tottenham Court Road, London W.1 Tel.: 01–387 0501 Chairman Director Secretary Chief Training Officer	29,750	3,600,000	2·5% £86,000,000
— 1.3.65 —	FOUNDRY INDUSTRY TRAINING COMMITTEE (Established as a committee of the Engineering I.T.B.) 50–54 Charlotte Street, London W.1 Tel.: 01–580 0575 Chairman Director Principal Training Officer Secretary	Number operating the Committee's recommendations approx. 3,000	approx. 250,000	Covered by Engineering I.T.B.s
5 19.11.64 S/I 1782	SHIPBUILDING I.T.B. Raeburn House, Northolt Road, South Harrow, Middlesex Tel.: 01–422 9581 Chairman Chief Executive Secretary Chief Training Officer	1,170	136,000	1·55% £1,940,000
6 24.6.65 S/I 1256	ELECTRICITY SUPPLY I.T.B. 30 Millbank, London S.W.1 Tel.: 01–834 2333 Chairman Secretary Chief Officer Training Officer	Electricity Council, Electricity Boards, London Transport Generating Board (17)	250,000	0·025% £65,000

Table 2.1 (b) (continued)

Serial Number *Initiating Date* *Statutory Instrument Number (S/I)*	*Board Information*	*Order of Magnitude (1968)*		
		Number of Establishments	*Number of Employees*	*Levy and Yield*
7 24.6.65 S/I 1257	GAS I.T.B. 17 Grosvenor Crescent, London S.W.1 Tel.: 01–245 9811 Chairman Director Secretary Senior Training Advisers	Gas Council and Area Boards (13)	120,000	1·25% £1,520,000
8 24.6.65 S/I 1258	WATER SUPPLY I.T.B. 104a Park Street, London W.1 Tel.: 01–499 2754 Chairman Chief Executive Officer Chief Training Officer Secretary and Accountant	570	44,000	1·1% £425,000
9 23.7.65 S/I 1391	CERAMICS, GLASS and MINERAL PRODUCTS I.T.B. 1st Floor, Bovis House, Northolt Road, Harrow, Middlesex Tel.: 01–422 7101 Chairman Director Secretary Assistant Director and Training Controller	4,140	415,000	Sub-Boards 1 and 2 0·75% Sub-Board 3 1·5% £3,440,000

10 9.12.65 S/I 2028	FURNITURE and TIMBER I.T.B. York House, Empire Way, Wembley, Middlesex Tel.: 01–902 1046 Chairman Director Two Joint Secretaries (Training, Administration) Chief Training Officer	4,770 (firms)	215,000	0·9% £1,750,000
11 23.2.66 S/I 143	MAN-MADE FIBRES PRODUCING I.T.B. 5th Floor, Bowater House, Knightsbridge, London S.W.1 Tel.: 01–589 9008/9 Chairman Secretary	38 (in three major firms)	50,000	0·2% £100,000
12 18.3.66 S/I 245	CARPET I.T.B. Evelyn House, 32 Alderley Road, Wilmslow, Cheshire Tel.: 0996 4 27118/9/0 Chairman Director Chief Executive Officer Chief Training Officer Secretary and Administrator	264	44,000	0·5% £172,000
13 18.3.66 S/I 246	KNITTING, LACE and NET I.T.B. 4 Hamilton Road, Sherwood Rise, Nottingham Tel.: 0602 61075 {Chairman {Director Secretary Chief Training Officer	1,500	160,000	0·4% £400,000

Table 2.1 (b) (continued)

Serial Number *Initiating Date* *Statutory Instrument Number (S/I)*	*Board Information*	*Order of Magnitude (1968)*		
		Number of Establishments	*Number of Employees*	*Levy and Yield*
14 20.7.66 S/I 823	COTTON and ALLIED TEXTILES I.T.B. 10th Floor, Sunlight House, Quay Street, Manchester 3 Tel.: 061 832 9656/8 Chairman Chief Executive Secretary Chief Training Officer	1,600	200,000	0·4% £560,000
15 15.8.66 S/I 969	AGRICULTURAL, HORTICULTURAL and FORESTRY I.T.B. Bourne House, 34 Beckenham Road, Beckenham, Kent Tel.: 01–650 4890 Chairman Director Secretary Chief Executive Officer Chief Training Adviser	140,000	500,000 (estimated)	£3 p.a. per whole-time employee £940,000
16 15.9.66 S/I 1112	ROAD TRANSPORT I.T.B. Capitol House, Empire Way, Wembley, Middlesex Tel.: 01–902 8880 Chairman Director General Director of Administration Chief Executive Officer Director of Training	100,000	900,000 (estimated)	1·6% £11,500,000

17 7.11.66 S/I 1347	HOTEL and CATERING I.T.B. Ramsey House, Central Square, Wembley, Middlesex Tel.: 01–902 8865 Chairman Director Secretary Assistant Secretary Chief Training Officer	200,000 (approx.)	1,000,000 (approx.)	1·0% £3,000,000
18 8.3.67 S/I 263	CIVIL AIR TRANSPORT I.T.B. Staines House, 158-162 High Street, Staines, Middlesex Tel.: 79 57171 Chairman Director and Chief Training Officer Secretary Chief Executive Officer	340	60,000	£8 p.a. per employee £480,000
19 18.5.67 S/I 648	PETROLEUM I.T.B. York House, Empire Way, Wembley, Middlesex Tel.: 01-903 4161 Chairman Secretary Chief Executive Officer Chief Training Officer Director of Training	1,200 (estimated)	84,000 (estimated)	
20 9.8.67 S/I 1062	RUBBER and PLASTICS PROCESSING I.T.B. Brent House, 950 Great West Road, Brentford, Middlesex Tel.: 01-568 0731 Chairman Director Secretary Chief Executive Officer Chief Training Officer	4,000 (estimated)	250,000 (estimated)	

Table 2.1 (b) (continued)

Serial Number *Initiating Date* *Statutory Instrument Number (S/I)*	*Board Information*	*Order of Magnitude (1968)*		
		Number of Establishments	*Number of Employees*	*Levy and Yield*
21 9.10.67 S/I 1386	CHEMICAL and ALLIED PRODUCTS I.T.B. Staines House, 158-162 High Street, Staines, Middlesex Tel.: 79 51366 Chairman Director Secretary Chief Executive Officer Chief Training Officer	3,000 (estimated)	500,000 (estimated)	
22 29.5.68 S/I 786 (1968)	PRINTING and PUBLISHING I.T.B. 3-4 Chester Gate, Regent's Park, N.W.1 (temporary) Tel.: 01–486 3521 Chairman Director Secretary Chief Training Adviser	4,000	400,000	
23 29.5.68 S/I 787 (1968)	PAPER MAKING and PAPER PRODUCTS I.T.B. 3-4 Chester Gate, Regent's Park, N.W.1 (temporary) Tel.: 01–486 3521 Chairman Director Secretary Chief Training Officer	1,300 (estimated)	250,000 (estimated)	

DISTRIBUTIVE I.T.B.			
FOOD, DRINK and TOBACCO I.T.B.			
FOOTWEAR, LEATHER and ALLIED PRODUCTS I.T.B.			

Note.—The LOCAL GOVERNMENT TRAINING BOARD—A unique Training Board outside of the provision of the Industrial Training Act was established in September 1967. It was set up by the Local Authorities Associations and the Greater London Council in co-operation with the employers' sides of the negotiating machinery and the trade unions. The Board proposes to operate a group scheme for training to be paid out of a 10*d* per head of population levy to be apportioned between the various levels of the local authorities. Grants will be payable for all approved training of employees not already within the scope of I.T.B.s, for example, employees not within the scope of the Engineering I.T.B., the Construction I.T.B. and the Hotel and Catering I.T.B.; or employees not already provided for by national pooling arrangements, for example, not teachers, fire service personnel, police, health visitors and midwives.

3

Industry and the Technical Colleges

H. A. NOBES, F.C.I.S., M.B.I.M.

Head of the Department of Management Studies, Portsmouth College of Technology

INTRODUCTION

Memorandum No. 1 of the Central Training Council stresses that further education and training are complementary aspects of a single process. Co-operation between industry and technical colleges of various types is therefore vital. This aspect of industrial training is considered at length during the Portsmouth courses given to training officers from industry and to those given to training advisers of industrial training boards. Most of the main subjects covered during those courses find a place in this book. This chapter outlines the opportunities and services which the technical colleges can offer to industry. It is not intended to trace here the history of college schemes, within the National Certificate framework, nor to give a comprehensive account of the work of the colleges. The aim is to indicate mainly those activities relevant to training officers, and the three main headings will be:

Types of technical colleges.
The range of general courses.
Courses in management studies.

Probably the best approach to the subject is implicit in an industrial training manager's plan for the development of staff from the age of 15 to the end of useful planning. He requires first-class skill among his operatives and craftsmen. He must also plan for all round ability in the managing director designate. As the impact and impetus of the Industrial Training Act, 1964 increases, future managing directors will recognise how much they have to gain by ensuring that there is adequate and efficient training not only of craftsmen but of employees of every other grade.

TYPES OF TECHNICAL COLLEGES

Broadly, the types and functions of the various technical colleges are as follows:

1. *Local colleges.* It is here that a great range of work for industry exists

to meet local needs for courses for operatives, craftsmen and technicians. These courses may lead to qualifications such as City and Guilds Certificates and Ordinary National Certificates (O.N.C.). Many are keyed to local schemes for apprenticeship, by day-release study, block-release or, in the case of certain apprenticeships, through full-time first year apprenticeship study.

2. *Area Colleges*. These make similar provisions, and in addition offer advanced courses of two to four years duration from the age of 18 for Higher National Certificates (H.N.C.), Higher National Diplomas (H.N.D.) and Final City and Guilds Certificates.

3. *Regional colleges*. Some 25 regional colleges have been in existence since 1956. These vary greatly in size and tend to concentrate their courses at post A-level standard, for H.N.C., H.N.D., external degrees of the University of London and more recently for degrees of the C.N.A.A.[1] For the latter degrees, colleges perform very much the same function as universities and similar high regard exists for the safeguarding of standards. There is, for example, external assessment of internally set and marked examinations. C.N.A.A. degree courses have now replaced the former four-year Diploma in Technology sandwich courses. The C.N.A.A. has also broadened its scope to include many three-year full-time courses. Most of the regional colleges are now proposed in the 1966 White Paper *A Plan for Polytechnics* as 'Polytechnics' and are to continue their work by means of the new C.N.A.A. degrees and other advanced qualifications.

4. *National colleges*. There are also four national colleges for advanced work related to particular industries, namely:

National College for Heating, Ventilating, Refrigeration and Fan Engineering.
National College of Agricultural Engineering.
National College of Rubber Technology.
National Leathersellers College.

THE RANGE OF AVAILABLE COURSES

Craft and operative skills are given much emphasis in local and area colleges. The first year of apprenticeship itself may in some trades be completed within a college and be recognised by the trade unions and

[1]C.N.A.A. is the Council for National Academic Awards. It was established in 1964 by Royal Charter as an autonomous body with powers to award Honours and Ordinary Degrees comparable in standard with those awarded by the universities. C.N.A.A. courses are on the whole more orientated to the requirements of industry than most university courses. It is likely, therefore, that in the future, industry will become increasingly aware of the significance of this great change, especially for the development of staff of unusual potential who otherwise might remain unqualified.

The C.N.A.A. is also empowered to award research degrees on the same basis as that previously reserved for the universities. This has removed the major handicap from which colleges of technology had suffered in developing research activities of high calibre. During the period January–September 1966 (the last for which figures are available at the time of writing), the registration of 73 candidates for research degrees was accepted. There are indications that this number is likely to rise rapidly.

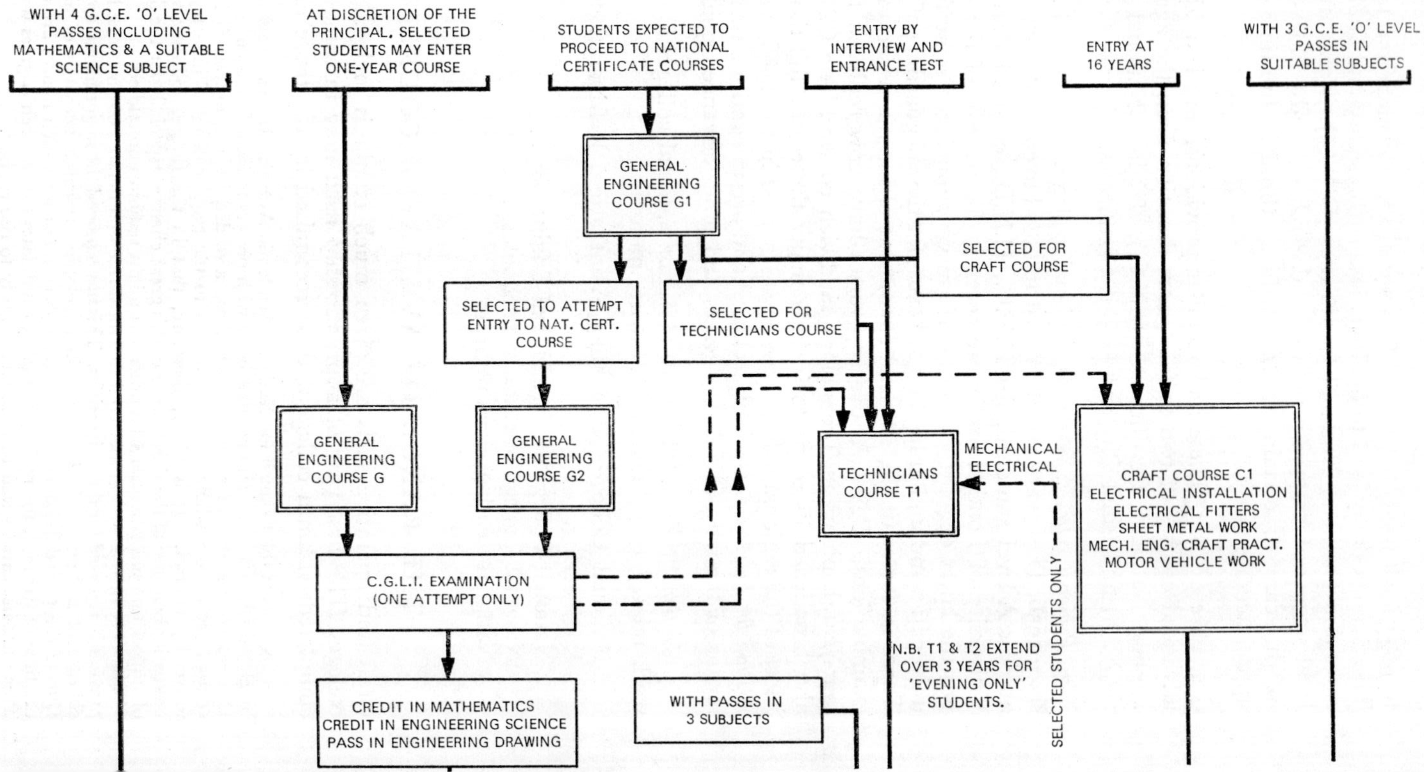

WITH 4 G.C.E. 'O' LEVEL PASSES INCLUDING MATHEMATICS & A SUITABLE SCIENCE SUBJECT
AT DISCRETION OF THE PRINCIPAL, SELECTED STUDENTS MAY ENTER ONE-YEAR COURSE
STUDENTS EXPECTED TO PROCEED TO NATIONAL CERTIFICATE COURSES
ENTRY BY INTERVIEW AND ENTRANCE TEST
ENTRY AT 16 YEARS
WITH 3 G.C.E. 'O' LEVEL PASSES IN SUITABLE SUBJECTS
GENERAL ENGINEERING COURSE G1
SELECTED FOR CRAFT COURSE
SELECTED TO ATTEMPT ENTRY TO NAT. CERT. COURSE
SELECTED FOR TECHNICIANS COURSE
GENERAL ENGINEERING COURSE G
GENERAL ENGINEERING COURSE G2
TECHNICIANS COURSE T1
MECHANICAL
ELECTRICAL
CRAFT COURSE C1
ELECTRICAL INSTALLATION
ELECTRICAL FITTERS
SHEET METAL WORK
MECH. ENG. CRAFT PRACT.
MOTOR VEHICLE WORK
C.G.L.I. EXAMINATION (ONE ATTEMPT ONLY)
SELECTED STUDENTS ONLY
N.B. T1 & T2 EXTEND OVER 3 YEARS FOR 'EVENING ONLY' STUDENTS.
CREDIT IN MATHEMATICS
CREDIT IN ENGINEERING SCIENCE
PASS IN ENGINEERING DRAWING
WITH PASSES IN 3 SUBJECTS

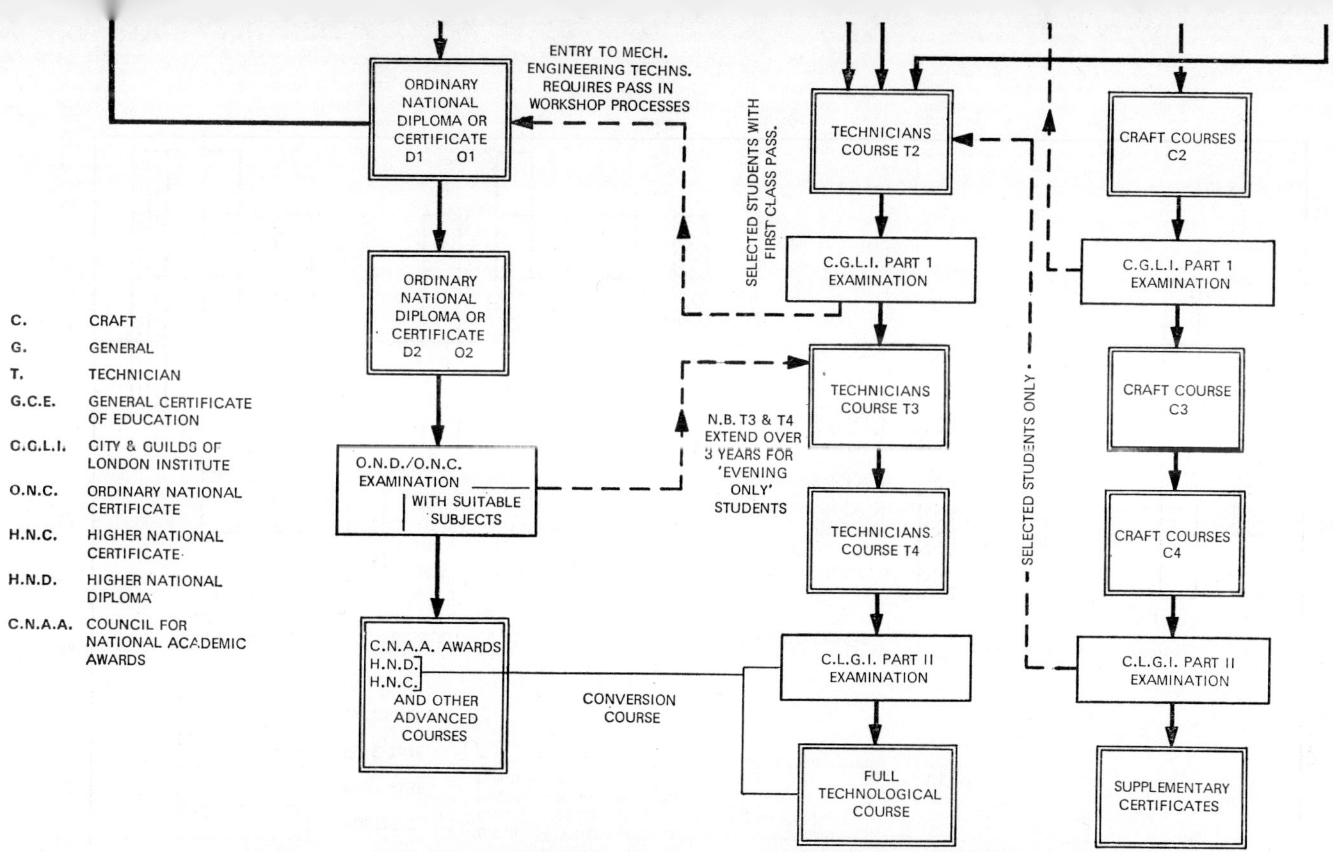

Fig. 3.1 *Courses Chart for the Faculty of Engineering. The Chart shows the different courses available in Engineering and the transfer loops that exist between them.* (*With acknowledgment to E. D. Rees, B.Sc.(Eng.), M.I.Mech.E., M.I.Prod.E., M.Inst.F., M.B.I.M., Head of Department of Mechanical Engineering and Naval Architecture at Portsmouth College of Technology*)

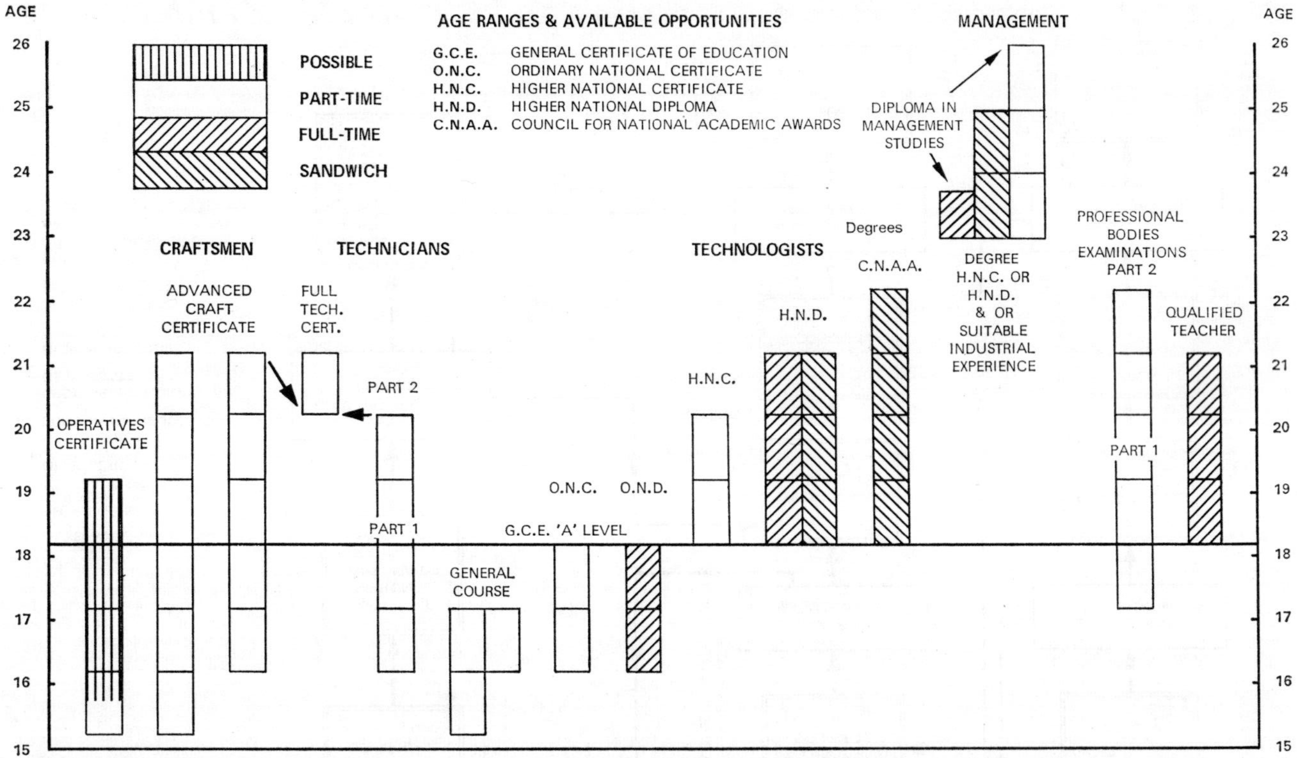
AGE
AGE RANGES & AVAILABLE OPPORTUNITIES
MANAGEMENT
POSSIBLE
PART-TIME
FULL-TIME
SANDWICH
G.C.E. GENERAL CERTIFICATE OF EDUCATION
O.N.C. ORDINARY NATIONAL CERTIFICATE
H.N.C. HIGHER NATIONAL CERTIFICATE
H.N.D. HIGHER NATIONAL DIPLOMA
C.N.A.A. COUNCIL FOR NATIONAL ACADEMIC AWARDS
DIPLOMA IN MANAGEMENT STUDIES
Degrees
PROFESSIONAL BODIES EXAMINATIONS PART 2
CRAFTSMEN
TECHNICIANS
TECHNOLOGISTS
C.N.A.A.
DEGREE H.N.C. OR H.N.D. & OR SUITABLE INDUSTRIAL EXPERIENCE
ADVANCED CRAFT CERTIFICATE
FULL TECH. CERT.
H.N.D.
QUALIFIED TEACHER
PART 2
H.N.C.
OPERATIVES CERTIFICATE
PART 1
O.N.C.
O.N.D.
PART 1
G.C.E. 'A' LEVEL
GENERAL COURSE
26
25
24
23
22
21
20
19
18
17
16
15

Fig. 3.2

the industrial training boards. There are strong indications that these provisions will be expanded and used more in the future. The training and concurrent education in general subjects are of a high standard.

The stages of technical education which are today open to the youth and the young men in industry are varied but even among industrial training officers (I.T.O.s) they are not always clearly understood. Fig. 3.1 gives an example of the complex network of courses available in engineering.

It is useful for the I.T.O. to have, in addition, a summary of the level of the types of courses and qualifications available. Some information is given in Fig. 3.2 and indicates the age at which various examinations may be taken. It gives an outline structure to which the I.T.O. will wish to add detailed information concerning local courses. There is, too, a large amount of valuable information in the prospectuses which the colleges issue periodically.

Many firms operate their own schemes for rapid training and retraining of operators and other grades of their employees. Supplementary college courses can often be provided from national schemes. Furthermore, most of these courses can be 'tailored' to meet local needs. Colleges may provide lectures and even complete courses within the premises of firms, e.g., where shift classes can be run conveniently only 'on site'.

In addition to the wide range of courses offered for City and Guilds and Ordinary National Certificate examinations, there are many others also subject to the regulations of the professional institutes. Recognition by an appropriate professional body is of increasing importance not only to the trainee, but to the supervisor, and to the manager. The institutions set out clearly their requirements which are reflected in college syllabuses.

Many professional and degree courses demand study of 'Industrial Administration'. By way of illustration of the amount of this work, the Department of Management Studies at the Portsmouth College of Technology devotes some 1,700 hours per annum of teaching time to introducing students of other departments to the elements of industrial administration. More than half of the students who receive this instruction are on mechanical engineering courses. The remainder are in electrical engineering, civil engineering, chemistry, mathematics and architecture. As well as instruction in industrial administration many colleges are giving increasing attention to supervisory and management education. For many years, colleges have provided courses to meet the examination requirements of the former Institute of Industrial Supervisors, now re-named the Institute of Supervisory Management.

Similar provision will now be made for the New Certificate of Supervisory Studies, awarded by the National Examinations Board on Supervisory Studies (N.E.B.S.S.). Another recent development has been the provision of short courses for shop stewards, derived from an agreement between employers and trade unions. Finally, the colleges now offer to industry a wide range of management training courses, mainly in the regional colleges. This work will now be described in more detail.

COURSES IN MANAGEMENT STUDIES

Whilst there remains some controversy about education for management, there is increasing agreement on the value of some training of the individual for management and of providing specialised assistance of high standard to the practising manager.

Management situations (like people) differ, and good management entails the taking of correct decisions in these situations. In this sense it may be doubtful whether 'management itself' can be taught. On the other hand, potential managers can usually learn from persons who have already had practical experience in this field and by discussion with personnel drawn from different sectors of industry. Management training can do much to prepare the individual to meet novel situations.

Much work on education or management in the U.K. is carried on particularly in the regional colleges and there is no doubt that the new polytechnics will give great attention to it.

THE DIPLOMA IN MANAGEMENT STUDIES

The most intensive scheme for junior and middle management is the Diploma in Management Studies (D.M.S.). It is based upon pioneering work of Col. Urwick and of the British Institute of Management. In 1961, it became the new 'post graduate' diploma course offered only by college departments which are approved by the Department of Education and Science. The examinations for the scheme are externally assessed each year, but all teaching and examinations are the internal responsibility of each college. The total number of diploma students enrolled each year is now over 1,000.

An essential preliminary is that students should be qualified and should have industrial experience. (Applicants for full-time teaching posts on the staff of Management Studies Departments are appointed for their relevant industrial experience as well as their academic qualifications.) At Portsmouth, there are four streams of students for the D.M.S.—(*a*) the full-time course which lasts for eight months, (*b*) the block-release course which has full-time students three weeks per term for two years, (*c*) the part-time day-release course on one afternoon and evening per week for three years and (*d*) the evening course of two evenings per week for three years.

Periodically these groups are brought together for residential weeks and weekend conferences. This enables the students to benefit from lectures given by distinguished guest speakers, and from discussions on experiences in industry. On these occasions the students also take part in business games and in involved case studies which simulate real situations and which can then be given the time, preparation and briefing to ensure their effective use.

The Diploma course consists of two parts. The first is designed to provide students with a reasonably uniform background of knowledge on some of the basic functions of management. The theme is 'operating management' and this is divided into separate but connected parts as follows: Production organisation and control: Marketing management: Finance and cost: Statistical control.

In the second half of the course the aim is to study management 'principles' and to develop through practice a skill in dealing with problems and situations. The subjects are: Organisation in management: Financial management and control: The background of business (economic, social, legal): General management (principles and practice): Quantitative analysis: Practical project work.

The Diploma course is an excellent preparation for anyone about to assume responsibilities in management.

FUTURE D.M.S. TRENDS

Today there are firm indications that new trends may develop. The advent of the Industrial Training Act cannot cause changes overnight in existing Further Education provisions. Nevertheless its implementation over the years is certain to cause important repercussions especially in the field of management training. Already, where appropriate facilities exist, requirements have been voiced for the Diploma in Management Studies course syllabus to be modified so that it may meet the more specific needs of particular industries. It would be expected that this type of requirement would stem solely from industries covered by the Act. But the yeast, it seems, is also beginning to work elsewhere and public social services like the hospitals are also beginning to ask for similar provision.

TRENDS IN OTHER MANAGEMENT COURSES

The trend for special provision courses for the D.M.S. is likely to be typical of many changes which may very soon develop as a consequence of enlightened integration between, on the one hand, clearly formulated industrial training requirements, and on the other, the modifying of existing facilities and syllabuses by colleges within the Further Education system.

Clearly it would be unrealistic for colleges to lower standards or to be prone to change existing facilities and provisions too swiftly. The process of integration between training and education may be complex but at least it is increasingly being recognised. It was stated at the January 1968 Coombe Lodge conference on 'Educational Aspects of Industrial Training Boards' that training and education 'are clearly inter-related' . . . and that 'By careful integration a course can be designed to achieve the maximum amount of progress in both directions in the time available, since each is part of the learning process.'

One of the problems in the past has been a frequent failure by industry to be sufficiently specific and, at the same time, adequately forthcoming in support of its own requirements of the educational system. Once both of these aspects are tackled with increasing vigour there would seem little reason for Further Education not to make an equally lively response.

Meanwhile it is evident that the pattern and standard of residential management training courses which has prevailed successfully at the main colleges in the past will tend to change in emphasis; and in so doing perhaps change very subtly in content, in that emphasis will be not so much on general principles, but rather more on particular applications and interpretation. In other words the trend will be towards analysis of specific needs and problems.

Nevertheless it is inevitable that in order to supply general demands many general courses will continue to be offered. These will doubtless range over such subjects as Introduction to Management, Executive Development, and functional techniques such as Management Accounting, Company Finance, Statistics, Operational Research, and Office Management. However, if the first of these courses is examined it should not be too difficult to contemplate modified versions to meet specific needs.

At the Portsmouth College of Technology, for example, this course and the longer Executive Development course have been running at least once a term for the last five years. The 'Introduction to Management' divides neatly into three separate weeks of residential study which run consecutively. The first week is on effective communication and deals with topics such as report writing, effective speaking, efficient reading and chairmanship. The second week on management organisation includes business administration, financial control, marketing, and background economics. The third week concentrates on problems of human motivation and the structure and practice of industrial relations. Clearly all of these topics could, if necessary, be orientated towards the needs of particular industrial training boards or their industries. Similar modification could be applied to 'Executive Development' type courses.

There is little doubt that eventually management training and education will be developed to C.N.A.A. postgraduate level and will then comprehend all general and functional aspects. Meanwhile its very robust offspring—industrial training technology—is developing so rapidly that it is likely to emerge ahead of its parent into first degree, and probably postgraduate diploma if not postgraduate degree, status.

TRAINING OF INDUSTRIAL TRAINING OFFICERS

One of the most convincing indications of the capacity of the technical colleges to adapt to new needs of industry for management training is the development of the type of courses on which this book is based. There are now some thirty colleges offering introductory courses for industrial training officers. These courses are based on the recommendations of the first report of the Central Training Council's Training of Training Officers Committee. Most of the courses involve four weeks' full-time study plus additional time, usually two weeks, on project work in the field.

Portsmouth courses consist of eight weeks' residential work plus project work in the field. All told Portsmouth have now run nine of these—five general and four specially designed for training advisers of the Hotel and Catering Industry Training Board. The aim is to run a one-year diploma course in Industrial Training Technology, and ultimately a postgraduate degree course.

INDUSTRIAL LIAISON CENTRES

One other unique provision to meet the needs of industry must be stressed. Recently, industrial liaison centres have been set up as a joint venture between the Ministry of Technology and the technical colleges. In about two years over 50 centres were established throughout the country. The main function of a centre is to act as a source of information for firms, and to serve as links between industry and the educational system. The service offers information on management techniques and many centres are now located in departments of management studies.

At Portsmouth two liaison officers, a chemist and an engineer cover an area 50 miles in radius which contains 2,000 firms. About 300 visits are made each year and 600 enquiries are received. Practical help and advice are given by different departments of the College.

CONCLUSION

It is a far cry to the back-room days of the early 1920s when many technical colleges were thought of as dingy establishments in which dedicated evening institute students worked under very difficult conditions to secure 'further education'. The range of facilities has now been vastly extended and re-designed to meet the needs of the future. Today the work of the technical colleges integrates as never before with the needs of industry. These facilities aim to anticipate the important principle behind the Industrial Training Act—and urged upon the industrial training boards in the first Memorandum of the Central Training Council—that training should include associated further education as an essential element.

BIBLIOGRAPHY

Publications by the Department of Education and Science:

A Plan for Polytechnics and Other Colleges, H.M.S.O. (May 1966)

Trends in Education, H.M.S.O. (Oct. 1966)

Management Studies in Technical Colleges (Second Report of U.K.A.C.E.M.), H.M.S.O. (1965)

General Studies in Technical Colleges, H.M.S.O. (1962)

Further Education for School-leavers, H.M.S.O. (1966)

On Course. Quarterly journal of Further Education—news over the whole field of education for industry and commerce

Careers Booklets:

MINISTRY OF LABOUR, *Choosing Your Career*, H.M.S.O. (1963)

CENTRAL YOUTH EMPLOYMENT EXCHANGE, *Careers Guide*, H.M.S.O., 6th Ed. (1964)

General:

MINISTRY OF LABOUR, *Memoranda Nos. 1, 4 and 5 of the Central Training Council*

RANKIN, J. A., *Industr. Train. Int.* 'Guide Lines for their Sandwich Degree Courses' (Feb. 1967)

4

The Training Function in the Working Organisation

A. O. MARTIN, B.A., PH.D., DIP. PSYCH., A.B.PS.S., F.W.I.

Principal Psychologist, Training Department (Industrial Training Research and Information), Ministry of Labour

INTRODUCTION

The main purpose of this chapter is to suggest a systematic way of thinking about 'the training function'. It will not be concerned with any specific level or form of training, but with training as such, and, in particular, the manager's responsibility for it. By looking at training as a function of management based on a system of operations, information can be derived about the steps which the line manager and the training specialist can take with their colleagues in the working organisation to ensure that training is effective.

GENERAL PRINCIPLES

A systematic analysis must be based on sound general principles. Three basic principles are suggested as a starting point for this chapter:

1. Training is the responsibility of the line manager.
2. The aim of training is to improve job performance by extending knowledge, inculcating skills and modifying attitudes, so that individuals can work in the most economical, efficient and satisfying way.
3. Knowledge can be extended, skills inculcated and attitudes modified most effectively when good conditions for learning are provided.

THE MANAGER'S RESPONSIBILITY

The first responsibility for the performance of any job belongs to the individual himself. The supervisor has a double responsibility: he is responsible not only for his own job performance, but also for the performance of his staff. This pattern of 'shared responsibility' continues up through the working organisation at all levels. The principle 'the ultimate responsibility for the performance of staff rests with the manager' is an empty one unless it can be accepted that responsibility must in fact be shared. It is clear that many factors will influence job performance; training is only one of the many methods by which performance can be improved. The manager must, therefore, be as responsible for training as he is for improvements in methods, materials, equipment, lay-out, communications and relationships, and all the other activities of his working organisation.

This does not mean that every manager should be a training specialist, any more than he should be a work study specialist or a designer of equipment and machinery. But it does mean that every manager must know enough about training to ensure that the training function in his organisation is being carried out effectively. He must know whether the knowledge, skills and attitudes which have to be learned reflect the true needs of both the individual and the organisation as a whole. He must know whether the best conditions for learning them are being provided. Above all, he must be able to check whether the learning was effective. He cannot assess whether aims were correct, nor whether learning conditions were right, until he is able to apply the yardstick of effectiveness to the performance of those who have been trained.

The amount of detailed training knowledge required by a manager will depend upon the size of the firm, the variety of its activities and upon the number of training specialists working in it. In a small firm, the line manager may have to oversee every aspect of the total training effort. He himself may have to draw up plans, carry them out and then find out how near he has come to his training target. On the principle of 'shared responsibility', he will be wise to have his supervisors trained in methods of instruction so that they can do most of the face-to-face training as part of their general duties. He may also employ a full-time instructor who can provide guidance to these supervisors and assist in such tasks as the preparation of training manuals. All other training activities may be the manager's own personal concern. The bigger the organisation, the more varied the types of training to be done, the more difficult it is for the manager to gain specialised training knowledge—and to have the time to use it.

There will be a point, according to the size of the organisation and according to what that organisation does, when the manager must delegate most training duties to a specialist above the instructor level, either inside the organisation or outside it through a group scheme. But although he delegates duties, he can never delegate his final responsibility, particularly for training policy. Training effort is essentially team effort. *All* members of the working organisation are members of the training team and they will all have something to contribute. Because of this, training specialists—even training departments—should be regarded as providing services for the whole organisation on behalf of managers, and managers must know enough about training to ensure that these services are effective.

AIMS AND EFFECTIVENESS

If the *general* aim of training, quoted earlier, is recognised and accepted, it follows that training must satisfy real needs. The modern manager is not content to accept training as an act of faith, based on somebody's assumptions about what knowledge, skills and attitudes are appropriate. He has to ensure that actual needs have been identified and analysed. Furthermore, he will know that there is little point in trying to measure the effectiveness of training unless he has a sound base-line from which to work. This means that the *specific* aim of every training scheme, course or programme should be described in such a way that later measurement of effectiveness is possible.

This is a lot easier said than done. Sometimes it is easy in theory to spell out specific aims, but difficult in practice to find suitable yardsticks. Sometimes yardsticks are available, but there are reasons, rational or irrational, why they cannot be used. Sometimes, particularly where training and education merge, the aims are so long-term that one cannot be sure whether one is measuring the effectiveness of the training scheme, or measuring something other than training which has had an influence on performance. However, the existence of difficulties does not make the task impossible; difficulties simply present a greater challenge.

THE TRAINING FUNCTION

Training may now be examined more closely as a function, and it may be analysed in terms of what is actually *done*. Beyond stressing the team approach, this chapter is not concerned with who carries out specific duties, or with who is being trained in what spheres of knowledge, skill or attitude. It is concerned with the generality of the training function itself, which should be capable of application equally to managers learning stock control, to engineers learning how to diagnose faults on machines, or to operators learning how to produce goods or services of the right quality and in the right quantity. There is a common logic to be found and a common chronology to be followed.

Since a start should be made from aims, the first general task is to *identify training needs*. When this has been done, a *policy* has to be formulated to meet them. If this policy is acceptable to all concerned, detailed consideration must be given to its *implementation*, with special reference to providing optimum conditions for learning. Finally, methods must be found by which the *effectiveness* of the training can be measured. This test of effectiveness will also be a test of how well training policy was formulated, and of the accuracy of the earlier identification of training needs on which that policy was based.

It is important to remember that these four main tasks or steps will not be carried out *in vacuo*; they will be done in the context of what may be called social factors, which impose certain limits or constraints. Constraints will be imposed by the policies of the industrial training board(s) concerned with the particular working organisation, as well as

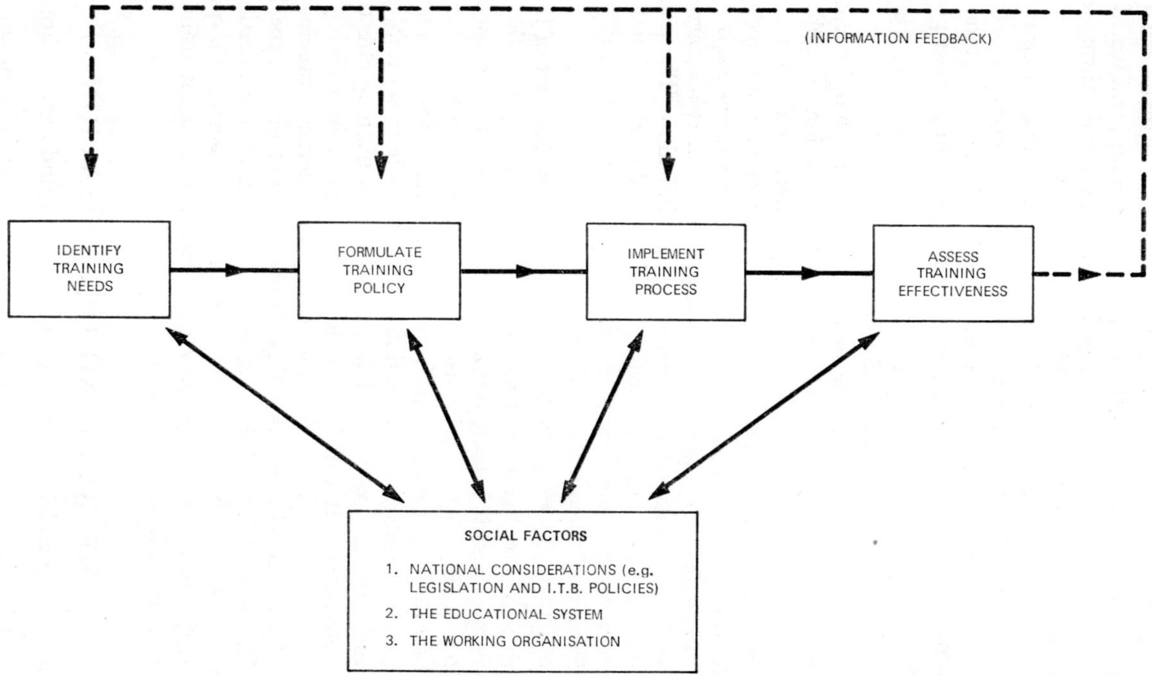

Fig. 4.1 The Training Function

by national legislation. Another factor is the existence of the educational system. To integrate training with further education, the working organisation must make full use of the educational system and dovetail its efforts with what that system can provide. The last main social factor is the working organisation itself; we do not live in an ideal world where the training and education needs of the individual will always exactly match those of the organisation which employs him.

The four main steps in the training function, and the three broad types of social factors which provide the context of its operation, can be set out as shown in Fig. 4.1. This 'information loop' way of looking at the training function serves to draw attention to several important points:

1. The assessment of training effectiveness is seen as a crucial stage which provides 'feedback information'.
2. With 'feedback' the training function becomes dynamic, leading to constant re-examination of needs, re-formulation of policy and review of the training process itself. Without feedback training is static—a once-for-all exercise which can never be evaluated.
3. There is a two-way interaction between the various steps and the social factors which provide constraints on how they are carried out. These constraints are not permanent and inflexible. The policies of the industrial training boards, provisions of the educational system and working organisations themselves will change as more 'training information' becomes available.

Other chapters on 'The Training Framework' in this part of the book deal with two broad groups of social factors—legislation and the educational system—and it is therefore appropriate to look next at *the working organisation* itself. Then a brief list will be made of some of the topics covered by these four main steps in the training functions: (*a*) identification of training needs; (*b*) formulation of training policy; (*c*) implementation of the training process; and (*d*) assessment of training effectiveness.

For convenience, it will be assumed that a specialist training officer exists who acts as a focus for questioning and decision-making about training in the working organisation. However, the team approach must be kept in mind. The training officer will not ask all the questions and make all the decisions; he will try to ensure that appropriate questions are asked by other people and that relevant decisions on training are made at the right levels in the team.

THE WORKING ORGANISATION

The questions to which the training officer must find answers are:

1. What is the organisation seeking to do—what is its production or service policy?
2. What are the present work processes and systems by which it seeks to reach these objectives?
3. What technological and other developments are foreseen which may change these processes and systems?

4. What is the organisation's personnel policy—what statutory or other conditions of employment may affect training?
5. How is the firm organised and controlled—what functions are given to managers at different levels and what formal and informal systems of communication have been built up?
6. How far has joint consultation developed—what local and national trade union agreements exist?
7. Is there a training committee—if not, would one be helpful?

These are all background questions of considerable importance. Until the answers are known, there is a danger that short-term training needs may be identified at the expense of the longer-term and that policy may be formulated in such an artificial way that it could never be successful.

THE FOUR MAIN STEPS

Every step in the application of the training function is limited in some way by the social factors which have been listed. Every step taken will provide 'feedback information' which will in turn modify these factors. Sound appreciation of limits and mature judgment of the extent to which they can be modified are the hallmarks of the expert training specialist.

The detail of the four main steps will be examined by again listing questions which have to be asked and answered. It will be sufficient, in this chapter, to have answers of the type which will lead to further questions—a check-list approach. The training specialist and his colleagues must find the detailed answers themselves. They will differ not only from organisation to organisation, but also from time to time in the same organisation if the training function is a dynamic one.

1. THE IDENTIFICATION OF TRAINING NEEDS

Two broad groups of basic questions are implicit under this heading. What are the *jobs* for which training is necessary, and who are the *people* who have to be trained?

Jobs

What are the training priorities?

This question will lead to an examination of measures of operational efficiency. If the training officer asks further questions about manpower turnover, accident and sickness rates, casual absenteeism, restrictive practices and labour disputes, quality and quantity standards (e.g., experienced worker standards), wastage and scrap rates, down time and waiting time, customer complaints and other opinion pressures, he should be able to identify barriers to improved performance which training might remove.

What are the circumstances in which the jobs are performed?

It is not uncommon to find that a so-called 'training problem' could be solved more readily or more economically by a change in job circumstances, methods of work, or design and layout of equipment. Whether or not this is so, safety considerations are always important in analysing a job and devising a training programme for it. Special hazards may exist when the job is paced, dirty, hot, cold or noisy, or when methods, design and layout are poor.

What are the techniques by which these jobs may be analysed?

There is a wide range of analytical techniques available according to the type or importance of the job: trained observation, questioning, job breakdown (as developed by T.W.I. Job Instruction), formal job analysis, skills analysis, work study and the statistical techniques often required to complete a large scale analysis.

What levels of required performance does the analysis reveal?

Every job can be analysed in terms of the attitude/knowledge/skill pattern required.[1] It is often useful to express the results in terms of levels of abilities and interests, following the National Institute of Industrial Psychology (N.I.I.P.) Seven Point Plan.[2] The identification of training needs on this basis carries the great advantage that the resulting 'specification' can also be used for selection purposes.

People

What people have to be trained to these levels of performance; and where will they come from?

The analysis should reveal whether the job needs a manager, supervisor, professional or technological specialist, technician, craftsman, commercial or clerical worker, operator, sub-operator or some other category—and how many of these are required. Are they to be recruited from outside, selected from within, or is the problem one of retraining existing workers? Depending on the answers to these questions, the training officer may have to ensure that adequate selection methods are used. He may himself be responsible for selection as well as training.

2. THE FORMULATION OF TRAINING POLICY

With all these facts and estimates at his disposal, the training officer is in a position to formulate a policy for consideration by line managers. In doing so, he must look ahead to the implementation of the training process and the assessment of the effectiveness of the training, and draft his proposals within the limits imposed by legislation and I.T.B. policies, the educational system and the working organisation.

He will not draw up plans in isolation. He will have done his analysis and made his projections as a co-ordinator of a team effort, taking care to keep all the 'key people' informed as the outlines of policy develop. The end product of the team effort may be a well-written and well-presented report, but its value will be limited if it is not based on proper consultation. The training officer must make use of all his social skills and be highly critical of his own work. Questions he should be asking are: 'Is there anything I have left out? Is there *anyone* I have left out? Is there anything I could explain more simply? Am I being realistic? Is there anyone likely to oppose this general policy?' Even if he is satisfied that he has the right answers to these questions, the training officer will be well advised to circulate his final report in draft form before submitting it for formal discussion and approval.

3. IMPLEMENTATION OF POLICY—THE TRAINING PROCESS

Once again, there are two broad groups of questions to be asked and answered. These cover *training administration*, and *training methods, aids and devices.*

Training Administration

What type of training is to be given?

The earlier identification of training needs will indicate whether priority should be given to induction and background training, initial job training, progressive training, refresher training, specific retraining or 'appreciation' training. The term appreciation training—an important but often neglected area—refers to training, particularly for supervisors and managers, in a related area of work about which they need a broad knowledge in order to perform their own work more satisfactorily (e.g., computer appreciation courses and brief courses leading to a better appreciation of the training function itself).

Where is training to be carried out, and what form should it take?

Training may be on the job; in a training bay or training position; in a works school, centre, workshop or works staff college; at a centre run by the industry or by the appropriate training board; at a Government Training Centre; at an educational establishment; at a privately run centre; or even by a home-based correspondence course. Training can be wholly within the firm or be supplemented by day release, block release or sandwich arrangements; by full-time educational courses; by evening courses or correspondence courses. Quite often, the needs of the firm and the individuals concerned would best be catered for by group schemes (involving a group of firms).

What is the length of the training period?

Usually it is better to judge the length of training by reference to the type of performance necessary for the job, rather than have as the starting point 'how much time can be spared and what can be done in that time?'. However, this is not an invariable rule. Where a well-defined criterion of job performance exists (e.g., experienced worker standard), there is no virtue in any other approach. Supervisory, managerial and many specialist jobs of a 'social' nature are different—they are 'expanding universes' where criteria of performance are difficult to establish. For such jobs, a start is often made from the 'how much time can be spared' principle.

How should programmes and courses be designed?

As a subsequent chapter will show, course design is an art in itself. In the present context, it is possible only to point to the need to specify the aims, and to appreciate the problem of reconciling the needs of the trainees with those of instructors and others who have to fit training commitments into their other jobs.

Training Methods, Aids and Devices

It is not possible to design a training course without reference to the methods, aids and devices which will be used. Working from the general to the particular, the training officer and his instructors will need a broad appreciation of learning and teaching theory, including aspects such as motivation, incentives, attitudes to learning, creativity, individual differences, group processes and the design of instruction. They must also take into account the learning problems of particular age groups and particular categories of trainees.

What are the methods to be used?

Methods may be classified broadly into those suitable for *group* instruction, for *individual* instruction or for *general* use with groups or individuals. The choice will depend on such factors as subject matter, the location of the training (e.g., on or off the job), the numbers of trainees available at any one time and their age and previous experience.

The range of *group* methods is biggest, and it is convenient to list them in approximate order of the degree of active participation they demand from the trainees. Unless it is vivid, interesting and supported by visual aids, the lecture tends to be the least demanding method. At the other end of the 'participation scale' can be placed 'sensitivity training' (T-groups). It is possible to list between these extremes, lesson method (built on questioning technique), discussion (with or without chairman or expert witness), case study (including in-tray exercises and business games) and role play (including role-reversal).

Purely *individual* methods are correspondence and private study courses, but even these are often supplemented by discussion or similar group exercises. Among *general* methods are programmed instruction, Training Within Industry Job Instruction, practical demonstration and project work. All these may be used with groups or with individuals, depending on the medium employed. Thus, closed circuit television or film can make available to a group a practical demonstration normally observable in detail by one individual only. This raises the next question.

What learning aids and devices are available?

It is essential that aids and devices are labelled 'aids to learning' rather than 'teaching aids'. The training specialist is concerned with helping the learner rather than himself. It goes without saying that the correct use of aids and devices also simplifies the task of the teacher, but his needs should be subordinate to those of his learners. The correct description of aids is thus a point of some importance, since it will reflect the training specialist's approach to his work.

Aids and devices can be grouped according to the senses through which the learner is getting his information. Although in some circumstances the senses of taste and smell can be used (and opportunities for doing so should not be neglected), the four main groups are visual, audio, audio-visual and 'multi-sensory'.

In terms of *visual* aids the range is very wide—from chalkboards to fixed models. It includes magnetic boards, 'flannelgraphs', silent projectors of various kinds, textbooks, notes and other printed handouts. Purely *audio* aids include tape recorders, discs, signal generators, and are not restricted to use of the human voice. *Audio-visual* aids, which employ both 'sensory channels' at once and thus increase the amount of information, include sound filmstrips, sound film, direct closed-circuit television, video tapes and broadcast television.

The *multi-sensory* group includes all those aids which make use of the sense of touch as well as sight and hearing. Real machines can be placed in this category, as well as working models and simulators.

The general principle should be to secure maximum participation of the learner by using methods, aids and devices which make use of all the relevant senses, and to reduce straight lecturing to the minimum.

4. THE ASSESSMENT OF TRAINING EFFECTIVENESS

Without training records there can be no test of training effectiveness. The training officer must be conversant with the general principles of the design of both instructors' records of trainees and trainees' records of their own progress.

What is meant by 'assessing effectiveness'?

'Assessment' is used as a neutral term to cover the linked but separate processes of validation and evaluation.[3] The *evaluation* of training seeks an answer to the question 'is this training programme worthwhile?'. It has to do with 'cost-effectiveness'. But before a programme can be evaluated in this sense, it is necessary to ask whether the training programme is valid.

In doing so, it is useful to distinguish two elements: there are two distinct questions to answer. Firstly, does the training do what it set out to do? Secondly, is the design right—was the training designed with the real requirements of the job clearly in mind? It is not unknown for a training course to aim at the wrong target and it is certainly not uncommon for the purpose of a course to be vague.

The training officer is, therefore, concerned with two types of validity. When he asks whether a course or programme achieves the specific objectives of the training, he is asking about its *internal validity*. When he asks whether these specific objectives were the right ones he is asking about its *external validity*.

It is only when these two questions have been answered satisfactorily that the training officer and his colleagues can move on to *evaluation*—the analysis of the value of training in the 'cost-effectiveness' sense. Training costs must be weighed against increased efficiency, both short-term and long-term, and the possibility of any alternative methods of achieving the same short-term and long-term results should be considered. Evaluation of training is not a matter for the training specialist alone. He will work closely with the firm's accountants, whose costing techniques can be applied to training as they are applied to other areas of the firm's activities.

What methods of assessment should be used?

There are two broad systems—continuous assessment and terminal assessment. In continuous assessment, tests are applied throughout the whole training process and are designed to measure *developing* performance, over the whole period of training. Traditional terminal assessments have the disadvantage of testing the trainee at only one point in time when 'mood of the moment' or 'state of health' might be significant. They also need frequent review to check whether they are really examining what the trainee needs to know or to do.

The training officer will naturally use objective methods as far as he can, but he will frequently have to rely on his own judgments and those of others. Although objective measures are to be preferred, they cannot be used unless they meet three requirements. They must be scientifically sound, they must be administratively convenient and they must be 'politically' acceptable. Administrative convenience and political acceptability will often dictate continuous assessment rather than terminal assessment methods. Tests given as *part* of the training process can be fitted into the programme, and individuals will accept them as natural to the scheme of things. The political acceptability of objective testing is often very doubtful in training for managers, supervisors and other older people, who might not take kindly to the formal examination atmosphere so difficult to avoid in terminal testing.

Objective knowledge tests, tests of skill and attitude scales are available or can be constructed with expert assistance. When they cannot be used, subjective methods should be made as objective as possible. This is usually achieved by ensuring (*a*) that the observer or interviewer is well trained, (*b*) that his subjective judgments are based on as much *factual* material as he can collect, and (*c*) that wherever possible, his findings are compared with those obtained by another *skilled and independent* observer.

The need for research

It is not easy for the training specialist and his colleagues to *identify* variables within and outside the training situation which may have a bearing on performance, let alone control them. For example, the measures of operational efficiency examined in the original identification of training needs may change during the training process for reasons which have nothing to do with training. There may have been changes in work methods, type of supervision, quality of materials, sales policy or service policy. If the relative effectiveness of different training methods is being examined, it is sometimes possible to equate training groups (or allow for differences by statistical means) according to age, sex, intelligence and specific abilities. It is not so easy to make allowance for influences such as instructor ability, location of training, and health and personality characteristics of trainees. Large samples, and random allocation of trainees to different methods and different instructors, may minimise the influence of these variables—but it is a very fortunate training officer who can design his 'research' to the satisfaction of a critical statistician. Particularly in the small firm, he simply has to do his best to assess effectiveness as objectively and disinterestedly as he can.

He will be well advised to make a modest start with a form of training which is readily measurable—and not, for example, plunge straight into an attempt to test the effectiveness of manager training. He may also consider the possibility of linking up with other training officers in the locality, under the general guidance of a trained research worker in the neighbouring university or technical college, so that 'group' schemes of testing effectiveness of training could be worked out. There is wide scope for co-ordination of research; the bigger the samples and the better the overall design of experiments and investigations, the more likely it is that 'feedback information' for more general application will emerge.

REFERENCES

1. FRYER, D. H., *Personn. Psychol.*, **4**, 19 (1951)
2. RODGER, A., *The Seven Point Plan*, N.I.I.P. London (1952)
3. MARTIN, A. O., *Personn. Mgmt*, **39**, No. 340, 88 (1957)

BIBLIOGRAPHY

D.S.I.R., *Training Made Easier*, H.M.S.O. (1960)

WELLENS, J., *The Training Revolution*, Evans Bros. (1963)

KING, S. D. M., *Training Within the Organisation*, Tavistock Publications (1964)

SEYMOUR, W. D., *Industrial Skills*, Pitman (1966)

PART 2

PREPARATORY CONSIDERATIONS

5

Effective Recruitment and Selection

A. O. MARTIN, B.A., PH.D., DIP.PSYCH., A.B.PS.S., F.W.I.

Principal Psychologist, Training Department (Industrial Training Research and Information), Ministry of Labour

INTRODUCTION

The term 'selection' covers several distinct processes which are related to the 'social factors' mentioned in the last chapter as the context in which the training function is carried out. The main purpose of this chapter is to identify these processes and show how good training should be based on good selection. The following will be considered: pre-employment selection through the educational system; manpower availability; job analysis within the firm; and the use of evidence gathered by reports, tests and interviews. Special attention will be given to psychological tests, where they may be obtained, and the training which is necessary for their proper administration and interpretation.

BACKGROUND TO SELECTION

The aims of selection are broadly the same as the aims of training—greater efficiency, economy and satisfaction. Both selection and training are aspects of the general problem of fitting the man to the job; both seek to achieve the best possible match between the talent and potential of individuals and the requirements of the organisation.

It is useful to distinguish between different types of selection. The training officer and his colleagues may be concerned with the following items:

1. Recruitment and selection from *outside* the firm.
2. Selection from *within* for new or different jobs.
3. Examination of potential for retraining when existing jobs change in their nature.

These types of selection usually *precede* training and thus become foundations on which it can be built. But it is important to note that selection may also be concerned with the employment of people already trained elsewhere for whom little or no training is required beyond 'induction' and background courses. Professional and technical people are often in this category.

SELECTION THROUGH EDUCATION AND TRAINING

In passing through the educational system, everybody is trained to some extent for work in industry or commerce. Everybody possesses certain basic attitude/knowledge/skill patterns on which vocational training may be built. The educational system, through examinations and other means, is also selective; recruits needed by industry and commerce have been partially selected through steps in the educational ladder. In the same way, subsequent industrial training is also selective; people who fail to reach the standards required are normally diverted to lower level jobs.

MANPOWER AVAILABILITY

In areas of manpower shortage, it may not be possible to refine the 'educational selection' process much further, except that larger firms with a range of vacancies may be able to ensure that the available people are placed in posts most appropriate to their aptitudes, abilities and interests. However, manpower shortage should never be taken for granted. Firms should as far as possible budget their requirements in advance and base them on accurate job analysis. This may often reveal that some jobs need not be done at all, or that changes in methods can reduce the number of people needed for them. Firms should also check whether an apparent manpower shortage is real, or whether there is anything wrong with the processes by which they attract potential staff. It is rarely sufficient to hang notices on the gate (a last resort rather than the first step to take). The many sources of supply should be tapped directly. Employment exchanges, youth employment officers, careers masters and mistresses and other agencies should be informed of vacancies in good time. These direct sources as well as press publicity should be tried before any assumptions are made about 'manpower shortage'.

Firms naturally prefer to select the best from an abundance of applicants, but, when manpower shortage is real, they may have to accept a lower level of talent. If this is so, the training process must be made even more efficient. Lower quality people will need better and longer training. As a corollary, if talent is abundant, care must be taken not to select at so high a level that people have to work below their capacities. This could result in costly labour turnover and absence rates. The general principle should be to select the lowest level capable of doing the job at the required standard after proper training, without excluding a leaven of higher potential for subsequent promotion to supervisory jobs.

JOB ANALYSIS

Careful job analysis is the only secure foundation on which to build proper selection and training procedures. Neither can be built solely on *personal opinions* of what a job demands in the way of individual characteristics, and of what training is necessary to bring individuals to the standards required. As already indicated—and it bears repetition—job analysis *may* reveal that the job either does not need doing at all or that it should be radically changed. Ideally, it results in itemised *job specifications* (and/or *job descriptions*), and *personnel specifications* for selection purposes. It is advisable to use a single analytical system to which both can be related.

THE SEVEN POINT PLAN

This system[1] is the most widely accepted in vocational guidance and personnel selection. Since youth employment officers and staff of employment exchanges are trained in its use, any personnel specifications based on it will be readily understood by them. A firm can thus make better use of their services in finding the recruits it requires. The main headings of the Seven Point Plan are:

1. PHYSIQUE	Items relating to health, hearing, eyesight, voice, disability, etc., which may be of occupational importance.
2. ATTAINMENT	Measures of educational and occupational achievement or failure.
3. INTELLIGENCE	Measures of the intelligence or general ability level required.
4. SPECIAL APTITUDES	For example, mechanical aptitude, verbal or figure fluency, manual dexterity, drawing ability.
5. INTERESTS	Analysis of relevant interests (e.g., intellectual, practical-constructional, physically active, social). There may be interests useful in performing the job; alternatively, good job performance may be inconsistent with certain personal interests.
6. DISPOSITION	Qualities of temperament, general 'character traits', e.g., acceptability, influence on others, steadiness, reliability, self-reliance.
7. CIRCUMSTANCES	There may be domestic circumstances to be considered, for example in selecting for a job involving travelling or awkward shifts. This heading also concerns the physical environment of the job (e.g., paced, dirty, hot, cold, noisy work and similar features).

COLLECTING THE EVIDENCE

Some of the items in the Seven Point Plan can be measured objectively by medical and other tests; others by reports, such as headmasters' reports and employers' references, and educational and vocational training certificates; some may be arrived at only by questioning or by judgments of a more subjective character. The best practice is to use the interview as a means of supplementing the factual evidence collected from various sources and for personally assessing the individual as a whole person in the light of this evidence.

APPLICATION FORMS

A considerable amount of information can be made available to a firm at the *application* stage. It is important not to wait for the interview stage to collect it. The time of skilled interviewers is costly, and should not be wasted in doing a job which can be done by a well-designed application form. Investment in the proper design of such forms—perhaps by a consultant—can pay quick and substantial dividends. The form should reflect the system (e.g., the Seven Point Plan) by which the *personnel specification* has been drawn up. Care should be taken to differentiate the factual information which can be supplied by the candidate from that which will be obtained elsewhere (e.g., school reports, employers' references and evidence of educational achievements). Though reports and references may be subjective, they can often be used as the first 'sieve' in the selection process. However, it must be ensured that the 'labels' attached to potential employees by others are still valid. Reports and references should be recent enough, and their sources sufficiently respectable, to ensure that the information is reasonably accurate.

OBJECTIVE TESTS

Tests are of value only in the hands of people who have been trained to administer and interpret them. This is already well accepted in respect of 'medical' tests (e.g., eyesight, colour vision and similar items covered by section 1 of the Seven Point Plan), but still insufficiently appreciated in respect of objective psychological tests (e.g., of general intelligence, special aptitudes and other personality characteristics). Some tests may be administered to one individual only at a time; others are specially designed for use with groups. The test user must thoroughly understand their purpose, how they were designed and standardised, and how scores may be interpreted.

The design of tests of this kind must be left to the experts; 'home-made' tests, which cannot be standardised on large populations, are usually valueless. Worse still, they will frequently have a 'face validity' (i.e., they will be superficially sound) but in fact bear little relationship to the qualities and attributes that they are designed to measure.

The concept of *validity* is a difficult one which cannot be discussed here. The reader should go direct to the recent account of psychological tests given by Anstey.[2] Much of this book relates to the technicalities of designing, validating and standardising objective tests, but each chapter contains a summary specially designed for the layman who may not need detailed technical knowledge. See also Vernon.[3]

Psychological tests must be distinguished from devices and techniques requiring subjective interpretation, which are better described as 'clinical instruments'. Examples are Rorschach 'inkblots', Szondi photographs, Thematic Apperception picturecards and 'graphology'. Psychological tests may be classified in two broad groups: *cognitive* tests (of ability or knowledge relating mainly to intellectual processes) and tests which are basically *non-cognitive* and relate mainly to the more emotional aspects of personality such as temperament, interests and attitudes.

COGNITIVE TESTS

While cognitive tests do not fall into watertight compartments, it is permissible to draw a distinction between tests mainly of *innate ability* and tests mainly of *acquired knowledge*. Ability tests can be further classified into three broad groups (which overlap to some extent), namely: tests mainly of *general intelligence*; tests mainly of *special aptitudes*; and tests mainly of *creative originality*.

Tests of acquired knowledge and skill are usually known as *attainment tests*. While these usually measure specific and limited areas (e.g., arithmetic and spelling), trade tests and 'work sample' tests may often belong to this category. Work sample tests may measure proficiency *after* training, or they may take the form of trials of elementary tasks, performed early in a training course, which seek to predict success in training. Current research on work sample tests used for prediction suggests that such tests may come to be known as 'trainability' tests for certain occupations. They may be of special value in selecting for retraining older workers who are often anxious when faced with paper and pencil and similar tests.

NON-COGNITIVE TESTS

These are often referred to as 'personality tests' but they are better described as non-cognitive tests, since most psychologists prefer to use the term 'personality' to cover *all* the attributes of an individual. Non-cognitive tests may be grouped into tests of *general personality factors* (e.g., extroversion–introversion), tests of *specific temperamental traits* (e.g., sociability) and tests of *special interests*.

It is worth pointing out that if physical tests are included, the objective tests listed above could also be classified to correspond roughly with the first six headings of the National Institute of Industrial Psychology (N.I.I.P.) Seven Point Plan. (Vernon[3] lists many well-known tests on a slightly different classification.)

1. Physical tests (i.e., relating to physique).
2. Attainments.
3. Intelligence or general ability.
4. Special aptitudes.
5. Interests.
6. Temperament.

Another point to be borne in mind is that although a wide range of tests is available for use in vocational guidance, to assist in identifying many attributes of an individual, it does not necessarily follow that the same tests can also be used for selection purposes. It is very difficult indeed to *validate* selection tests against subsequent job performance. Satisfactory criteria of performance have to be identified, and other variables affecting performance must be taken into account.

USE OF TESTS

Broadly speaking, unless a firm employs a professional psychologist on a permanent or consultancy basis, the only tests available for use in industry are those of general ability and special aptitudes. The general conclusion of an enquiry made by the N.I.I.P. (Frisby[4]) into the use of such tests for engineering apprentices was that they could be of value. A marked relationship was shown between performance on the tests and academic achievement at the technical college. No adequate criterion of workshop performance was available and it was thus impossible to assess the value of the tests from this point of view; but evidence from supervisors' ratings pointed to the conclusion that the tests were useful in this respect also. However, even general ability and special aptitudes were not enough. Useful as these tests were, the report on the enquiry stressed the importance of *other* factors in the degree of academic success achieved by the apprentice. It is therefore quite possible that if other, non-cognitive, tests had also been used, prediction of both academic success and job performance would have been better.

SELECTION

Since the results of the N.I.I.P. enquiry were expressed in cautious terms, it follows that in no circumstances should a general ability or special aptitude test (or a battery of tests covering a range of measures of ability and aptitude) be used as the *sole* criterion on which a selection decision is based. Though the tests are as objective as possible, results are still to be interpreted in probability terms. The general lesson of most enquiries into the use of selection tests in industry (see Bibliography) is that tests are of value as the *final refinement* of a selection system which is in all other respects as perfect as it can be. Even then, psychological tests should be used only by people who have been trained in their administration and interpretation.

TRAINING IN THE USE OF TESTS

The two main suppliers of tests in this country are the National Institute of Industrial Psychology (N.I.I.P.) (14 Welbeck Street, London W.1), and the National Foundation for Educational Research (N.F.E.R.) (The Mere, Upton Park, Slough, Bucks.). Both organisations, in common with most other publishers of tests, lay down stringent requirements for test purchasers and test users. For example, tests distributed by N.F.E.R. are categorised according to the minimum qualifications necessary for their administration and interpretation, as follows:

LEVEL A Tests require only the minimum technical knowledge.

LEVEL B Tests can be used only by people qualified by successful attendance at a recognised course of training, or otherwise qualified.

LEVEL C Tests are for the most part 'clinical instruments' or non-cognitive tests for use only by qualified psychologists.

Tests which are most suitable for use in industry are generally included in levels A and B.

The N.F.E.R. does not provide training courses in the use of tests, but may be prepared to advise on suitable courses elsewhere. Courses in personnel assessment and selection are run by the N.I.I.P., which also offers an advisory service on the installation of tests. The Institute of Personnel Management (I.P.M.) (5 Winsley Street, London W.1) may also run courses and will be able to advise generally on testing and other aspects of selection procedures. Advice is also available from the British Institute of Management (B.I.M.) (Parker Street, London W.C.2).

GROUP SELECTION PROCEDURES

Staff at operator level (broadly defined as those who have no responsibility for the work of others) are normally selected by a single interviewer who has collated the 'paper' information available about the candidates. He may also have administered objective tests. At higher levels, the decision is more often made on the basis of the findings of interview boards or 'panels'. Although it is selection *by* a group, this procedure is not to be confused with *group selection*, a system which may often be found useful in choosing senior staff whose successful job performance depends largely on social skills and similar leadership qualities.

Group selection procedures invariably include situations specifically designed to test these qualities. The two most frequently used situations are a group discussion and a group exercise or project involving about eight candidates at a time. The activities of the members of the group are observed by trained interviewers using rating scales of quality and quantity of participation. Individual interviews, most often by a 'panel', supplement information obtained by these methods. Further information may be obtained from the N.I.I.P. (Group Selection Procedures, N.I.I.P. Paper No. 5, National Institute of Industrial Psychology, London).

THE INTERVIEW

By way of introduction to the next chapter, some brief remarks are given about the interview. As with objective tests, the interview is a valuable tool in the hands of those trained in its use, and used in the right conditions. It is a means of supplementing the evidence supplied by the application form, by reports, references and certificates, and by objective tests. It will supplement not only in the sense of establishing new evidence, including the subjective impressions and judgments of the interviewer, but also in the sense of relating and explaining any discrepancies between items of evidence. The interviewer(s) will seek to gather up all the parts into a coherent whole, so that the individual being assessed can be looked at side-by-side with the *personnel specification.*

Selection is essentially a *prediction* process. It may be short-range, for the current job, but quite frequently it must also be concerned with longer-range judgments relating to promotability and future manpower succession. Short-term or long-term, selection decisions are always based on judgment of probabilities from the evidence available. There will inevitably be margins for error, but they can be reduced significantly if the collection of evidence is skilled and thorough and the judgment is well-informed. Thus, staff employed on selection work—those who collect the evidence and make the judgments—themselves need proper training.

REFERENCES

1. RODGER, A., *The Seven Point Plan*, N.I.I.P., London (1952)
2. ANSTEY, E., *Psychological Tests*, Nelson (1966)
3. VERNON, P. E., *Society: Problems and Methods of Study* (Ed. A. T. Welford), 61, Routledge and Kegan Paul (1962)
4. FRISBY, C. B., *et al.*, *Tests for Engineering Apprentices in British Industry*, N.I.I.P., London (1959)

BIBLIOGRAPHY

ALBRIGHT, L. E., *The Use of Psychological Tests in Industry*, Munksgaard, Copenhagen (1964)

BRITISH INSTITUTE OF MANAGEMENT, *Survey of Selection Methods in British Industry, B.I.M. Information Survey 108*, B.I.M., London (1964). (This is a confidential document available to collective subscribers only)

EYSENCK, H. J., *Uses and Abuses of Psychology*, Pelican (1953)

FRASER, J. M., *A Handbook of Employment Interviewing*, MacDonald and Evans Ltd. (1950)

INSTITUTE OF PERSONNEL MANAGEMENT, *Selection and Placement*, MacDonald and Evans Ltd. (1949)

LOCK, H. F., *Use of Tests in Selection Procedures*, B.I.M., London (1956)

MANDELL, M. M., *The Selection Process*, American Management Association Inc., New York (1964)

ORGANISATION FOR EUROPEAN ECONOMIC CO-OPERATION, *Job Analysis—A Tool of Productivity*, H.M.S.O. European Production Agency, London (1956)

NATIONAL INSTITUTE OF INDUSTRIAL PSYCHOLOGY, *Group Selection Procedures* (*N.I.I.P., Paper No. 5*), N.I.I.P., London

RODGER, A., and CAVANAGH, P., *Society: Problems and Methods of Study* (Ed. A. T. Welford), 217, Routledge and Kegan Paul (1962)

ROFF, H. E., and WATSON, T. E., *Job Analysis*, MacDonald and Evans Ltd. (1961)

SIDNEY, E., and BROWN, M., *Skills of Interviewing*, Tavistock Publications (1961)

SUPER, D. E., and CRITES, J. O., *Appraising Vocational Fitness by Means of Psychological Tests*, Harper and Row (1960)

VERNON, P. E., *Personality Tests and Assessments*, Methuen (1953)

6

Interviewing Techniques

A. G. BEVERSTOCK, B.SC., M.A., PH.D., C.ENG., M.I.E.E.

Education and Training Officer, Southern Electricity Board

INTRODUCTION

Interviewing is a difficult process—partly a science—partly an art. It is a combining of the subjective and the objective.

Subjective analysis, from the impact of the individual personality at the interview, has to be co-ordinated with the objective facts obtained by other means. These have to be selected, amplified or modified and synthesised. In this way the attempt is made to arrive at a balanced judgment on the suitability of a candidate for an appointment, or on the potentiality of an individual for training for a particular job. Alternatively, the objective may be to give guidance on career choice or on some related development activity for the individual.

The purpose of this chapter is to show that successful interviewing cannot be achieved without considerable care and planning, and that it should never be treated in a casual way. 'To spend a lot of money and time choosing a machine of which the average working life is, say, ten years, and little of either on selecting men and women who may be with the firm for forty years is short-sighted, to say the least, of industry and commerce.' (Sir Miles Thomas)

THE NEED FOR TECHNIQUES

The interview, therefore, demands a systematic approach, not only to achieve the best judgment possible, but for economic and scientific reasons. The interview itself costs money, as do the preliminaries, such as press advertisements. In these days of severe competition, substantial sums can be involved.

Secondly, studies of the unsystematic interview indicate that its dependability is little better than would be expected from chance considerations. The reliability of the unplanned interview, that is, the degree to which the same results would be repeated under as nearly similar conditions as possible, is low. The validity is the degree of correlation between the results of the interview and the subsequent performance of the individual in the tasks for which he has been selected. It is limited by the 'dependability', and is usually lower. Hence unplanned interviews have low validity.

Thirdly, the interview is regarded as an important aspect of the modern concept of 'communication'. It can be regarded as an upward form of communication as, for example, in the employment interview where information is obtained from the candidate. It is an example of downward communication when a manager discusses progress with a trainee. Moreover, the interview is often the first 'live' contact a person has with the organisation, and the applicant forms an impression of the firm which he is seeking to enter by the way in which the interview is conducted, and the bearing and performance of the interviewers themselves.

The importance of the interview, together with the essential aspects of the interview process itself—the corroborating and supplementing of facts, the checking of basic facts, the linking of these with the personality of the interviewee and the co-ordination of the whole in a final judgment —requires, therefore, a system which enables the purpose to be achieved in an efficient and effective way. Such a system demands the application of techniques. The techniques, however, need not necessarily be involved.

THE PLANNED INTERVIEW

Proper planning is the key to successful interviewing. The planned interview includes:

1. A carefully selected series of questions. Where there is a selection panel, each member should be allocated a group of questions appropriate to his special knowledge and experience.
2. A list of items on which judgments are required.
3. A planned assessment procedure of the type outlined below.

Clearly, the range, level and content of questions will be designed according to the particular requirements for which a person is being interviewed. The following scheme illustrates, for example, the type of questions which might be asked of applicants for craft and engineering apprenticeships.

(a) Arising out of school experience

1. Size of school, i.e., numbers.
2. School or class positions held.
3. Amplification of application form detail given, concerning games played.
4. Subjects in which most interested and why.
5. Subjects in which most successful and why.
6. Short description of science experiments done by candidate.

(*b*) *Arising out of leisure pursuits*

7. Books, papers, magazines read during last six months.
8. Hobbies—care is needed here to ascertain degree of interest and success in the hobby given.
9. Visits to exhibitions or places of interest.
10. Club and social activities.
11. Holidays—visits to other towns, abroad, etc.

(*c*) *Arising out of technical knowledge*

12. Simple mechanical and electrical terms, e.g., watt, horse-power, force, ohm etc.
13. The use of simple tools.
14. Properties of materials.
15. Information concerning mechanical devices.

(*d*) *Arising out of career literature*

16. Questions to test the extent to which the candidate has obtained information about the firm or organisation.

A similar question plan can be devised for selection purposes at any level but must be drawn up in relation to the assessments required. The ultimate aim of questioning is to assist with the problem of making judgments and a primary task, therefore, should be to decide the items on which assessments are to be made.

ASSESSMENT PROCEDURES

It is common practice for such items to be arranged within the design of the Seven Point Plan devised by the National Institute of Industrial Psychology and dealt with in some detail in Chapter 5. It has been found practicable, however, to use three main groupings which have certain advantages in day-to-day work, where, as with most interviews, time is an essential factor. These are:

A. Acquired knowledge and experience.
B. Personality features including disposition and character.
C. Special assets which include intelligence, aptitude and interests.

The assessments required in Group A can be made with reasonable accuracy and special tests assist to aid the assessment of items under Group C. Group B presents the greatest difficulty. Although personality tests exist, their validity has yet to be proved, and they are still regarded in this country with justifiable suspicion.

It is helpful to an interviewer when making personality assessments to know a little about certain aspects of individual psychology.

Personalities can be grouped under four headings—extroversion, introversion, aggression and sociability. Some people will be especially strong or weak in one or more of these groups. Others will be composed of shadings from all or some of the groups.

Each group contains strong and weak features. The extroverted individual will be enthusiastic, friendly, self-confident and persuasive, but tends to be impulsive, impatient and lacks analytical powers. The introverted person is imaginative, meticulous, patient and has powers of analytical thought, but usually lacks confidence, and may be unduly sensitive and self-conscious. The strongly aggressive person will be dynamic, decisive, extremely self-confident, but may be ruthless, egotistic, intolerant and tactless. The person with positive sociability is usually selfless, unassuming and fond of people, but may lack mental toughness and the power of critical thought and may be impractical and even naive due to a strong desire to see only the best in people.

In making personality assessments, it will aid the interviewer if the above headings are borne in mind. In drawing up the specification for a job or a training scheme, an attempt should be made to weight the groupings in relation to the personality features required. Consider, for example, interviewing for a management training scheme. Restricting considerations to personality features, those that are being sought are those that are considered to be basic personality features required in a general manager, some of which may be developed under the scheme.

Opinions will vary but the following list of management skills in rank order of importance is given for purposes of illustration:

1. Ability to understand and handle people.
2. Good powers of analysis.
3. Ability to make rapid decisions.
4. Decisiveness.
5. Ability to inspire others.
6. Fair-mindedness.
7. Good organising ability.
8. Tact and tolerance.

The feature of high intelligence, and technical and verbal ability are assumed.

Features 1, 6 and 8 are characteristic of a person with strong social drive; 3 and 4 are characteristic of a person with a positive degree of aggression; 5 and 7 are possessed by the extrovert, and feature 2 is a quality of the introvert.

In selecting the manager an individual is being sought who has strong social drive yet possesses the aggressive qualities and strengths, but not the weaknesses, of the extrovert.

While any personality assessment is open to a considerable margin of error, a planned approach to the problem similar to that given is likely to yield results nearer the mark than overall impressions or 'hunch' judgments.

RATING SCALES

Rating scales can be of value, especially when the time available for interviewing is limited. The relative merits and profiles of candidates can be read at a glance and the scales help to channel the thoughts of assessors along clearly defined lines. By this means the interview is made a little more objective. A very useful form of rating scale has been recommended by the Interview Study Group convened by the Royal Institute of Public Administration. Ratings are made on a five point scale as in the following example for two items:

PLACE A TICK IN THE APPROPRIATE BOX

X	*X applies*	*Tendency to X*	*Average*	*Tendency to Y*	*Y applies*	*Y*
1. Good bearing and address						Unimpressive in bearing and address
2. Shows strength of character						Easily swayed

Rating scales should be simple in design, easy to fill in and the items should be clearly defined without ambiguity and overlap. They can be used for qualities which can be assessed at the interview, or for drawing together in a compact form assessments made from other data, including testimonials, interview form information, aptitude and intelligence tests.

Where rating scales are used for performance appraisals of existing staff, assessors should be left in no doubt about:

1. The special significance of the task.
2. The need for a constant effort to achieve objectivity.
3. The ways in which poor assessments highlight the assessor.

NARRATIVE REPORTS

When time permits, a report based upon notes made at an interview, and on information obtained from the application and interview forms, and any other assessments, can be an aid in reducing the subjectivity of the interview. The narrative report should be written, or a pen-sketch made, immediately after the candidate has been seen. There is no rigid pattern; an arrangement under the following headings has been found to be helpful in selection for training schemes:

1. Previous record—comments on degree of knowledge, attainments and outstanding achievements.
2. Personality—a brief description of disposition and character.
3. Potentiality—an assessment of the degree to which the candidate is suitable for, and likely to benefit by, the training scheme, coupled with his potentiality for the job for which he will be trained.
4. Profile—a brief note of assets and liabilities, i.e., strengths and weaknesses.

INTERVIEW METHODS

There are three main methods of arranging interviews—the panel or board method, the group method and the rotation method.

PANEL OR BOARD METHOD

The Panel or Board method is long established and is still widely used. Research studies have shown, however, that its reliability and validity are not high. It is time consuming and expensive, especially so for senior members of staff. For the candidate, the panel constitutes an artificial situation. It can, however, lessen the subjectivity of a single interviewer and it does allow different departmental interests to be represented. The most effective number of members is three to five.

GROUP METHOD

'Group Selection Technique' is the procedure in which group exercises and projects take place for all candidates. It is supplemented by intelligence and other tests, coupled with normal panel interviews. It can also refer to the interviewing of a number of applicants at the same time. Group sessions are useful for assessing a candidate's ability to handle ideas, his power of logical thought and expression, leadership tendencies and degree of reaction to stress conditions. The method usually yields a greater correlation between assessors' ratings than in the normal panel interview.

Group techniques have been found to be of value in selection for senior training schemes, e.g., staff development, managers, sales executives and administrators.

ROTATION METHOD

In this method, a single candidate is interviewed in rotation by two or three single interviewers. Each interviewer deals with a selected area of questioning, e.g., if there are three interviewers, one might deal with technical matters, the second with personality assessments and the third with special aptitudes. The method has the advantage of being natural and informal, the risk of personal prejudice on the part of a single interviewer is reduced, and the influence on judgment by other interviewers, as in the panel method, is avoided. The candidate has a more favourable opportunity of presenting his claims than in a single interview before a panel.

THE CARDINAL POINTS OF INTERVIEWING

The actual process of interviewing is highly subjective and there cannot be any rigid rules for a process of this kind. The following points will help in making the interview more effective and valid.

1. The main aim of an interview is to secure the co-operation of the candidate and he should always be shown consideration and courtesy. Avoid cross-examination. This is particularly important in the interviewing of apprentices and young people.
2. Always check identity and domestic state of the candidate.
3. Use a carefully planned procedure.
4. Do not be biased by preconceptions or general assumptions. For example, there is no statistical evidence that a jutting chin indicates determination, or a high forehead denotes high intelligence.
5. Always have the 'halo effect' in mind, i.e., the overall assessment of personality may be dominated by one outstanding and strong characteristic.
6. Good questions are direct, short, simple and open-ended. They are not ambiguous or vague. They do not give a clue to the answer expected. They invite the candidate to talk.
7. Always retain humility in interviewing. The interviewer who says he can 'pick his man' purely on the basis of experience is likely to be the one whose judgment is to be mistrusted.

BIBLIOGRAPHY

ANSTEY, E. and MERCER, E. O., *Interviewing for the Selection of Staff.* Allen and Unwin (1956)
MUNRO FRASER, J., *A Handbook of Employment Interviewing.* MacDonald and Evans (1950)
FEAR, R. A., *The Evaluation Interview.* McGraw Hill (1958)

7

Job Analysis

G. A. RANDELL, B.SC., M.SC.

Senior Lecturer in Psychology at the Management Centre of the University of Bradford

Many management and personnel problems arise because jobs at all levels have not been carefully defined and had their work objectives clearly stated. Frequently such problems can be alleviated by creating a purposeful, comprehensive and precise description of the work to be done, and communicating this information to the member of staff who has to select and direct the person to do the work, or passing it directly to the person appointed. In practice, it is not usually as simple as that sounds, for it is a demanding task to *create a purposeful, comprehensive and precise job description*, and then to develop this into systematic selection, training and management procedures.

TYPES OF JOB ANALYSIS

The analysis of a job to provide a useful description of it is a key technique in the world of work. There are different approaches to the methods of job analysis depending on the purposes to which the resulting information is to be put. These purposes include:

1. Helping the job choice of an individual—in a vocational guidance situation.
2. Deciding whom to choose—during a personnel selection procedure.
3. Developing a training scheme.
4. Defining how work has to be done after applying methods study and equipment design.
5. Depicting the type and form of the structure within which the work will take place by means of organisation analysis.
6. Determining the worth of each part of a job by job evaluation.

The main difference between the various approaches is one of emphasis. The inter-relationships between the resulting job descriptions are demonstrated when a systems approach to industrial behaviour is taken[1] when all procedures are seen to be aspects of the process of arranging for the optimum use of manpower, for maximising both output and job satisfaction. This chapter deals with the methods and techniques for the analysis of jobs for the purpose of providing a description of the work to be done so that the individual knows how and what he has to achieve. If he already has the necessary knowledge and skills then he should be able to do the work once he has read the job description. If he has just the aptitudes and inclinations for the work, then from the job description appropriate training can be planned. The preparation of this type of job description is one of the most important tasks of a training officer. It is also a demanding and useful training exercise for management trainees.

The essential point to note here is that it is implicit that considerable prior knowledge exists about the job that has to be done, either by the individual concerned or by the person responsible for training him. This means that for training purposes the job description need not go into the considerable detail that a methods specification requires, nor have the form or inferences of a job specification for selection purposes. Although a job summary is required, what is needed in a job description for training is emphasis on those aspects of the job that are unique or not necessarily noticed by those ordinarily concerned with the job. Emphasis should be on those aspects of the job which are known to be critical to success or failure, or to enjoyment or frustration.

The underlying principle behind writing a job description for training purposes is emphasis of those aspects of the job which are trainable and which are critical determinants of success or satisfaction. When this information is combined with an outline of the scope and objectives of the job, a purposeful, comprehensive and precise description should result. Many methods have been developed to achieve these aims. There are available check-lists and statistical approaches that have the advantage of comprehensiveness, but lack purpose and precision. There have been notable attempts at adding purpose, by Flanagan,[2] Rodger and Cavanagh,[3] and Miller.[4] The following method is a synthesis of previous approaches; it attempts to be comprehensive as well as realistic, and it lends itself to adaptation to the specific needs of different organisations.

A METHOD OF JOB ANALYSIS

STAGE 1: Observations of a fact-finding kind

Find out and record:

1. Exact job title, and number of people doing the job.
2. Main routine tasks, and occasional tasks, using a 'what he does', 'how he does it' and 'why he does it' approach, with an indication of the amount of time spent on each task.
3. What the work should achieve, its main and subsidiary objectives.
4. Minimum standards of work, and desired standards, including details about acceptable and unacceptable faults.
5. Responsibilities for the work and safety of others, and for equipment and materials.

6. Aspects of the job that cause difficulty or distaste.
7. Ways in which job is likely to change, in particular, the possible use of new equipment or work methods.

STAGE 2: Inferences by training and subject matter experts

Infer and record:

1. Essential knowledge required to do the job.
2. Background knowledge useful for providing a broadly based competence, or creating and maintaining interest in the work.
3. Skills necessary to do the work. Such skills could be motor, perceptual or conceptual and can only be inferred from detailed observation and skills analysis, for example, of the kind described by Seymour.[5]

STAGE 3: Assumptions made during stages 1 and 2

Make explicit and record:

1. Assessments made about the people entering the training in terms of their attainments in knowledge and skills, aptitudes, and ability and willingness to learn.
   This information is usually arrived at in consultation with the staff responsible for selecting individuals to be trained. The importance of such information emerges when decisions have to be made about the length of the training course, and the need for special training aids to facilitate learning. Also the type and level of abilities of individuals to be trained does not remain constant over time so it is often necessary to change the length and content of training in accordance with changes in the abilities of the entrants.
2. Details of the facilities, tools, equipment and materials available to do the work. This information is usually arrived at in consultation with line or production management. It is important to observe and record this so that the tools, equipment and materials used in training match those to be used on the job. If trainees learn on equipment different from that which they are going to use they may find it difficult to transfer their new skills from the training to the on-the-job situation. Also the asking of questions about facilities, tools, equipment and materials often focuses attention on the need to apply ergonomic principles to the design of jobs, so making it easier to train people for them.

When following the above method, the job analyst has the problem of judging how much detail he is going to record. If it is essential for the job to be performed in a standardised way or to a high level of precision, then considerable detail is required. If the work is not so well-structured or rigid, then little detail is required, or even desirable.

To follow the method described, various techniques need to be applied. Such techniques are not only necessary to get at the required information, but also are part of the need to gain support and acceptance for the methods and outcomes of job analysis. Co-operation is essential if training is to be a success within the organisation, and this is needed not only from management, but also from the employees involved in the job study and their supervisors and representatives.

TECHNIQUES FOR JOB ANALYSIS

Various techniques can be used to gain the information necessary for writing a job description for training purposes. The obtaining of information is not the only purpose behind these techniques; they are also aimed at gaining the co-operation of all those who are likely to be concerned with the outcomes of the training scheme. To achieve this may mean collecting more information than is really necessary for the immediate purpose of the job analysis. For the sake of good relations within an organisation this burden should be accepted by the job analyst, even though it may add considerably to his task.

The techniques that are used in any particular job analysis depend very much on the type and level of the job being studied and the precision of training required. For reasonably rare jobs, for example, of a management or supervisory kind, painstaking probing techniques are applicable; for well-populated technical jobs, like assembly operators, systematic surveying techniques can be used. The main techniques are dealt with below.

INTERVIEWING

The collection of data by purposeful face-to-face conversations is a skilled task, and requires training and practice to gain competence at it. Although approach and manner are of importance, the essential difference between an effective and ineffective interview is the point and form of the questions asked. Knowing what to look for is the prime requisite of a good interviewer. Stage 1 of the method described earlier can be used as the framework of job analysis interviews to give form and purpose to the questioning.

QUESTIONNAIRES

When a large number of people are involved, it is not always possible to interview them all, but it may be important to give all of them the opportunity to provide information to the job analyst. In such a situation, a questionnaire is a useful tool. This should be developed from a sample of pilot interviews, structured on the basis of the method described, and containing questions found to be the most relevant and revealing about the job being studied. Once again questionnaire design is a skilled task and considerable care has to be exercised to use this technique effectively.

GROUP DISCUSSIONS

A quick way of obtaining information from a group of employees is to get them together in a discussion about the job. Such an approach is particularly useful with a group of leavers for gaining insight into the difficulties and distastes of the job. If a controlled and congenial climate is established, it is possible to obtain considerable detail about critical aspects of a job in a relatively small amount of time by this technique.

WORK STUDY

For highly repetitive jobs, the various techniques of methods study can be useful. Process charts, pre-determined motion time systems, activity sampling and high-speed cine films have their uses, but if used indiscriminately, they provide a large amount of relatively irrelevant data for the training job description. Of course, if it has already been collected for the purposes of work study, then it can be used to provide useful detailed information.

SKILLS ANALYSIS

The more appropriate of the detailed techniques of job analysis for training purposes are those of skills analysis. In the hands of specialists, these are powerful techniques, and it is often essential that they are applied before really effective training can be developed, especially in the case of machine shop and assembly tasks. They entail a detailed analysis of all the information that an experienced worker uses when carrying out a task, using a sensori-motor process chart rather than just a methods analysis for a two-handed operator process chart. The importance of these techniques to job analysis has increased following the trend in industry towards skills with a high perceptual or decision-making content. Also, research has shown that even in apparently completely motor tasks the difference between a skilled operator and a learner is the ability the former has to carry out his work with the minimum of information. A skilled man learns to ignore information, as well as to attend to it.

SOURCES OF INFORMATION

Techniques of information collecting can, and should, be applied to a wide range of sources. It is not advisable to base a job description on information from a single source, even though there are frequent pressures for this to be so. The job analyst must decide how much and how wide his observation should be. The following are the main sources of information that can be tapped.

MANAGERS

An interview with the departmental head is often the start of the data collection stage of a job analysis. The information collected should indicate future changes in the job, and the manager's expectations about how the work is to be done and his ideal standards. Essential information about objectives should be obtainable from the management.

SUPERVISORS

The main source of information about the acceptable ways of doing the job, and about standards, is from those members of staff directly responsible for getting the work done. Essential background information, an overall account of the job and data on the equipment and facilities can also be best obtained from the immediate supervisors. Care should be taken when noting the supervisors' accounts of methods of doing the work, for it is probable that one of the reasons why they were appointed supervisors was because they were experts at the work, hence their ways of doing it may well be different from the ordinary workers. Also methods may well have changed since they last performed the tasks.

INCUMBENTS

The job as performed by a range of existing employees should be analysed. Essential differences in methods, as well as in difficulties and distastes about the job, may be found to exist between the experts, the experienced and the recently trained. It is also useful to study workers judged 'good performers' by the supervisors, and compare their methods with those judged 'less good'. At all levels of employment, the whole range of techniques of job analysis are applicable to collecting information from the incumbents of the job. Thus job analysis techniques accord in this respect with recent trends for target and objectives setting in consultation with the individual who carries out the job.

SPECIALISTS AND MEMBERS OF ASSOCIATED DEPARTMENTS

If other specialists have also studied the job then their reports are a useful source of information, e.g., such specialisms as methods study, organisation and methods, operations research, rate-fixing, job evaluation and computer systems analysis. Also any information about the job from any associated inspection, quality control or test department sources should be utilised. These departments will be able to provide details of faults in job performance and so enable a faults 'museum' to be built up.

LEAVERS

One of the most fruitful sources of information about the difficulties and distastes of a job is from those just about to leave it. A carefully controlled, but relaxed, group discussion held with them by a perceptive job analyst, in a comfortable room away from the office or shop floor, can provide critical information for the job description and can lead towards more purposeful training being given. (Thus reducing labour turnover.)

PERSONAL EXPERIENCE

It is often useful for the analyst to attempt to do the job himself. This has limited value but it may ensure a better understanding of what is involved and bring to his attention things he may have missed out in his other observations.

WRITING THE JOB DESCRIPTION

Once the analysis has been completed and all the items of information have been assembled, there remains the task of writing the job description. Various attempts have been made at short-cutting this vital and onerous part of the study, either by check-lists or other pre-planned note-taking forms. These are not recommended for it is rare to find any job fitting neatly into a general purpose scheme, and usually it is neither possible nor advisable to assemble all rough notes on a single form. Within any organisation there is a need for standardisation of procedures, as an aid towards comprehension and communication with colleagues. Of course, the final description can conveniently be made on pre-printed forms, as long as these forms allow for flexibility rather than encourage rigidity of description. But if training is to be successful, every job should be regarded as unique, so for training purposes any procedure which makes jobs appear similar, rather than accentuates differences, should be guarded against.

The form of the written document for each job should ensure that the resulting job description provides a whole picture of the job and is readily comprehensible. Tasks can be conveniently listed in chronological order of performance if they are sequential or cyclical in nature. Alternatively, they can be listed in order of importance, by subject matter or in any other way which accords with logical arrangement.

The amount of detail to be written depends on the type and level of the job studied, and the precision of performance required. For example, if programmed instruction techniques are to be applied considerable detail will be required. Mager[6] makes some useful comments about the need for using precise expressions when describing jobs prior to writing programmes. Gagné and Fleishman[7] set out an approach for recording detail about technical jobs, under the headings 'activity', 'performed on', 'tools' and 'specific skills'.

In general, short, crisp expressions are the most useful kind of prose to use when writing a job description, preferably each phrase should begin with a verb in the third person singular of the present tense. Verbs should be selected with care, so that the maximum amount of specific meaning is conveyed. Qualifying clauses beginning with 'may' or 'occasionally' can be used to indicate that not all employees doing the job perform that task, or a task not customarily performed by any particular employee.

The outcome should be a written description of a job with emphasis appropriately placed on training aspects. From this document a syllabus can be derived, a course structure and time-table planned and the use of training aids decided upon. The document should also indicate what terminal training tests would have to be developed to enable the attainment of the individual trained to be measured and the effectiveness of training evaluated.

CONCLUSIONS

MEASURABLE CRITERIA

One of the main outcomes of the approach to job analysis described in this chapter is the opportunity it provides for developing measurable criteria of performance, both for end-of-training reporting and for subsequent job performance. Such criteria are implicit in the data collected during the analysis. It would be an essential responsibility of those concerned with training to check the validity of the job description by developing an attainment test based on the information contained within it and applying it both to samples of job incumbents and to people who have just completed their training. It is also important to repeat this exercise at regular intervals of time. If this is done, not only will evidence accumulate as to the effectiveness of the training, but also about its continuing relevance to the job that has to be done. It is highly probable that the job description will have to be changed frequently, to keep it in line with current practice.

FLEXIBILITY

The procedures for job analysis described in this chapter may appear to be over-elaborate and unnecessarily tedious. It is, of course, not always necessary to follow all that has been suggested, for it is part of the skill of a highly experienced job analyst to be able to exercise judgment about the amount of detail required. Nevertheless, job analysis is a demanding and time consuming task. As job descriptions accumulate, common features between jobs will emerge, making it possible for a job classification scheme to be developed. Although such a scheme will have limited use for reducing the work of job analysis, for all jobs are unique, it would help in the preparation of job descriptions for jobs that do not yet exist. Such 'synthetic' information, coupled with data from predetermined motion time analysis, would make it possible to make a start on a training procedure for the new jobs. However, a full job analysis should subsequently be undertaken for the job, as soon as it has been going long enough to provide the necessary data.

THE CONTINUING NEED

Clearly there is work for full-time job analysts in every medium or large firm. Some firms employ full-time specialists, others rightly consider that rotational training as a job analyst is itself an invaluable training exercise. Firms attempting to implement management by objectives usually find that full-time analysts are an early priority. Programmed learning, by its very nature, presupposes a thorough and detailed job analysis. Other personnel management and manpower planning functions require job analyses to be undertaken before job or person specifications can be prepared.

Job analysis is an essential function of management. It is a prerequisite not only for training in manual skills, but also for training in all occupations, including selling, supervisory and managerial posts. Without it there can be little hope of ensuring high standards of training or valid criteria by which to measure and evaluate subsequent job performance.

REFERENCES

1. RANDELL, G. A., *Occup. Psychol.*, **40**, 115 (1966)
2. FLANAGAN, J. C., *Psychol. Bull.*, **51**, 327 (1954)
3. RODGER, A. and CAVANAGH, P., *Society: Problems and Methods of Study*, (Ed. A. T. Welford), 217, Routledge and Kegan Paul (1962)
4. MILLER, R. B., *Training Research and Education*, (Ed. R. Glaser), 31, John Wiley, New York (1962)
5. SEYMOUR, W. D., *Industrial Skills*, Pitman (1966)
6. MAGER, R. F., *Preparing Objectives for Programmed Instruction*, Fearon, San Francisco (1961)
7. GAGNE, R. M., and FLEISHMAN, E. A., *Psychology and Human Performance*, Henry Holt, New York (1959)

8

Initiating, Planning and Time-Tabling of Training

JOHN W. BARBER, M.A. (OXON.)

Senior Lecturer in Management Studies, Portsmouth College of Technology

INTRODUCTION

The aim of this chapter is to present in practical terms the basic essentials for ensuring and initiating effective planning and time-tabling of training. Before the 1964 Act, even in the best of companies, training has sometimes had little more than lip service. Even today it is often only the men at the top who agree among themselves about the importance of training. General policies are agreed. People are appointed to implement the policies. The word is passed down the line that training is essential and must be carried out. Yet when production pressures build up, training often suffers. The immediate needs of production are understandably considered by line managers to be more important than the implementation of training.

It is not the purpose of this chapter to propose new solutions to such old problems as management organisation and communication. Conflicts and pressures between departments may be inevitable. Line management and staff functions which ought to be complementary often do seem to work for their own ends. If this is taken as the starting point, then the real problem of initiating, planning, and time-tabling of training, becomes a question of how best to operate *within* this kind of situation.

The topic will be discussed under the following main headings:

The need for a practical training policy.
Old concepts and new.
How to diagnose needs and priorities.
How to initiate and co-ordinate.
Assembling a master plan.
Time-tabling.

THE NEED FOR A PRACTICAL TRAINING POLICY

AS YET MANY FIRMS HAVE NO POLICY

If the great majority of British firms had had effective training policies, it is almost certain that the Industrial Training Act would never have reached the Statute Book. Now, as a direct result of the training levy (the 'blunt instrument'), many firms have suddenly had to face the need for a training policy. If an engineering firm employs 2,000 people whose total emoluments average £1,000 per head per year, it will be confronted with an annual training levy of £50,000. Any managing director faced with an extra charge of £1,000 per week will quickly awaken to the need to secure something in return for it. If he already employs a personnel officer he will probably delegate responsibility for training to him.

Overnight the personnel officer is expected to become expert in a new field. If he is lucky, he may secure an assistant who will become his training officer. If there is no personnel officer, the firm may either decide to give someone else (not always the most suitable person) a part-time responsibility for training, or it may appoint him on a full-time basis. Alternatively, the firm may appoint someone from outside. Not even the last eventuality is, however, any guarantee that a genuine training policy exists. Hitherto, senior management has frequently not been sufficiently concerned about the implications of a training policy. No one has had time to think through the situation. This is because it is difficult to know where to begin, and because expert knowledge is scarce.

POLICY MUST BE REALISTICALLY FORMULATED AT BOARD LEVEL

In many firms, training has never achieved independent status as a staff function. The reason for this is simple. The policy makers often say that if the personnel manager is responsible for industrial relations, welfare, recruitment and training, he is then in a position to integrate them effectively. This is the theory. The vital question is—does he do so? In practice, industrial relations usually dominate. Wage negotiations, terms and conditions of employment, rate fixing, disciplinary cases and threatened disputes cannot wait and are usually given priority. Training *can* wait.

Faced with choosing between immediate and future needs, the personnel manager like the production manager usually (and naturally) puts immediate needs first. Furthermore, he is often consoled by the thought that no one will or would challenge the rightness of his decision. In these circumstances a good case can often be made for combining recruitment and training into a separate function. The fault in all this is not with the personnel manager but with policy makers at board level. It is there that separation of functions and equality of status should be determined more realistically.

Many of the less large firms are only now beginning to realise the true importance of training. Their directors are often both bewildered and impressed by the documentation emerging from the industrial training boards. The impact of this is having an effect and many directors are at last realising the tremendous possibilities of a far-sighted training policy. They are beginning to see too that policy on training must be realistically formulated at board level, and then clearly communicated throughout their organisations.

POLICY MUST BE BASED ON AN ACCURATE ASSESSMENT OF TRAINING NEEDS

The difficulty about any attempt to formulate training needs is that this demands a searching analysis. This analysis is not one which can be undertaken by a functional planning expert. It is one which demands the attention of all the functional experts and all the top managers as well. It requires a detailed investigation of the future development and production targets of the entire organisation. This forward planning is not something to be tackled spasmodically. It is a process which ought to be going on all the time. The extent of forward planning will vary according to present product, envisaged product, new techniques, required changes, modification or renewal of existing equipment, new materials to be used and the extent of new capital investment. All of these factors have to be studied, assessed, valued economically and synthesised into a master plan for future action. To decide on such a master plan for immediate and long term strategy is the function of every company's board of directors.

Yet greater in importance than all of these considerations is the need to analyse and evaluate the extent to which the firm's human resources will be likely to match the needs of the changing situation. This activity should provide the dynamic of every master plan. It is the main ingredient of an effective manpower policy. From it stem many questions. What new tasks and what new responsibilities have to be undertaken? What old skills have to be disregarded? What new skills have to be learnt and deployed? In the past, far too little effort has gone into finding accurate answers to these questions. It is not enough to agree manpower statistics in round figures and leave it at that. Someone at board level must be responsible for an accurate assessment of training needs.

INITIATING PLANNING AND TIME-TABLING MUST BE BASED ON PRIORITIES

One of the problems in training is to establish the priority of specific needs. Who, for example, is to say that before everything else all the foremen must be trained in good supervision? How many other needs have been identified which are equally important and equally pressing? If new skills, new methods, new techniques and new appliances are all to be employed in the next two years, will this affect people at all levels, and if so, where in this area do priorities lie and do these have preference ever specific training in foremanship? How are the trainers to be trained; where are they to come from? If they are to be seconded from their normal work, what is to happen to it in their absence? What about middle management training? What about management development? What about safety?

There are scores of questions like these to be answered before priorities can be given to training needs. Add to this the conviction of nearly every supervisor in every factory that he is understaffed, and the fact clearly emerges that decisions on training questions are essential at the highest level of management.

POLICY MUST BE FIRMLY BASED ON THE PRINCIPLE THAT TRAINING IS LINE MANAGEMENT'S RESPONSIBILITY

Assuming that senior management has identified training needs, has formulated its training policy and has endorsed priorities, it may nevertheless happen that training itself either breaks down or limps along. The explanation is almost always the same . . . 'Sorry but the job must come first . . .' Once this behaviour is allowed to go unchallenged, it may quickly deteriorate into regular practice.

If, however, senior line management lays down that absence for training purposes is to be treated in the same way as absence due to sickness, then middle management and first line supervision will respond accordingly. Where, for example, senior management itself takes an active part in the training, it is comparatively rare for absenteeism to occur.

Conversely, if senior line management is not convinced that training is of vital importance, then this attitude will quickly be sensed by those at a lower level. It will be taken as endorsement of the belief by first line supervision that the immediate task is more important than training for the tasks ahead. Few line supervisors ever want to send a valued member of their team on a training course. It is therefore essential to adopt practical measures which place clear responsibility for training upon line management at every level. Before considering in more detail what these measures should be, it is necessary to re-assess two traditional concepts by comparison with the modern concept of training.

OLD CONCEPTS AND NEW

TRAINING THROUGH EXPERIENCE

Experience teaches. Practice makes perfect. These adages are partly true. Certainly we learn from experience, and on balance experience probably is positive in its impact. But for those who learn incorrectly in the first place experience may simply confirm errors. Similarly, practice, which is so important for the consolidation of learning, can also deepen the ingraining of bad habits.

Learning by experience has one very commendable feature. The individual faced with the task has to master it himself. At best, therefore, learning by experience can be a challenge—full of job satisfaction. Unfortunately this is not the normal outcome.

From this it is evident that though training by experience alone goes on all the time, it produces very uneven results. Experience alone is usually an uneconomic way to meet training needs, especially those deriving from rapid changes in technology. Planned training becomes essential.

The first challenge for the training officer and for line management is then to discover how to motivate individuals positively towards changes brought about in their day to day experience by new methods, new techniques and new technologies. The second is to design and implement effective retraining of such a quality that it converts reluctance to train into high motivation.

TRAINING BY 'SITTING ALONGSIDE NELLIE'

In many firms, the only method of training is to put the learner alongside an experienced worker. If the learner remains at the experienced worker's elbow long enough, it is believed that he will learn the job. It is also believed that all that is then necessary is for the learner to be given a reasonable period of trial by exposure to the job itself. In a changing technological situation, this type of training cannot be said to be efficient even where an experienced worker is available. Results would depend upon many factors such as complexity of the work, the learner's inate capacity, his interest in the work, his confidence, his previous experience and his attitude to the experienced man.

Equally significant are the attitude and ability of the experienced man. He may resent having to teach his skill to another man, particularly if the learner is much younger. He may jealously guard his own skills and avoid passing on the more subtle tricks of his trade. Usually he has been selected to teach the new man because he is considered good at the job. Even this is not always true. He may in fact be inferior in some aspects of the work. Finally, he may not only be reluctant to teach but may also be incapable of doing so.

The environment itself can also be a serious handicap. Bad lighting, noise, interruptions, pace of the work, all affect the learner. If the immediate work group is unsympathetic, then once again the chances are that the learner will not succeed. Finally, when he is put on his own, if there is no one to whom he can confidently turn to for help and advice, he may at that late stage decide to look for a job somewhere else. In practice, this is often a peak period of labour turn-over.

Despite these factors many firms still train in this way. For the training officer of the future, one problem will be to discern the best circumstances in which 'training alongside Nellie' may be economically justifiable, e.g., where skill content is small and numbers are few. For some firms, it will then be an essential training need for the Nellies to be trained in instruction techniques.

PLANNING OF INDUSTRIAL TRAINING

The modern concept is for training to be planned. Wherever new methods are developed, and whenever labour turnover, movement of staff or changes in responsibility occur, training is essential. There are very few firms which do not have training problems arising from these sources.

In planned training the basic concept is different from that of learning by experience or learning by working beside an experienced person. Planned training must be thought out beforehand. It culminates in a discipline imposed from outside and may for this very simple reason be resented by the individual who receives it. That is why, therefore, the first and most essential planning task is for the trainer to devise training which from the outset will capture and maintain the interest of the learner.

The training planner has to identify, define, analyse and make explicit, exactly what it is that has to be learnt, whether it is a mental skill, a manual skill or a combination of both. Only after careful analysis of the content of the task can he effectively prepare a syllabus. For this he usually needs the assistance of line experts in the subject. Together they then draft the contents of the course at a level which will match the capacity of the intended learner. Thoroughness in this will determine the important question of training standards. If standards are not high, time, money and patience are sure to be wasted and the interest of the learner will probably be lost.

HOW TO DIAGNOSE TRAINING NEEDS AND PRIORITIES

DISCUSSION AT ALL LEVELS

Many firms begin by assuming that immediate training needs are all important. The far-sighted firm recognises that future training needs are also equally important and lays down its policy accordingly. It may even place priority on future needs. If it does this, it cannot succeed without support for its policy at all levels of line management. Gaining this support is a question of good communications and strong administration. For planned training to be effective, management must ensure that regular discussions take place at all levels down to first line supervision. At every level meetings should be informal and brief. They should be sufficiently regular (at least once a week) to ensure that two-way communication is always maintained on a defined list of priority items. Training must be kept on the priority list for regular review under the following headings.

MANAGEMENT DEVELOPMENT

In this special area, whatever the size of firm, discussions on succession to senior posts should be from board level down to departmental head. The intermediate should be the local factory manager. There should be a continuing informal flow of information concerning recommendations on individuals to be trained for future development and promotion. For less senior posts, confidential lists should frequently be discussed and reformulated by the factory manager and his heads of department. Central and local training officers should co-ordinate this work.

Control and planning of normal development is often associated with individual staff appraisals. These appraisals should measure the extent to which the individual has achieved work performance targets. For everyone they should highlight particularly the areas for development training in the existing post. In addition, and separately with individuals of outstanding potential, recommendations should indicate special and immediately desirable management development training. The training officer should co-ordinate educational assignments and special courses to be attended.

STEERING COMMITTEES

The factory manager should also have a special meeting each month for an overall review of training items discussed at his weekly executive meetings and for progress reports on all training. Members of this steering committee on training should be the factory manager, his heads of department and the training officer. Each departmental head should be fed with information on training needs and priorities by each of his section heads and by any training advisers in his department.

This committee should review and endorse all training plans. Time-tables should be the subject of earlier scrutiny and agreement within each department. Advice on all outside courses should be provided by the training officer.

The committee should also assess and review management development recommendations.

DEPARTMENTAL RECOMMENDATIONS

In order to stimulate and develop training as a line responsibility, each departmental head should request specific training recommendations from his section heads and line supervisors. The procedure would achieve two ends. It would intensify each supervisor's awareness of the performance and job content of each of his workers, and it would also bring to light weaknesses in production methods or procedures. Where a department has a special training adviser of its own, he should be responsible for co-ordinating its training recommendations with particular emphasis on content duration and possible dates of proposed training courses.

In effect, this would amount to a form of departmental steering committee controlled by the head of the department and his training adviser.

FACTS ABOUT EXTERNAL EDUCATION AND TRAINING FACILITIES

Every training officer should have a thorough knowledge of the educational system and in particular the facilities offered by the local and regional technical colleges. He should also have a working knowledge of other facilities. He must advise his training steering committee on these aspects.

It would be ludicrous if a firm went to the expense of designing a course syllabus, preparing testing and evaluating the content of the course, and then running it, if all this had already been done and was available locally.

Most of the industrial training boards provide information about appropriate courses already in existence and of how far these are eligible for grant awards.

The field is bewilderingly wide. There are many hundreds of basic information course syllabuses already in existence and selection is therefore difficult.

Apart from producing graduates, the universities do not as yet make extensive provision to satisfy specific industrial training needs, although some university extramural departments are beginning to work in this field. Hitherto, provision made specifically for management training had been so slight that industry itself donated large sums towards the provision of the two new business schools. However, even when the Manchester and London business schools are fully operational they can only supply a fraction of the demand that will then come from industry and from the industrial training boards.

In the past, local and regional technical colleges have been the main sources of supply for industry (see Chapter 3). For many years, these colleges have provided local courses to satisfy national examining institutions such as the City and Guilds which itself offers syllabuses by the hundred for craft, operative and technician qualifications. There is also massive provision by the colleges for Ordinary and Higher National Certificates and Diplomas. From now onwards, in their own right, degrees too will be offered in certain colleges through the Council for National Academic Awards; and in some of the regional colleges, post graduate degrees in subjects such as management studies and industrial training technology will also become available.

Short duration courses designed specifically for industry are also provided by a variety of national organisations such as the following:

*British Association for Commercial and Industrial Education (B.A.C.I.E.).
*British Institute of Management (B.I.M.).
*Industrial Society (I.S.)—formerly Industrial Welfare Society.
Industrial Training Service (I.T.S.).
Institute of Production Engineers.
*Institute of Personnel Management (I.P.M.).
Institute of Office Management.
*Institute of Supervisory Management—formerly Institute of Industrial Supervisors.
National Institute of Industrial Psychology (N.I.I.P.).
Royal Society for the Prevention of Accidents (Ro.S.P.A.).

Other typical bodies which provide (mainly management) courses specifically for industry are foundations such as:

Ashridge College of Management.
Cranfield College of Aeronautics.
Henley Administrative Staff College.

*These are described in Part 5.

Roffey Park—particularly for human relations courses.
Production Engineering Research Association (P.E.R.A.).

Many local education authorities also provide further educational residential centres, at which short courses are available for industry on subjects such as communications and human relations in industry.

Finally, an increasing number of companies, private consultancies and agencies, is now coming into the field. All offer training courses and industrial training services to industry. Of these some of longer standing have already received recognition for grant award purposes by some industrial training boards. Certain private firms which run special training courses to meet customer requirements have also had their courses recognised.

Eventually many of the industrial training boards offering advice to firms through regional and district staff officers will themselves offer specific training courses to firms in their own industry. Meanwhile, their grants schemes as they emerge often provide a good indication of courses which are considered to be of the required standard.

THE CO-ORDINATING AND INITIATING ROLE OF A TRAINING DEPARTMENT

From the approach already outlined, it is obvious that training officers ought to be specialists in their own right. They need to know about all forms of existing training facilities. They must know also about the latest techniques and be able to distinguish them from the latest 'gimmicks'.

They must above all make themselves knowledgeable about every aspect of their firm's activities. Whenever the need arises for a tailor made course, it is they, through their capacity to analyse the content of a new or old task, who will be expected to advise the line expert on how to give instruction on it. Peripheral interest will not be enough. Wise management insists that their training officers are given every possible opportunity for deepening their in-company knowledge and experience.

Throughout the approach to planning and time-tabling the basic theme has been that line management supported by the training officer is directly responsible for training. Unless, however, the training officer can communicate with equal weight with line management, shop stewards and senior management, he cannot adequately fulfil his role. He has to be so expert at convincing other people of the value and importance of training that he must finally listen to them telling him what training (he has convinced them) they want. This is the point at which his detailed planning starts. If he is to place himself in a position where his help and advice are sought after and respected, he must:

1. Keep all training records up-to-date.
2. Assist line management to select appropriate individuals for training (by reference to records).
3. By discussion with line supervisors and section heads, determine the

training need and with their assistance draw up syllabus outlines.

4. By discussion, agree which of their specialists can be freed to prepare and give particular sessions.
5. Arrange with lecturers-elect the details of their sessions (including visual aids).
6. Where necessary, advise line management of where, what and when training priorities should emerge.
7. On management and supervisory training, be prepared to lecture or take discussion sessions.
8. On apprentice and semi-skilled job training, be prepared to initiate and carry out intensive job analysis with the assistance of line experts.
9. Be a source of information and advice on all appropriate industrial training boards.

It is this solid work of the training department which puts it in a position to draw up a master plan of training for the year ahead.

THE ANNUAL TRAINING PLAN

It is only by consistent, conscientious and integrated work in the line departments that a worthwhile master plan can be created. The training officer's function is to co-ordinate the plan. At all levels of line management, he must also act as a permanent but unobtrusive catalyst for stimulating ideas that contribute to the creation of the master plan.

Before the plan is ever 'submitted', the training department will know whether it is acceptable at all levels including the board of directors. Before it is submitted to the factory steering committee, relevant parts of it will have been discussed within each of the departments. Without this prior acceptance within the departments, the chances of ultimate implementation would be remote.

When finally drawn up, the plan will consist of a summary of agreed proposals for courses or other activities in terms of:

Reason for the proposal and its priority.
Anticipated numbers and level of those involved.
Costs and expected results.
Total duration.
Number and duration of sessions.
Outline of course or conference content.
Proposed dates and location.
Number of times to be repeated.

The assembled plan is often presented under headings such as courses for all departments:

1. Induction and information courses.
2. Management and supervisory courses.
3. Specific job training courses (for technical specialists and for administrative, clerical, maintenance and production staff).

In large firms, these headings where applicable are repeated under the specific training listed for each departmental course.

Combined bar chart diagrams, showing the total training plan in terms of each department's commitments, are usually included and enable the year's plan to be seen, at a glance, in terms of duration of courses, their possible dates and for whom they are intended. When the plan has been agreed at factory level, final sanction for its implementation is seldom withheld unless in content, manpower and costs it is out of line with any new and undisclosed overall policy.

With management development schemes, a separate plan is usually made, with the factory manager and training officer acting as co-ordinators. This is then discussed and sometimes amended at board level before being finally sanctioned. When both plans have been agreed, the heads of department would be fully informed. In turn, they would pass appropriate details of the master plan to their section heads and line supervisors for eventual implementation.

Assignments for management development training would normally be passed by the training co-ordinator to the head of department for implementation.

TIME-TABLING

After the master plan has been agreed, the training officer would normally carry out most of the final time-tabling and course programming. The master plan itself should state that agreed approximate dates should be kept whenever possible and that one month's clear notice of final details should be given to all lecturers and line supervisors and course members. With training courses for specific skills, if the task analyst were the training officer, he and the line expert alone would know exactly what the training would involve. It is, therefore, they who normally should agree the final time-tabling. With more general courses, involving several speakers, there are many questions to consider.

In general, time-tabling involves much more than the mere placing together of a topic, a time and a date. The content, form and administration of a course all subdivide into numerous issues some of which are:

(*a*) *Course content*

What are the precise objectives?
What is the overall strategy?
Will final concepts be defined at the outset?
Who will write the introductory outline?
How will progressive learning be achieved?
What will be the pattern and structure?
What methods and techniques will be used?
How will maximum course member participation be achieved?
How and when will speakers and instructors be briefed?

Should a synopsis for each session be given or agreed?
What further help will session leader need?
Should a synopsis be displayed at the start of each session?
How can scope be left for amendments?
How is it possible to ensure flexibility, continuity and integration?
Should a course tutor always be present?
How best can a tutor act as a catalyst?
What provision should be made for free discussions?
Is it necessary to have prepared handouts?
Is it necessary to have note-taking exercises?
When and what periods should there be for private study?
How is it possible to ensure optimum balance of activity?
Is it necessary to rehearse or have 'dummy runs'; if so when?
What is the purpose and what should be the frequency of outside visits?
What period should be allowed for final preparations?
What new sessions should be included?
What revised timings should be made?
What time-tabling of open-forums for course members are necessary?

(*b*) *The Form of the Course*

What length will the course be?
Should it be part-time or full-time?
Should it be residential or non-residential?
How many sessions per day?
What kind of sessions each day?
What should be the balance of lecturing, instructing, demonstrating and discussion leading?
Should unbroken lecturing last not longer than twenty minutes?
Should unbroken instruction last no longer than five minutes?
Is it necessary to review each morning the previous days work?
What use should be made of appropriate visual and/or audio aids, materials and models?

(*c*) *Course Administration*

Ensuring initial discussion of details with all concerned.
Drafting the initial programme.
Securing line management's approval of the draft.
Producing a final well laid out programme.
Passing final confirmations to all speakers.
Securing agreed lists of course members.
Notifying course members well in advance.
Ensuring adequate accommodation.
Arranging adequate seating, lighting, heating and ventilation.
Giving clear instructions on car parking and telephone contact.
Ensuring appropriate breaks and refreshments.
Having adequate power-points and (if needed) black-out.

Securing on time, required equipment, models, materials.
Arranging for any required bench space or outdoor work areas.
Arranging for required wall display facilities.
Making contingency arrangements for 'stand-in' sessions.
Organising payment of visiting speakers' fees and expenses.
Making arrangements for meeting and introducing visiting speakers.
Coping with last minute changes to the time-table.
Checking and distribution of handouts.
Arranging for typing of any required copy.
Following up with 'feedback' comments to speakers.
Doing open-forum report and course report for line management.
Arranging follow-up discussions with supervisors of course members.

Most—if not all—of the above check points can be tackled by the majority of training practitioners. On the thoroughness of this attention to detail depends the quality of the time-tabling. The final advertisement is the course programme itself. In quality of layout, printing and paper, the programme should reflect the high standard of work which has preceded it.

In larger firms, in addition to the issue of detailed course programmes, a useful method of communicating overall time-tabling is by the issue to all supervisors of a foolscap sheet once a month giving bar chart time-tables of all training activities for the following month.

Detailed time-tabling is tedious work but it is vitally important that it should be efficiently carried out. 'Dummy runs' and careful re-assessment of time-tabling should precede the launching of new courses. In this way major difficulties can be avoided.

No training plan is better than the people who have created it. To a large extent, their participation, their co-operation and their interest guarantee it against failure. Training must be undertaken vigorously and with maximum help from line management. Half-hearted support is demoralising for everyone concerned.

CONCLUSION

In Great Britain, the Industrial Training Act has given an impetus to the initiating, planning and time-tabling of training which had hitherto been thought impossible. The consequence is that in practice as well as in theory, training is assuming its true position of importance in industry. In-company courses must, therefore, now reach higher standards than ever before achieved. If this happens industrial training can produce increased dividends for industrialists and greater job satisfaction for their employees at all levels.

BIBLIOGRAPHY

MINISTRY OF LABOUR, *Glossary of Training Terms*, H.M.S.O. (1967)
MINISTRY OF LABOUR, *Industrial Training Research Register*, H.M.S.O. (1967)
MINISTRY OF LABOUR, *Training Abstracts Service* (including *Notes for Users*), Training Department, Ministry of Labour (1968 onwards)
MINISTRY OF LABOUR, *Training Information Papers*, H.M.S.O. (1968 onwards)

9

The Costing and Evaluating of Training

DOUGLAS GARBUTT, A.C.I.S., A.C.W.A., DIP.ED.

Senior Lecturer in the Department of Industrial Management, College of Aeronautics, Cranfield

INTRODUCTION

The first newsletter issued by the Shipbuilding Industry Training Board in 1966 bore, in very large print, this message:

> 'Note to Company Executives. This document concerns the Industrial Training Act and has financial implications for your Company.'

Industry is now finding that the cost of training is an important problem. How can the costs of training be found and what is the significance of costing data? These are the problems discussed in this chapter.

WHAT IS COST?

The accountant's approach to costs is fundamentally simple. First, he establishes, within a period of account, those expenditures which can be associated with the activities and functions carried on by the firm. The basic period of account is one year and it is sub-divided into shorter periods of a week, month or year for purposes of control.

Some costs are direct: they vary in proportion to production activity. If production rises, they rise. If production falls, they fall.

Other costs such as training costs are indirect to production activity. These are overheads. Although they vary, they do not have a directly proportional relationship to the output of the firm.

Accountants have developed various procedures for associating indirect costs with products and these rules (of apportionment and allocation) are based on the idea of sharing indirect costs equitably.

The costs of training can be built up by following the same approach, and this is dealt with in some detail later.

OPPORTUNITY COSTS

Economists define costs in a broader sense as being a measure of monetary sacrifice. In these terms, the opportunity foregone may be a better measure of costs than the accounting expenditures. However, measures of opportunity costs are hard to establish, especially where conflicting interests are involved.

Whilst the accounting information is not ideal for many purposes, it is at least measurable in broad terms, and it must be the main source of information about training costs.

STATISTICAL TECHNIQUES

The task of the training board is to establish the costs of training in its industry *as a whole* and then to distribute its grants on a reasonably equitable basis. This means that the board may use a statistical sampling technique for establishing training costs and then make grants on a standard scheme.

This gives an accurate picture of the amount of training given for the industry as a whole within stated limits of confidence, but it could of course be highly inaccurate in measuring the costs of an individual firm.

NEED FOR ACCOUNTING OF TRAINING COSTS

Before the Training Act, few firms made a systematic effort to ascertain training costs. The Engineering Board attempted in its early days to establish the amount of costs in the engineering industry. That industry covers 29,750 firms and the Board made a survey of about one hundred firms which were known to have effective and extensive training schemes. They found that in very few cases were training costs analysed as part of the normal accounting process and attempts to establish a cost of training an apprentice were unsuccessful. Firms simply did not have the information and were unable to produce it in a standardised way.

In the past, industry has been 'losing' its training costs but it is now clear that the skeleton must be taken out of the cupboard so that the true costs of training may be established.

Industrial training boards may not require firms to keep records of training costs. Should these be kept by the firms themselves? The answer to this question is undoubtedly yes.

NEED FOR A UNIFORM DEFINITION OF COSTS

Since many firms are subject to more than one board, it will be extremely difficult for them if boards evolve radically different definitions of training costs. With this in mind, a Ministry of Labour Working Party prepared a comprehensive list of definitions. At the time of going to press, this had not as yet been published.

THE EXTENT OF TRAINING COSTS

By January 1968, there were 21 industrial boards and they will be imposing levy amounting to about £140 million. The Engineering Industry Training Board will be responsible for more than half of this. Some of the best known engineering firms will pay £2 million or more in levy and similar amounts in training costs.

Once training begins to cover all the employees of a firm and once the need for retraining is recognised, then, clearly, the costs of training will rise and will need to be controlled. There is no inherent difficulty in establishing training costs, any more than in establishing the costs of any other service, such as accounting or production control.

ACCOUNTING FOR COSTS

Cost accounting requires three things:

1. A *list of costs* likely to be incurred by the undertaking. Some system of primary classification is needed if all costs are to be ascertained.
2. The identification of *cost centres*—functional or technical locations or divisions for which costs may be ascertained and used for the purpose of cost control.
3. The adoption of *cost units*—units of quantity of a product or service in relation to which costs are expressed.

The essence of cost accounting is that expenditures are related to work done.

THE LIST OF COSTS

The items of cost will vary considerably, according to the type of manufacture and the raw material and the method of distribution, and very few firms have a similar cost structure when the matter is investigated closely. Some costs are incurred by all businesses—personnel costs, occupation of buildings, administrative and financial expenses and depreciation. Others, such as loose tools and consumable supplies, are found in manufacture but not in trading.

Some industries, such as printing, have a uniform cost accounting system which can easily be modified to allow for the identification of training costs.

The following is a suggested list of all costs for a firm. From this it will be necessary to select those items and amounts which relate to training.

Group 1—Personnel costs and payments to personnel.

Wages—ordinary time.
Wages—overtime.
Salaries—ordinary time.
Salaries—overtime.
Wages—casual and part-time labour.
Commission.
Bonuses.

Group 2—Personnel costs, other payments.

National insurance.

Luncheon vouchers and gross cost of meals.
Pension fund contributions.
Fees to consultants, auditors, counsel, etc.
Accommodation (hotels and hostels).
Miscellaneous.

Group 3—Fees paid for external training.

Fees paid for co-operative training.
Fees paid for industrial training.
Fees paid for group training.
Fees paid to research establishments.
Fees paid to Ministry of Labour schools.
Fees paid to consultants.
Education fees to local authorities.
Education fees to universities and other non-profit-making institutions.

Group 4—Building costs and services.

Rent.
Rates.
Water.
Repairs and maintenance.
Security.
Fire prevention.
Electricity.
Gas.
Oil.

Group 5—Production costs.

Machine hire and rental.
Outwork and rental.
Royalties.
Machine repairs and maintenance.

Group 6—Administration costs.

Printing and stationery.
Postages and telegrams.
Telephone.
Newspapers and periodicals.
Subscriptions.
Furniture, office machine repairs and maintenance.
Meetings and excursions.

Group 7—Insurances.

Employers' liability.
Accident.
Fire.
Burglary.

Fidelity bond.
Cash in transit.
Vehicle.

Group 8—Transportation and travel.

Vehicle licences.
Vehicle repairs and maintenance.
Petrol and oil.
Travel expenses and fares.

Group 9—Depreciation and other provisions.

Land and buildings.
Plant and machinery.
Vehicles.
Furniture and office machines.
Patents and trade marks.

Group 10—Purchasing and selling expenses.

Advertising for products and staff.
Commission other than staff.
Carriage inwards.
Customs duties.
Carriage outwards.

Group 11—Adjustment or reciprocal accounts.

Group 12—Materials.

Production materials used at actual cost.
Production materials used at standard cost.
Loose tools.

This list is based on R. J. Ryall's list (see *Dictionary of Costing*, Pitman 1946) quoted by K. S. Most in *Uniform Cost Accounting*, Gee & Co. 1961.

The list must be modified in the light of the requirements of each type of industry. It is a subjective list of expenses and in order to establish the cost of training it is necessary to allocate the items in relation to their objective. Those costs which relate to training must be separated out and then associated with training activities.

Problems of identifying costs

The costs in the list require careful definition if accurate and comparative data is to be produced, and it is worth considering one or two of the items.

Personnel costs

All items should be taken into account, including bonuses and commission, so as to charge the gross cost of wages and salaries. Accounting procedures vary and some firms may include overtime and other premium payments as overhead whilst others include it in direct cost. Similarly, employee related payments such as Selective Employment Tax (S.E.T.) may be included in overheads. Where meals are provided through ordinary company facilities and at company cost during training, the costs should be the normal price charged to employees plus a proportionate share of the subsidy (if any) on canteen meals taken within normal company canteen facilities.

All payments for costs of travelling made to trainees during training, irrespective of whether these are normally borne by the firm or not, should be included.

Fees paid for external training

The list given is fairly long. In most cases it may be possible to classify course fees under two headings:

1. Fees paid for industrial training and approved by industrial training boards.
2. Fees paid for industrial training and not approved by industrial training boards.

Since some fees are subject to a percentage grant, this will involve splitting the amount between the two categories.

Production costs

Machines may be used wholly or partly for training purposes. Where the equipment used is expensive, the normal time-booking system in the firm may be used to record the time taken by trainees. If this is charged at normal rates, a charge for general overheads may be recovered from training, which seems justified.

Administration costs

These costs may not arise entirely within the training function, and in this case training should be charged with a fair proportion of the facilities provided.

Materials costs

Materials used by trainees may be specially purchased or drawn from stores and should be charged in accordance with normal company procedure. This may be:

1. Actual cost of materials, including direct expenses such as carriage and freight.
2. Actual cost plus a charge for handling or storage costs (possibly calculated as a percentage of actual cost).
3. Standard cost. In this case, the standard cost will be calculated in relation to the normal product in which it is incorporated. Since the trainee is likely to be training for jobs in relation to the normal product, it is reasonable to use the normal standard cost.

This implies that a proportion of the price variance on materials may be included in indirect costs.

Material costs may also include the cost of small tools when this is in accordance with normal procedure within the company.

TRAINING COST CENTRES

In order to control training and costs, it is necessary to identify certain functional departments. These are cost centres. In most firms, there will be a cost centre which will be the training manager's department. To this will be apportioned and allocated all the costs of running that department. Another cost centre which may be identified is the training school.

The size of the firm and the nature of its products and technology will influence the size of the training department. In a large firm with a large training programme, the training activity will tend to divide into specialised functions. Ultimately, each function might be a cost centre. The following is a suggested list of training functions.

Recruitment and Selection of Staff.	
Establishing Training Needs.	
Designing Courses.	
Arranging Courses	Recruitment and selecting of trainees. Organising course material, aids, etc. Training and programming of instructors.
Running Courses	Course Direction. Instructor's activities. Aids in use. Trainees, ancillary activities. Examinations and inspection. Ceremonies, etc.
Servicing Courses	Administering. Providing aids.
Assessment of Training	Validation. Evaluation.
General Training Overhead	Management of the training function. Provision of general services.

The cost of training may be established in the first place on a functional or cost centre basis because this is the most convenient basis for establishing budgetary control.

The manager of each section will be responsible for the costs incurred in that section.

TRAINING COST UNITS

If the output of training is regarded as being the trained or retrained person of a particular type, e.g., a five-year-trained apprentice, then it is desirable to know what is the cost of producing one such apprentice as a cost unit.

The costs may be:

(*a*) *Direct to the cost unit.* This is mainly the cost of wages and salaries actually incurred whilst the trainee is undergoing training.

(*b*) *Indirect to the cost unit.* These are the costs arising within the cost centre. They include such items as payment for provision of training facilities, buildings, rent, heating and lighting, administrative charges and so on. These costs may be incurred whether training takes place or not. Although it may be decided to associate these general costs with particular trainees, it has to be recognised that they do not vary in relation to the amount of training actually done.

There is no generally accepted classification of training 'cost units' beyond the vague distinctions between skilled, semi-skilled and unskilled labour.

In the petroleum industry, the following categories are used:

Operative
Craft
Technician
Technologist
Commercial
Graduate trainee
Supervisory
Management
Others

A similar classification is used in engineering but it also includes draughtsmen and scientists. The commercial group is called 'administrative clerical and office workers'. Supervisory grades may include foremen and the management category may include works superintendents and department managers. 'Other workers' may include canteen workers.

Other categories may be based on the function within which an individual works, e.g., sales, personnel, transport and so on.

Training is often identified by the type of qualification obtained. It can also be sub-divided into initial, refresher and retraining. It is often classified according to whether it is full- or part-time, day or evening, sandwich or block release, residential or non-residential and so on.

The categories generally accepted in an industry will be used for costing purposes. These are usually a mixture of the above categories.

The categories generally do not seem to have an objective basis since they differ markedly from one country to another.

Confusion arises between occupational classification and the classification of trainees: these are not synonymous. An apprentice will only be a trainee when he is undergoing training whether this be on or off the job.

ESTABLISHING THE NUMBER OF TRAINEES

The correct measure to take is the number of trainees completing the course. Where training extends over more than one year, the number still in training at the end of each year may be taken.

Comparisons may be facilitated by expressing these units in terms of trainee years. One trainee taking a one-year course is the equivalent of 13 trainees taking a four-week course. These units may also be used in apportioning general training overhead to each category of training.

GENERAL PROBLEMS IN ESTABLISHING TRAINING COSTS

OVERHEADS

Trainees make use of many of the general facilities provided by a firm such as, for example, the costs of payroll accounting. It is therefore legitimate to charge training with a fair proportion of these costs, using the overhead rates in use. Where cost rates are not used, a proportionate charge should be made.

CREDIT ITEM

Costs will be charged gross against training. It is important that any credit items should be deducted from costs. These may include fees recovered from local authorities, charges made for meals and accommodation, charges made to production departments for use of training equipment and so on.

The value of saleable work produced by trainees should also be credited to the training department at standard cost of the work done.

The net cost of training to the firm will be training costs plus levies minus grants but it must be stressed that levies and grants should not be included in breakdowns of cost.

ON-THE-JOB-TRAINING

The problem of on-the-job-training is determining what proportion of a person's time is training and what proportion production work. In practice, it would be difficult to do this on an individual basis. It follows that training boards may be ready to agree on rules that are generally equitable.

One approach is to value the loss to the firm which arises from the fact that trainees do not produce as much on the same equipment as trained workers. A valuation at average cost to the industry would provide an incentive to the more efficient firms. Strictly, allowance should also be made for loss of profits. The allowance is added to training costs.

Against this must be set the fact that trainees are paid less: the savings to the firm must be deducted from training costs.

An alternative suggestion is that the average value of a trainee's output should be expressed as a percentage to be deducted from their gross wages and related costs.

THE EVALUATION OF TRAINING COSTS

The assessment of the value of training includes all ways of testing whether training is worthwhile, efficient or effective.
Evaluation implies deciding whether training justifies its cost and is doing what it sets out to do better than alternatives such as increased investment in plant and machinery, changes in methods and so on.
Validation is deciding whether, and how well, training is doing what it sets out to do. Validation is not considered here.

ORGANISATIONAL GOALS

In evaluation, training methods are related to the organisational goals of the firm—are they improving profitability, increasing productivity or the quality and reliability of the product or service?

COSTS AND BENEFITS

Ideally, it is desirable to balance costs against benefits and to determine whether the benefits outweigh the costs. To the industrial firm, the benefits of training may be improvements in efficiency such as decline in absenteeism or labour turnover. These lead to identifiable cost savings. Training benefits, however, may also be less tangible as when a telephonist's training leads to a decrease in complaints. Benefits also accrue to the individual and to society from training.

Training leads to changes in the trainees; these lead to changes in job performance; these lead to a more effective realisation of company goals.

At each stage, it is useful to be able to make comparisons: the ultimate comparison—of goals to training methods is the most difficult to make.

BUDGETARY CONTROL

Day to day control of the training department and, to some extent, data to justify further spending on training can be obtained by means of judicious comparisons.

The data for many of these will arise naturally from the process of planning, co-ordinating and controlling the utilisation of resources in training.

Taking the general expenses of the training department they may be considered as a percentage of the firm's expenses, or as a percentage of the wage bill. What proportion of the firm's budget goes on training? How does this compare, for instance, with spending on such items as the social and welfare facilities?

If the training budget is growing, is that growth related to growth in training activities as measured by the number of trainee years recorded?

If capital expenditure is required on a new school or on plant and equipment, the amount requested may be compared to the spending on other capital projects.

Some organisations calculate a course cost, per member per day of course, and this makes it possible to get standards which exercise restraint over costs whilst allowing for increases in activity.

Comparisons of this kind are made of present and predicted costs with past costs, internally. They can also be made externally provided the data have been collected and analysed on the same basis. Some industries have inter-firm comparison schemes which allow comparisons of training costs to be made. Even unit costs per trainee-year may be calculated.

Essentially, these comparisons are useful provided it is realised that there can be no absolute standard. Rising costs may be accepted, for instance, if training activity is increasing or if the quality of facilities provided has been improved.

COST-BENEFIT ANALYSIS

Essentially this approach is the normal approach to the measurement of the profitability of projects. Expenditure on education and training, including income foregone, is treated as a negative cash flow and then compared with the positive cash flows subsequently generated.

These cash flows are, in fact, difficult to identify and measure. In the U.S.A., social scientists have compared the cost of acquiring higher education (including the income foregone during student days) with the lifetime earnings which follow for the individual. One study showed a return of 9% compared to an 8% return on all forms of business investment. Since the return excluded social benefits it was undoubtedly under-stated.

One English study shows that graduates here obtain, on average, a 20% return on the costs to them of their higher education.

It is possible to calculate the returns which various categories of employees are obtaining from their training and relate this to the costs of giving that training. The returns to the firm are not so easy to show since so many interrelated factors influence the ultimate profitability of an enterprise. However, it is sometimes suggested that a firm may save on the salary it pays to a skilled man of their own training as compared to the salary paid to someone recruited from outside. If this is so, the savings would indicate a benefit which could be measured and related to the costs of training less the costs of hiring the trained man.

COMPARISON OF COST SAVINGS

Where training is undertaken for operators, it is possible to compare the outputs of two groups, one of which has training (or a new method) to another group which has had none (or was trained under an old method).

Assuming that rates of pay are related to actual production and that other factors such as rate of waste of material and spoilage of production are unaffected, the increased production will show savings in the over-recovery of fixed costs. This gives a fairly accurate measure of the superiority (if any) of the method with the highest production levels.

Properly speaking, this method is a validation method since it indicates which method should be used if training is undertaken.

Unfortunately, it does not answer the broader question which top management may ask 'should we invest in training or are we better off investing in other directions—new plant, new methods or new products?'

Table 9.1 summarises the data in a comparison of this nature made by W. Douglas Seymour. (See *Industrial Training for Manual Operations*, Pitman 1954.)

Table 9.1

	Method 1	*Method 2*
Total production in year (in units)	275,000	375,000
Increase in production (%)	—	38
Labour costs	£	£
6 weeks at £3 6*s* 6*d*	20	20
44 weeks at £3 6*s* 6*d*	144	144
Bonus earned	16	56
Total annual earnings	£180	£220
Increase in earnings (%)	—	22
Cost per 1,000 units (in pence)	157	140
Cost saving (*d*/1,000 units)	—	17
Total economy on labour cost over year		£26
Increased overhead recovery		
Given by Production Increase × overhead recovery rate.		
If recovery 100% of labour rate { i.e., $\frac{100,000 \times 8}{1,000}$ shillings }		£40
Total Savings		£66

STATISTICAL ANALYSIS OF DIFFERENCES

In this method, data for an industry are collected on the size of firm, type of production, type of market and organisation, level of costs, profitability, turnover and capital investment. Taking, also, the training undertaken, the size and composition of the labour force, it is possible to analyse the data and to state the degree to which any one factor, e.g., spending on training, is related to any other factor, e.g., profitability.

Such a study would be suitable for a training board to undertake. It might also be done by an employers' association but it would not be practicable for anything but very large companies.

It might, however, produce data of high reliability.

MANPOWER PLANNING

This is the most useful approach to evaluation of training both nationally and for the individual firm. This approach evaluates training and education costs in the light of the actual 'stock' of trained persons in any category and the numbers needed to cope with future demands for their skills.

The benefits of training are the potential losses which are avoided by undertaking systematic training.

Forecasting future employment patterns is a Herculean task, for account must be taken of a host of factors most of them changing rapidly and many of them interdependent or themselves functions of what happens to employment.

In the Netherlands, for example, short-run employment projections are made with the aid of the following econmetric model developed by the Central Planning Bureau:

$$a = 0{\cdot}39v + 0{\cdot}76K + 0{\cdot}7Pm\text{-}v + 0{\cdot}66$$

where a = total employment,
v = total volume of sales excluding stock formation,
K = level of gross profits per unit of production,
$Pm\text{-}v$ = the ratio of the import price level to the internal price level corresponding to v.

For the medium and long term, projections from the past are often checked against the opinions of people in industry as to what will happen in the future and against any other information that may be available about future trends.

Manpower projections must be made not only industry by industry but for each occupation. However, standard occupational breakdowns are not sufficient in the view of the most advanced manpower forecasters, at least for educational planning: one occupation—say managers—may require very different kinds of educational preparation from one industry to another or even within the same industry.

For the individual firm, manpower planning involves four main problems.

Determining categories of training needed

This is not necessarily the same as determining the number of people to be recruited.

One factor that ameliorates problems of supply at lower levels is the hierarchy of skills—the *more* skilled men can perform *less* skilled operations.

On the other hand, occupational structure and the attitudes of many men on the shop floor work the opposite way, and men who could do a job refuse to do it or are not allowed to do it.

Determining cut-off points

These are the points at which firms will not train internally but will either train externally or recruit trained personnel. The scale of need may indicate these: training programmes should only be set up internally if there will be a sufficient demand for the training over a period of time. Up to a point, external training will be cheaper, since a higher rate of occupancy of facilities may be obtainable. Specialists are also likely to be more efficient.

Relating training to the present structure and future development of the company

Determining the labour needs and, thereby, training needs is a vital component of company budgeting which should be based on market forecasts and company policies.

Evaluating the risks in manpower and training requirements

All forecasts involve risks which can be quantified in the form of probabilities. This is especially true of training requirements which are derived from forecasts of business activity. A vital component in the prediction of training needs is the rate of labour turnover, a factor which training may be able to affect.

The point about manpower planning is that the firm which recruits and trains enough labour for its needs will be in a position to benefit by the availability of a trained labour.

The firm which does not train for its own needs is inevitably dependent on chance when it needs to recruit; chance in finding the right skills and chance in finding them close enough to the place of work.

CONCLUSION

The introduction, by the Industrial Training Act, of a system of financial burdens and financial incentives intended to encourage better industrial training has made, and must make, industry more aware of the costs and benefits of training. All interests involved in the industrial training system are concerned with obtaining more training at the lowest cost compatible with good standards. Firms, industries and boards must collect reliable cost data on which planning may be based.

Control of costs can be based on comparisons within the firm and externally using budgetary control procedures and inter-firm comparisons. Assessment of the benefits of training in precise terms is more difficult but can be done and there is ample scope for building up more and better information on which the economic evaluation of training may be based.

BIBLIOGRAPHY

ALLEN, K. R., *Tech. Educ. & Industr. Train.*, **5** No. 11, p. 516, Nov. 1963; **5** No. 12, p. 578 (Dec. 1963)

DAVIDSON, A. F., *Management Accounting*, p. 6 (Jan. 1967)

FORRESTER, D. A. R., *Tech. Educ. & Industr. Train.*, **9** No. 2, p. 60 (Feb. 1967)

GARBUTT, D., *Accountancy*, p. 784 (Dec. 1967)

10

Operational Research

A. W. INGRAM, C.ENG., M.I.E.E., M.B.I.M., M.O.R.S.

An Operational Research Engineer with the South Western Region of the Central Electricity Generating Board

INTRODUCTION

The very rapid growth in the size of industrial organisations over the last 100 years has led inevitably to an increase in the complexity of decision taking within the management function. The order of magnitude of the financial sums involved in the development of new processes, new products or new methods of working, which the rapid growth of science has forced upon industry, are sufficient to cripple a company if decision on them should be weak or ill-conceived.

In these circumstances of ever growing complexity and individual specialisation, it is essential that modern management be well equipped with scientific 'tools'. Problems to be tackled are not confined to the private sectors of industry and commerce alone; but are present in both local and national government, and within military organisations where perhaps there has been a greater awareness of the help that science can give to the decision taker. Undoubtedly the accelerating pace of scientific change has contributed largely to the complexity which surrounds modern management.

Operational research provides the modern manager with scientifically based information which enables him to narrow the unknown areas in which decision taking is to take place. Knowledge of operational research has become a necessity in management training.

The object of this chapter is to acquaint the reader with the purpose and scope of operational research and to mention some of the techniques which have been evolved to assist industrial managers in getting the best out of the human and material resources under their control.

ORIGIN OF THE TERM

The term operational research derives from the Armed Forces and in particular with the Royal Air Force. In 1938, the pioneers of radar were faced with the difficulty that although the equipment they had developed worked perfectly in the laboratory, the results obtained in the field when the equipment was in service with operating squadrons were, to say the least, disappointing. It was apparent that the apparatus was both too complicated and delicate to be handled as a matter of routine by regular R.A.F. servicing personnel. In consequence, a team of civilian 'boffin boys' who had been engaged on radar research were posted off on 'operational duties' with the R.A.F. Squadrons. Their duty was to work in the operational climate and solve the problems which arose there.

The success of this innovation is, of course, well known and the very practical achievements of Fighter Command and their high rates of enemy interception some two years later, in 1940, will for ever remain a testimony to the success of the methods employed.

The successful results in the case of radar led to the more widespread use of the technique of employing scientists in the field and ultimately to the complete acceptance by each branch of the Armed Forces of the operational research principle. Indeed, its development has been officially described as one of the chief scientific features of the war.

After the cessation of hostilities, attempts were made to convert the operational research sword into a ploughshare for use in the rapidly developing world. Progress was slow at first but the value of operational research had been proved and now most large business concerns, nationalised industries and government departments employ operational research groups. The problems tackled and the techniques used are in many ways different from the wartime examples, but the central theme remains—that of using the scientific method to assist industrial management to manage successfully.

DEFINITION

Inevitably any new branch of science gives birth to its own vocabulary and jargon. The experts claim that the existing vocabulary is inadequate to express their meanings and whole sets of new words find their way into the language. Often these serve only to bemuse and bewilder the uninitiated, and to confuse even the experts responsible for their introduction by the many and varied interpretations placed upon them. In this respect, operational research is no exception. There are for example dozens of definitions of operational research itself. The most universally accepted, is that of Sir Charles Goodeve, Fellow of the Royal Society and Director of the British Iron and Steel Research Association:

> 'Operational Research is the use of scientific methods in providing executive departments with a quantitative basis for decisions regarding the operations under their control.'

The use of each word in this definition is deliberate and important. The title 'Operational Research' is intended to convey that it does not include applied scientific research concerned with development, in the narrow sense, of new or improved products or processes. It is concerned with the study of the operational requirements of the actual functioning of the processes as they pass into use in the factory or of their possible development in that industrial environment.

The words 'scientific methods' are used to convey the concept that the true disciplines of operational research are observation and analysis implying that emotion and 'hunch' are not part of the stock-in-trade of operational research.

The phrase 'provide executive departments' implies that operational research acts as a service to management and is advisory in character and does not attempt to usurp the executive function. The term 'quantitative basis' expresses the need to present a *measurable* value in numbers and the logic of mathematics.

Although it has been stated earlier that the term 'operational research' as a specific description of this particular application of the scientific method only dates back possibly to 1938, the idea of studying the performance of men and equipment under normal working conditions is, of course, by no means novel. Engineers and professional managers have been doing it in one way or another for a very long time but they lacked the modern techniques of recording and analysing which have only become available in post war years and have only been given wide practical application during the last decade.

The modern concept of operational research is therefore the result of evolution over a great many years. Charles Babbage, Lucasian Professor of Mathematics at Cambridge invented his 'difference engine'—the first computer—nearly two centuries ago. The activities of Boulton and Watt laid the foundations of production control at their Soho Foundry in Birmingham also nearly two centuries ago. Many other pioneers followed them and related knowledge on this subject has continued to grow. The industrial revolution in due course gave place to the scientific revolution. Many facets of the rapidly developing situation have been gathered together and today, in this age of 'galloping technology', they have been formed into the systematic approach or method known as operational research.

THE PRINCIPLES OF METHODS OF OPERATIONAL RESEARCH

Today when the applications of operational research are increasing in variety, it is not easy to define either the scope or the techniques of the subject. Rapid growth is, of course, characteristic of nearly all the sciences at the present time. However, despite this rapid growth each branch of science retains readily identifiable features, which places any aspect of the particular branch of science into neatly labelled categories. As an illustration, two branches of science both of which have undergone very rapid expansion in recent years are chemistry and nuclear physics. Despite their rapid growth, there still remains a neat line of demarcation between them. Chemistry always deals with phenomena which can be traced to the configurations of the orbital electrons of the atom. Beyond that point, when science attempts to explore the secrets of the proton, then the subject changes from chemistry to nuclear physics.

No such clear line of demarcation can, however, be drawn around operational research nor can the future be expected to provide such boundaries. The very success of operational research methods and techniques depends upon ability to borrow methods from all the sciences in order to solve the problems confronting management. It is, therefore, misleading to speak of the 'methods' or 'techniques' of operational research since the methods used in O.R. investigations are those used in science in general.

The laws of hydrodynamics provide an example of the parasitical nature of operational research. It is widely understood that these laws are essential in designing the hydraulic systems in industry. It is also well appreciated by all engineers that no efficient circulating water system in a power station could ever be designed without an adequate knowledge of the principles of hydrodynamics. Operational research borrows these same laws to help solve such analogous problems as the flow of billets of steel into a rolling mill or the flow of a sliver of cotton through a doubling machine in a textile factory, or even to the flow of people into the Oxford Circus underground station during the rush hour. Thus operational research specialises in no one particular science but rather in the scientific method. The four fundamental aspects of operational research are:

1. Operational research is scientific in its methods—that is, it depends upon logical reasoning from observed facts to produce hypotheses for experimental verification in the factory.
2. Operational research is quantitative—that is, it relies on measurement and on expressing things in numbers. This does not mean that absolute precision in measurement is always essential. It may often be better to reach a decision quickly from data which is approximately accurate rather than lose the opportunity of making the decision effective through spending too long in achieving the last refinements of precision.
3. Operational research is concerned with decisions about the future and with the effects of change. Many of the techniques are used in order to predict what would be the consequences of making some change in a method of working. These predictions must be surrounded with some measure of uncertainty and it is largely for this reason that concepts of the theory of probability continually recur in operational research.
4. The function of operational research is, and must remain, advisory —that is, it is concerned with the measurement and prediction of the effects of decisions which management may, or may not, choose to make.

Although it has been stated that it can be misleading to speak of the methods and techniques of operational research, experience does show that certain types of situation occur again and again in industry and in their solution certain specific techniques are frequently used. These are usually known as the techniques of operational research. More specifically certain principles characterise all the many applications of operational research in its diverse fields.

These principles are:

The measurement principle.
The operational principle.
The feedback principle.
The co-operation or team principle.
The practical principle.

THE PRINCIPLE OF MEASUREMENT

Right from the outset, operational research has always placed great emphasis on the importance of measurement. Likewise it has always been the practice of science to measure those things with which it deals. The great physicist Lord Kelvin made the same point in the following words:

> 'I often say that when you can measure what you are speaking about and express it in numbers, you know something about it; but when you cannot measure it, when you cannot express it in numbers, your knowledge is of a meagre and unsatisfactory kind.'

In operational research, this concept of measurement has been extended even to those areas where it was customary to believe that measurement could be neither practical nor useful.

Operational research groups associated with certain industries have, for example, made attempts to measure such things as the flow of information or the average length of time an intending passenger waits for a bus.

THE OPERATION PRINCIPLE

In some fields of scientific research, it is possible to isolate a self-contained problem and to tackle it in one corner of the laboratory to the exclusion of all else.

This is not so in operational research for in the activities of an industrial concern a change in one sector often leads to repercussions elsewhere and to total effects which are often complex. These inter-actions between one part of the operation and another must be taken into account in the investigation.

In the solution of such problems, the use of a mathematical model of a whole cycle of operations has enabled predictions to be made of the full effects of several possible changes. By this means, a decision as to the best course of action can be more reliably made.

Operational research differs from most research in that it is not concerned with a single material, process or man, but with the way in which all these factors contribute to a complete operation or system of operations. In this context, the human operator is in many respects the least understood of all the factors.

THE FEEDBACK PRINCIPLE

Operational research must be thought of as a continuing process. It is very rare that a problem can be solved once and for all time. The very fact that an investigation has been made and that as a result the system has been modified means that conditions change and that a new situation arises.

In turn, this new situation will need to be examined to ensure that it is functioning correctly and has not of itself produced some further difficulties. Investigations cannot, therefore, be carried out on the basis that problems once solved can be dismissed. Operational research must be thought of as a communication process. It involves the condition known to electrical engineers and more recently to programmed learning educationists as feedback.

The feedback between stages of an operational research investigation involves an intensive communication between all those involved in the problem and its solution. One of the most vital stages in an O.R. project is that of defining the real nature of the problem and this requires an intensive feedback between the members of the O.R. team and their 'customers'.

THE CO-OPERATIVE PRINCIPLE

From the outset, the advantage of team-work in operational research has been appreciated. The stimulation of the mental processes resulting from the cross fertilisation of ideas between a group of individuals tends to arise in geometric rather than arithmetic progression. This tendency is further accelerated by the inclusion of additional professions and mental disciplines among the members of a team. Mixed teams comprising many professions are now a common practice in industry, not only for solving problems involving the operation principle, where many specialisms are included, but also in those problems which appear to belong to a more restricted field. A typical team used by Imperial Chemical Industries recently to investigate the future development of one of their factories consisted of:

An engineer with O.R. experience (Team Leader).
A technologist in the appropriate branch of chemical engineers.
An accountant/economist.
A mathematician.
A work study officer.
A statistician.

Teams frequently include engineers, physicists, chemists, biologists and where the behaviour of people is involved, social scientists.

THE PRACTICAL PRINCIPLE

Operational research must be practical in its objectives and although the concept of practicability is such an ambiguous one, what is really implied in this principle is that in any situation the correct criteria are examined and that no 'red herrings' have been introduced.

It is an important part of any operational research investigation to ensure that the improving of the criterion being used as a measure of performance will, in fact, improve the overall effectiveness of the operation.

A Service example of this will serve to focus this particular point. During the last war, one measure of Squadron efficiency as judged by Coastal Command Headquarters in the R.A.F. was 'serviceability percentage' which was a factor expressing the total number of aircraft either flying or serviceable and ready to fly, as a percentage of the total number of aircraft in the Squadron. This may at first sight appear to be a reasonably accurate indication of Squadron efficiency. However, if the system of scheduled maintenance used in the R.A.F. is examined, it reveals that the more an aircraft was flown the more frequently it was grounded for its periodic inspections which took place every 25, 50, 100 flying hours, etc. Thus, it will be obvious that the more an aircraft was flown the more often it had to be declared as being unserviceable while it underwent its maintenance inspection. One way in which a high serviceability percentage could be obtained would have been to get each aircraft in the Squadron serviceable and then NOT to fly it; in this way a report of 100% serviceability could be rendered daily to Command Headquarters.

However, if the function of Coastal Command were examined it would reveal that its main duty was the routine patrolling of the sea lanes approaching this country and to give early warning of the presence of enemy aircraft or submarines, and not to have serviceable aircraft sitting on the ground. Thus, the real criterion of a Coastal Command Squadron's efficiency was its flying hours and not its serviceability percentage.

In Fighter Command, however, such a criterion as serviceability was a very real yardstick of Squadron efficiency, since in this Command the real objective was to be able to put as many aircraft into the air as possible at once in an emergency, and thus a high state of readiness was imperative.

Hence, the most critical and often the most difficult stage in an operational research investigation is fundamentally one of defining the real nature of the problem—and it sometimes happens that definition brings its own solution.

THE TECHNIQUES OF OPERATIONAL RESEARCH

It has been stated earlier that operational research is not a single technique or body of techniques but that it exploits all the methods of scientific research. Nevertheless, certain basically similar problems yield to treatment from similar methods and in consequence certain methods do find frequent application. However, it must be remembered that operational research exists to offer a service to management in the solution of a wide variety of management problems, and not to tackle merely those selected problems which happen to fit established, and well ordered techniques.

These techniques are used for three main purposes:

1. The recording and obtaining of the necessary data surrounding the problem, with special reference to the difficulties and possible sources of error.
2. Reducing these data to a digestible form whereby they can be scientifically analysed and interpreted.
3. The development of new methods of obtaining and processing data on the basis of facts already obtained and theories established.

Most of the procedures for handling data are embraced by the science of statistics, the basis of which is the theory of probability which provides a degree of quantitative assessment in an area where otherwise only random and chance forces appear to exist. However, the application of the statistical method is very properly the concern of a trained statistician who should be consulted at the outset in all but the simplest of cases. As an illustration of the potential danger associated with an incorrect interpretation of statistical data, it might wrongly be construed when studying safety statistics that one of the most dangerous situations to life is lying in bed since it can be proved quite conclusively that more people die in bed than in any other place.

STATISTICAL METHODS

PROBABILITY

Many of the problems subjected to operational research involve measurement of frequency of occurrence of various events. The successful solution of the problem requires a correct interpretation of these occurrences. This class of problem thus involves the concept of probability. The probability scale enables data of this sort, in which exact statements cannot be made, to be processed in such a manner that definite probabilities can be assigned. For this purpose, use is made of the probability scale in which unity, or one, is assigned to an event which represents an absolute certainty. Similarly at the other end of the scale, zero, or nought, represents absolute impossibility. Between these two extremes, it is possible to assign a quantitative measure to the degree of probability. For example, the probability that a tossed coin will come down heads is 0.5.

With life in general and more particularly in the industrial environment, situations are seldom so simple and straightforward that it is required to deal with the probability of one event at a time. It is more usual that the probability of several independent events occurring at the same time needs to be known. Here the addition law and the multiplication law make it possible to calculate the overall probability. These fairly simple mathematical laws enable a quantitative assessment to be made of the probability of two or more mutually exclusive events occurring either at the same time or in a given order of succession. These laws can, for example, together with data on the reliability of certain plant and equipment, enable designers to take a calculated risk in either the provision or non-provision of standby equipment of important plant and machinery.

FREQUENCY DISTRIBUTIONS

In many situations, measurements are taken of a set of articles representing values of a variable. Where some hundreds or even thousands of these different values have been calculated, it is impossible for the mind to grasp their significance unless the data have been summarised in some way. This is usually done by working out an 'average value'. In doing this, however, much valuable information on the true behaviour of the 'population' (the statistical term for the variable being studied) can be lost.

Frequency distributions, which constitute the main method of dealing with extensive data of this type, enable the statistician to condense these values into a comprehensive form. This in turn enables various estimates and deductions to be made which it would not be possible to achieve from the mass of data in the raw state.

The most common of these various frequency distributions is known as the Normal or Gaussian curve of distribution. The logical understanding of this particular shape curve is attributable to De Moivre who first published his findings on this subject in 1733; but La Place and Gauss are other mathematicians whose names are associated with this law. This knowledge forms the basis of many techniques in operational research and is indeed the very foundation of statistical quality control.

Not all statistical data fit this particular law; there are many others among which perhaps the more well known are the Binomial and Poisson frequency distributions. Each have their appropriate part to play in the interpretation of large quantities of data.

CONGESTION OR QUEUE THEORY

Random occurrence of events can be usefully described in mathematical terms for solving problems of congestion arising from fluctuations in loading of a particular part of an industrial system. This is particularly true of any part of the organisation which offers a service, since the provision of too many facilities here can be very wasteful and the degree of utilisation would be extremely low on some occasions. On the other hand, the provision of too few such facilities might well represent a loss of efficiency in the delay which would inevitably occur in the queueing to receive the appropriate service. Work in this field was pioneered by a Danish engineer named Erlang who was investigating traffic intensity and the average duration of calls on the Danish telephone system. This type of problem is quite common in industry, varying from the canteen queue to the berthing of shipping through locks where the provision of unloading facilities commensurate with the job they have to do is essential.

This particular mathematical technique has been used by operational research units associated with the Esso Oil Company in the design of their Marine Terminal at Fawley, the Central Electricity Generating Board in the design of Shoreham Harbour Project and the Steel Company of Wales in the handling of ore ships at Swansea.

DECISION THEORY

Since decision-making is one of the major functions of management it follows that operational research workers should have studied the process and made an assessment as to how it might be improved. There are three particular techniques which have been evolved to assist in this process. They are the decision theory, gaming theory and operational gaming. In following the principle of rational decisions, a manager aims to achieve maximum expected use as the outcome of his decision. A formula has been evolved in which quantitative assessments of various courses of action can be evaluated thus:

$$\text{Expected utility} = (\text{Probability of success} \times \text{Value of success}) - (\text{Probability of failure} \times \text{Cost of failure})$$

Another technique is to follow the 'minimax' principle in which the almost intuitive action of a manager to select a course of action which would minimise his maximum losses is pursued. Each of these methods requires the evaluation of the probabilities of certain events and the values and costs associated with them. This is not an easy matter and can hardly be expected to be completely accurate. Nevertheless, the exercise is a useful one because it sharpens the manager's perception of the risks he is contemplating in preferring a particular course of action and the consequences he may have to face if that chosen should prove to be the wrong one.

OTHER OPERATIONAL RESEARCH TECHNIQUES

It has not been possible to describe all the very many techniques that are included in the general field of operational research. The use of matrix algebra and linear programming in the optimising of transport costs, the use of stock and production control models and more latterly the introduction of network analysis, are among the services offered by operational research groups. Among more specialist fields are the subjects of cybernetics and ergonomics both of which have a very real part to play in the more complex organisations of our industrial society.

Under one name or another, scientific methods are finding increasing application to administrative and policy problems. Although management appears always ready to sit in judgment upon any newly introduced technique, its great plea is that it has not the time to understand its complexities and technicalities. It is also almost true that every manager believes that his own particular sphere of operation is too complex, too much dependent upon human judgment, for formal scientific techniques to be successfully applied. However, it should be remembered that management science progresses mainly through developments originated from those who are not managers and this is the service that operational research endeavours to supply.

Management today is aware that however complex an operation—and however susceptible it may be to human judgment, there is nevertheless a measuring job to be done. It is in this field that the scientific techniques of operational research have an increasing contribution to make.

BIBLIOGRAPHY

DUCKWORTH, W. E., *A guide to Operational Research*, Methuen University Paperbacks (1965)

CHURCHMAN, C. W., ACKOFF, R. L., and ARNOFF, E. L., *Introduction to Operations Research*, J. Wiley (1957)

SASIENI, M., YASPAN, A., and FRIEDMAN, L., *Operations Research—Methods and Problems*, J. Wiley (1959)

PART 3

TRAINING METHODS AND TECHNIQUES

11

The Basic Concept Approach and the Film Loop

LEICESTER W. D. BALL, C.ENG., A.M.I.MECH.E., M.N.Z.I.E.

Visual Education Consultant, Author and Lecturer

INTRODUCTION

The aim of this chapter is to explain within the context of industrial training what is meant by the basic concept approach to learning and then to describe its practical application through the use of the basic concept film loop.

THE BASIC CONCEPT APPROACH TO LEARNING

The concept approach—just what does this mean? Whatever instructional method is used, the ultimate object is to implant in the mind of the student a clear concept or idea about the topic of instruction. But the term 'single concept' can be misleading. For, included in even the simplest topic of study are a large number of separate, if perhaps related, concepts. For example, behind the use of a pencil in making a drawing are all of these individual ideas and this is by no means a complete analysis:

1. The general idea of the pencil's purpose—to make meaningful pictures.
2. The specific idea of what the pencil does—makes a mark on material.
3. The specific idea of what produces this mark—the deposit of pencil material on the paper.
4. The ideas of what the pencil must do in order to deposit the mark material—(*a*) have its point make contact with the paper, (*b*) move in contact with the paper.
5. The ideas of what must be done to the pencil to get the right kind of mark—(*a*) it must be applied with a suitable pressure, and (*b*) it must be so controlled as to deposit uniform (*or varying*) amounts of its material throughout the movement . . . and so on . . .

Elaborate as this analysis may seem, all these ideas (with several more, of course) are embraced in the overall concept of using a pencil. And it is the proper appreciation of this mass of related information which enables the user of the pencil to perform the operation of drawing with it to that performance level which is called 'skilled'. Consideration will show that there is nothing difficult in any of the ideas, nor in their individual application, nor in the relationship between them. But the 'presentation' must be in the learner's own 'terms of reference'. That is, both the 'thing' being treated and the 'medium' used to inform about it must be meaningful to him.

Programmed learning has done much for education in establishing the importance of such an analysis as that above as a proper basis for sound teaching. However, programmed instruction is still basing its supply of information to the learner on conventional theory—the step-by-step method—which is believed to be the source of much of the reluctance or actual resistance to learning common to learners of all age levels.

The step-by-step theory rests on the obvious and demonstrably sound observation that in order to get from an initial point or state to a more (*or less*) remote objective, progress must be by steps or stages. The theory then assumes, and requires (*in my view falsely*), that the objective can only be properly achieved if the learner learns each step thoroughly before he progresses to the next. This has resulted in what for the student is often isolated and meaningless activity: no matter how relevant and important the activity may be to the teacher who sees it from his pinnacle of experience and achievement. Furthermore, the theory ignores a very fundamental psychological need. An objective must have sufficient attraction for people before they will strive to reach it. For most, and especially for learners, it is necessary to be able to see and appreciate not only the objective itself, but what steps are necessary to gain it, what is the nature of each step and what is the likelihood of the step being performed.

As a simple illustration, consider the process of 'learning to walk'. Few if any, would dispute the following statement—'A baby learns to walk by steps'—Does it?

For months, it observes what walks (*the parent*); for what purpose the thing walks (*to get something*); what device the thing uses to walk with (*legs*) and what these legs must do in order to walk (*make steps*). Then, and *only then*, after this long process of observation, does it try to make the steps. And always, the *whole* act, never a single step!—much less the separate operations of toes, foot, lower and upper leg, etc., that enable the step to be made.

In its mind, the baby has achieved an integration deriving from observation and a deeply felt purpose, it has a clear and attainable objective, and comprehends the *whole* sequence of the exercise, *before* it performs, indeed, before it will try to perform the detailed mechanics of the operation. Only after it has failed to perform the whole operation will it submit, and always impatiently, to any instruction about details.

Observation of any learner, animal or human will establish how instructive is this attitude and how strongly it is maintained throughout the whole of life. Yet conventional teaching practice is often based on just such detailed instruction of *isolated* operations and insists on too high level performance in each operation before permitting progression to the next.

The basic concept method reverses this procedure. It maintains the natural approach of the baby and of all young animals. It places considerable emphasis at the very outset of the study on inducing maximum involvement of the learner, on ensuring maximum clarity and appreciation of the overall essentials and of the wholeness of the exercise. This, of course, requires a rigorous analysis of the topic, a removal of inessential extraneous matter and a very close look at the mechanics of both learning and instruction. In the following remarks about the latter two topics, space permits only a cursory treatment.

LEARNING

Learning can be defined as—'effective change caused in the learner's behaviour', that is, a change produced in his recording apparatus (*by an 'input' from the topic of learning*) which causes subsequent change in his reactions to similar (*or related*) inputs.

THE BASIC ELEMENTS IN LEARNING

There are two fundamental components of the learning process:

1. The learner.
2. What he learns.

It is salutary here to recognise that the teacher is *not* absolutely essential to learning. This does not belittle the teacher, on the contrary, acceptance of this should enable him to enhance his true role.

LEARNING ENVIRONMENTS

There are two kinds of learning environment:

1. *The 'natural' direct situation*—where the learner gains the stimulus and information directly from the topic through his own unaided observations, experiments, assessments and subsequent practical action.
2. *The 'artificial' or instructional situation*—where the learner gains his stimulus and information mainly through an agent (*the teacher*) or a situation contrived by the teacher.

The degree of artificiality depends on how real the teacher makes the situation in the *learner's own terms*, that is, on how much the learner is *involved*, willy-nilly, in the situation. *Effectiveness of instruction is very dependent upon this involvement.* In general, the degree of student involvement depends on how powerfully the learning stimulus affects him: and/or how strongly he feels the need to acquire the object of the learning; and/or how capable he feels of performing the mechanics of the learning process. If the presentation does not attract him, as it were in spite of himself, and if it has no practical reality for him, he does little to try to acquire the necessary information. Furthermore, if he cannot understand the information in his own terms (that is, connect it with his existing background of knowledge as a reference) *or*, if the medium of communication is obscure, confused or meaningless to him, he will not and *cannot* learn.

In general, that appreciation of a subject which constitutes sound learning, that is, knowledge of, and ability to perform in the topic, necessitates the development of 'thought habits' and 'neuro-muscular habits'. The key word is *habits*, and these, in turn, are acquired by the repetition of actions until their sequence becomes automatic and reflexive.

Once a habit has been ingrained it is well known how difficult it is to eradicate it, or superimpose another on it. Hence the extreme importance of using instructional methods which permit ready adaptation to new processes and skills.

INSTRUCTION

THE PURPOSE OF INSTRUCTION

1. *To reduce risks, time and effort* otherwise involved in self-learning through trial and error.
2. *To reduce the time and effort* involved in learning successful, established procedures *from* an instructor and *through* repetitive manipulative practice.

THE FUNDAMENTAL AIM OF AN INSTRUCTOR

The fundamental aim of an instructor is to present the topic such that:

1. The learner's interest is so intensely aroused that involvement is inescapable and persistent.
2. The material is readily comprehensible.

HURDLES TO INSTRUCTION

Anything that comes between the learner and a clear communication of the fundamental facts and ideas of the topic is an *interface*. This may be a skirt on a lathe-saddle (which prevents clear observation of the transmission of motion from the lead screw gears to the rack); or it can be the *opinion* of a teacher (which leads to a distorted view of the facts and ideas); or it can be the ineffectiveness of the medium of communication (such as words and their usage producing a false image of objects and ideas).

MINIMAL REQUIREMENTS FOR COMMUNICATION

1. There must be for both instructor and learner a *common awareness* of at least the general nature and function of the topic of discussion.

(For example, if the subject of discussion is a cross-cut chisel, both the instructor and learner must know what the chisel looks like and what it does.)

2. There must be a common meaningfulness in the medium of instruction. (For example, if the medium is words, 'cross-cut chisel' must mean the same thing to both instructor and learner.)

To the extent that these conditions are not met, effective communication fails. A few careful observations in the course of any lesson will establish how common is this problem. It is no proper answer to say that the learner *should know* the medium of instruction.

METHODS OF INSTRUCTION

Conventional methods of instruction

These form habits of manual skill* which make it enormously difficult for a person both to accept the need for a change in the habit, or to change the habit itself. This is because conventional instruction concentrates, either through expedience or ignorance, on producing competence in a skill as the *end* in itself.

Basic Concept Instruction

This on the contrary, whilst always aiming at skilled competence with *a* tool as the practical *means* of obtaining an end result, concentrates *its initial* instructional emphasis on developing a thorough grasp of what the essentials of the exercise are, that is:

The purpose of the *object* of the exercise.
The critical characteristics of that object.
What sort of operating device is required to produce these characteristics.
What are the critical characteristics of this operating device.
What the device must do to produce the result—and so on.

For example, if the learner is required to produce a flat surface on a workpiece with a chisel he should thoroughly appreciate:

1. Why it is necessary for the surface to be 'flat' and 'smooth'.
2. What flatness and smoothness mean in practical terms, i.e., having all the 'high spots' on a *plane*, and a sufficient number of 'high spots' to give it 'planeness'; and high spots whose 'heights' are small enough to give it the required degree of smoothness.
3. That the chisel must have a tip of such a shape that it will remove chips.
4. That when removing chips the extreme tip of the chisel wedge must be *controlled* so as to move always on the 'flat plane'—and so on.

*Manual 'skill' can be defined as competent control of the movements a tool must make in order to perform the operation.

If the proper medium of communication is used, these ideas can be easily and rapidly conveyed.

The method puts its emphasis on result and *what* is involved in the production of the result. Whilst in no sense deprecating the importance of *how* (i.e., the skilled use of a *particular* tool), this nevertheless puts the *how* into perspective as only one of the several means. It thus develops mental attitudes highly receptive and adaptable to the rapidly changing processes and skills of today's world. The method speeds learning by providing rapidly the understanding that otherwise comes only from long repetitive practice. By removing unessential extraneous matter, it induces concentration and highlights the concept.

With its emphasis on simply described essentials, the method is applicable to any level of study. It minimises the effects of illiteracy and provides not only a common national, but an international, basis for equal understanding.

The basic concept method is not new, it is as old as learning itself. It has been employed by the best teachers through the ages. Long ago Pestalozzi, and before him Roger Bacon, deprecated our tendency to complicate learning by concentrating on unimportant details or by using unsuitable mediums of communication.

Today, when visual communication has, through the advertising screen, pictorial books, films and television, become so much more powerful than the written word, it has become possible to utilise the method to the full.

A recent development in the application of the basic concept approach to learning will now be considered.

BASIC CONCEPTS AND THE FILM LOOP

Within the context of the basic concept method, the daylight-projected film loop is an instructional device of great potential for both teacher and student. It will be seen that the loop requires not only very careful design of its content and treatment, but also equally careful application as a special technique for conveying concepts by visual impact. Content and treatment require high technical and teaching expertise, and material and facilities which cannot be expected of the ordinary teacher or film company. (This, of course, does not mean that a teacher cannot make a simple film loop for his own personal purposes as an aid to make his point.) The teaching technique, however, is simple, logical and easy to acquire and use.

SOME BASIC PREMISES ABOUT LEARNING

1. *The Role of the Senses*

The 'learning channels' through which all information is acquired are the senses. (Consider what it would mean to try to give instruction on even the simplest object or operation to someone who has no sense of sight, smell, hearing, touch or taste.)

2. *The Role of the Visual Sense*

The most effective of these senses for conveying the nature and meaning of a topic of instruction is the visual one. (Consider which of these senses a person would ask for if obliged to lose all except one by an operation.)

3. *Kinds of Appreciation*

In learning about a thing, it is necessary to acquire two associated but different kinds of appreciation:

(*a*) Understanding of the ideas in the thing and their relationships (such as words and their formation into sentences to make meaningful expressions; or such as the frame, cranks, sprockets, chains and wheels on a cycle and their 'linkage' to make a meaningful machine). Properly designed visual presentation provides a rapid and effective means of ensuring understanding of this kind.

(*b*) Manipulative facility with the thing itself must be achieved. In this area, very few people are unable to *perform* at all! The degree of competence with which they perform is another matter and depends to a greater or lesser extent on *feel practice* (*kinaethesis*). By showing clearly the movement or sequence of movements in a tool manipulation, as distinct from the actual physical 'feel', visuals can greatly accelerate the acquisition of this 'feel competence'.

4. *The Role of Movement*

All animals, including the human one, are intensely attracted by movement. Indeed, over the critical period of his evolution, man survived primarily because of his powers of observation of the moving object that might be, on the one hand, a source of food, on the other, a menacing seeker of it.

This instinctive interest in movement can, and should, be utilised to the full in teaching. It is an understatement to say that this is difficult to do with words.

It must be remembered, however, that this vital interest in movement has limitations. Like the interest stimulated by the other senses, it is readily and easily distracted unless the learner is highly sophisticated and disciplined. This resistance to concentration is quite natural and has a high survival value since, obviously, if an animal concentrates for more than a few seconds on any particular sight, it is likely to miss another which may provide an easier meal or sudden death!

Ability to concentrate consistently is an art acquired neither naturally nor easily, and yet it is fundamental to the learning of a topic of instruction. Intelligent use of movement, possible with the film loop, can help considerably in stimulating learner concentration.

REQUIREMENTS OF THE VISUAL TREATMENT

To be fully effective for a learner, the 'visual scene' must not only starkly expose, and thoroughly communicate a specific point, but also this must be comprehensible against a background of reality. In addition, the visual impacts must so combine as to produce a coherent story in their own right. In some pictorial textbooks, the object is to make the essential points and tell the essential story by means of sketches instead of words. Thus the sketch, or a sequence of sketches, forms the basic text with words used sparingly either as labels or to elaborate the sketches. A brief perusal of the *Pictorial Textbook of Engineering** will make this clear and will show how much more effective the treatment could be if the sketches were replaced by carefully designed models (or perhaps the real objects themselves if suitably modified) and placed on film. Today the film loop makes this possible. Not only can individual ideas be illustrated by a stop frame technique but so too can the sequence and coherence of these ideas be established thus conveying meaning and hence the opportunity for quick comprehension by the student.

BASIC CONCEPT TEXT LOOPS

As pointed out earlier, the 'single' concept approach can be misleading since there are many ideas embodied in even the simplest topic of instruction.† The basic concept loop accepts this. It analyses the topic into all its essential points and then selects those key points which are the real core of the lesson. By careful design and emphatic visual treatment of this essential core information, the basic concept is established and presents the fundamental message. 'Personal opinions' about theories, skills or techniques are either completely removed or kept to an unobtrusive minimum. Although the loops can be used as an aid they are much more than this. They are an authoritative text-core in coherent form.

Where these loops are to form the basis of a series of lessons, they must be similarly analysed to establish the essential core of each lesson and the sequence must be carefully selected and programmed in a way which allows maximum flexibility for the teacher or the industrial training officer. Integrated with this visual text, there is written text in the form of a text-booklet. Here the visual scenes of the film are set out in correspondence to the teaching point being made, together with brief notes and/or leading questions which involve the student in closer observation of the visual core of the subject. The object of this is to lead him to his own conclusion about the significance and importance of the point and its relationship with the other points which may have relevance.

**Pictorial Textbook of Engineering* by L. W. D. Ball and A. L. Odell, Cleaver Hume-Macmillan (1963).

†Accepting the term 'single' concept as descriptive of hitherto the great bulk of existing film loops, it is important to realise that the purpose of such a loop is to supply an 'aid' so that the teacher may clarify a part of his (*largely verbal*) exposition. Critical examination of such loops shows that although they may provide the opportunity for examining basic points, they are usually limited in their extent, descriptive in their treatment or include much extraneous or distracting material—or all of these things.

PLANNING A FILM LOOP APPLICATION FOR THE BASIC CONCEPT METHOD

This demands a thorough knowledge of the whole subject and of the pedagogical requirements involved. These require a highly qualified and widely experienced teacher. The planning rests on the proposition that for maximum communication both teacher and student must have a common, equally meaningful basis of reference. This proposition is attended to if the 'visuals' are designed and treated in the film with sufficient care and knowledge and ingenuity.

The plan calls for a background 'appreciation film', preferably, but not necessarily, with commentary. In this, the subject is treated in such a way as to provide a teaching vehicle which incorporates all the essential ideas in a general, semi-descriptive, semi-analytical fashion. Its two objectives are to provide a real life, functional and purposive pointer to essential points, and a common meaningful reference of objects, ideas and terms for both student and teacher.

From this film, each essential point is taken and used as the topic of a reference text-loop. These deal with the more abstract ideas and principles in the topic. If, for example, the topic (as in Measurement and Inspection) is Geometrical Shape—in particular, say, roundness, this reference loop examines the meaning of roundness in the simplest visual terms and establishes the practical problems involved in determining it. *The full series of such loops covers all the 'topics' of the background film.*

A subsequent series of loops stems from each of the reference topics and deals with the tools skills and techniques required to measure roundness. These skills are examined on the basis of skill-analysis rather than on the 'feel' aspect of the study. When the text-loops are shown, any physical objects or models of them, even if not identical in appearance, should wherever practicable be in view and available for reference. As soon as possible after the lesson, the student should engage in practical exercises associated with the study.

It should now be clear that the basic concept method involves complete planning and programming of a subject in terms of its essential topics, of each loop in terms of its essential points and of each point in terms of its essential elements.

Because these essentials are relatively few in each case, it is possible to treat them effectively in a short film loop. Since these essentials are interpreted fundamentally, there is little likelihood of offending against the instructor's treatment. The vital lesson core is provided with high impact and retention value and the instructor is relieved of the considerable problem of acquiring and preparing suitable material for a more tutorial role. He can choose the level and extent of his exposition to suit his audience and the local conditions, secure in the knowledge that the fundamentals of the lesson have been given maximum clarification. And he can feel happy that both he and the student are discussing, in the same terms, the same thing—the unequivocal image!

12

The Concept Approach and Programmed Learning

ERICA GLYNN, B.SC., PH.D., A.R.I.C.

Research Fellow in Programmed Learning, Nuffield Science Teaching Project, Science Education Centre, University of London

INTRODUCTION

This chapter raises certain basic questions for the training officer who may wish to initiate—or advise on—the use of programmed learning.

The aim of the first section is to examine the present position of programmed learning and the underlying principles. The second section provides a discussion of some aspects of the concept approach to learning and considers in particular to what extent an instructional programme provides a useful tool for applying this method.

PROGRAMMED LEARNING TODAY

Learning is not confined to the educational field, it is a universal and continuous process during human life. Nevertheless the need for efficient and fruitful learning is most apparent in scholastic and training situations. During the last 10 years or so, increasing demand for highly trained personnel has sharpened general awareness of this need. For instance, see the 'Resources for Learning' research project, set up recently by the Nuffield Foundation. This project, phased to proceed in six stages, hopes to indicate profitable directions for development of all learning resources such as programmed learning, team and group teaching, learning laboratories, audio visual aids, etc. The work of educational and other psychologists is providing techniques which are at least potentially capable of increasing learning efficiency.

Programmed learning—one of the techniques—is closely related to instructional methods used by good teachers for many centuries. A distinction is sometimes made between the terms 'programmed learning' and 'programmed instruction'. This distinction is gradually fading, and both terms, P.L. and P.I., will be used in this chapter. Socrates has been called an 'early programmer',[1] Pressey, Skinner and Crowder are among the most quoted modern pioneers.[2]

THE PRESENT POSITION

Programmed learning has in recent years become a firmly established instructional method. A National Centre for Programmed Learning has been set up by the Department of Education and Science under the direction of Professor Peel at the University of Birmingham. The Ministry of Labour has also initiated an Industrial Training Research and Documentation Centre for Programmed Instruction at Sheffield under the direction of Professor Kay. In general, the impact of programmed instruction on military and industrial training has been considerable.[3,4] Use in schools and establishments of higher education has been less widespread, probably for economic reasons. References to surveys of use of P.I. and lists of available programmes will be found in the Bibliography at the end of this chapter.

The following gives a measure of the growth rate of programmed instructions: United States industry was making almost no use of programmed instruction in 1960. A 1963 survey on the other hand indicated that 40 companies, in a sample of 277 large United States industrial concerns, had facilities for in-house production of programmes.[5] In this country, progress has been somewhat slower, but steel and engineering concerns, as well as the armed forces, have been particularly active in developing the use of programmed instruction.

The essential feature of programmed instruction is a properly constructed instructional programme of proved effectiveness. This programme can be presented in many ways, e.g., as a programmed book, or a film in a teaching machine. Most programmes are self-instructional devices, but instructor participation in some form is generally beneficial.

The instructional material, and in particular the instructional aims, are analysed in detail *before* the programme is constructed. The material is then broken down into steps (frames). The learner in most cases works through the programme individually, and at his own pace. He is required to make a response at each frame—answer a question, perform some operation, fill in a blank in a sentence, choose between alternative answers, etc. He is then immediately informed whether his response was the correct one.

Two main types of programming are used (they may both be used in a single programme): *linear*—all learners work through the same sequence of frames, irrespective of their individual responses, a low error rate is generally aimed at: *branched*—the learner's response to a given frame determines the subsequent sequence of frames which are presented to him; for example, he is asked to choose between a number of alternative answers, and is then directed to a remedial sequence of frames—'branch' —if he has chosen a particular type of incorrect answer.

Programmed instruction has entered a phase of intense evolutionary change, particularly with respect to format of the actual instructional programmes. The frequently quoted 'principles of programmed learning' have been subjected to critical investigation,[6] for example, subject matter breakdown into small steps arranged in logical order; active response by learner at each step; immediate confirmation of results; error rate kept below some arbitrary figure—5%. Group teaching machines have proved successful, thus indicating that some degree of pacing may be more beneficial than 'working at your own rate'.[7,8] It is difficult to forecast the ultimate conclusions to be drawn from some of this research on programmed learning.

It is possible to speculate about the future of P.I. in relation to some of the detailed points, but it is probably more profitable to consider which particular features and properties of programmed learning are likely to prove firm foundations for future success. These features can be summarised as follows:

1. Integration with other forms of training.
2. Precisely stated instructional objectives.
3. Assessment of effectiveness of training, by means of criterion test (validation).

These are discussed in some detail below.

1. *Integration* of the fundamentals of programmed instruction with other forms of instruction has helped to initiate a movement towards a generalised technology of education and training.[9] Procedures such as systems analysis, task and behavioural analysis, task specification, have been applied to programmed instruction as well as to other methods.
2. The construction of a programme requires a preliminary statement of clear cut *instructional objectives*. It is necessary to decide upon just what *the learner* is expected to be able to do *after working through the programme* that he was not able to do before. It is also necessary to be certain about *why* it is required that the learner be able to perform these tasks, i.e., *to apply systems analysis by viewing the training task in the light of general organisational problems*. This type of approach is being increasingly applied to all forms of instruction, partly because of the practical and research experience gained through programmed instruction. It is interesting to note the change of title of an excellent publication by Robert F. Mager. This was first published as *Preparing Objectives for Programmed Instruction*, subsequently as *Preparing Instructional Objectives—a book for teachers, student teachers . . . for anyone interested in transmitting skills and knowledge to others.* Fearon Publishers 1962.

   The necessity for preparing finite instructional objectives for programmed instruction can provide a useful new discipline for company personnel. It provides a challenge to examine deeply and analytically the tasks which managers and others are asked to perform. It can be of help in formulating detailed managerial strategy.[10]

In the same way, programme writing is a useful discipline in the educational field. It fosters a critical attitude to the purpose of immediate instructional data and method, as well as to the wider issues of course and curriculum content.

3. A programme is generally *validated*. The construction of a programme requires a searching preliminary enquiry into the tasks to be performed by the learner, leading to a *precise specification*—in performance terms—of these tasks, and of the *criteria* which will be used to assess whether the learner has in fact reached an acceptable standard of proficiency after working through the programme. A programme is not normally put into general use until it has been tested with a representative sample of learners. It could, for instance, be specified that 80% of trainees must score 70% in the final *criterion test*, which is taken after completing the programme. The programme is *revised* until this is in fact achieved.

   A properly validated programme offers an assurance of success in this way. The effectiveness of training can be maintained irrespective of changes in personnel, for as long as the overall training situation remains unchanged.

TEACHING MACHINES

In its initial stages, programmed learning was identified with the use of teaching machines. This gave rise to much hostility among teachers and instructors.

In fact, the *programme* is the actual instructional instrument, the machine merely provides one method of presenting a programme. The human instructor is still an essential part of most instructional systems.[11]

Much heat has also been generated in the 'programmed-book versus teaching machine' controversy. A book, like a machine, is merely a device by which the programme is presented to the learner. It is not surprising that experiments generally indicate that machines are not superior to programmed texts. Most machines are little more than 'page turning devices', which may in addition prevent the learner from cheating by reading the answer first. It is also by no means certain that 'cheating' in this way prevents efficient learning. The author has found that students in any case do not cheat, once they realise that they are entirely responsible for their own progress, while working through a programme.

A machine, however simple, may have intrinsic advantages in certain situations, for example, in a workshop or laboratory to keep the programme clean, or with apprentices who may, during their school days, have developed a mental block against 'book learning'. A book, on the other hand, can easily be used for work at home.

On the whole, machines present real advantages only when their mechanical or electronic potential is fully used. In simple cases, this can mean integration with audio-visual aids. 'Audio-programming', where the programme is presented on tape, sometimes in conjunction with visual information, has shown considerable promise. Computer-based instructional systems, using 'adaptive' teaching machines, have been used.[12]

One interesting device has been developed by Philips of Eindhoven. This is claimed to be a full 'training system combining theory and practice in physics and electronics education'. It consists of a book in which each left-hand page contains programmed theory and information, and each right-hand page the relevant circuit. A self contained laboratory test set and a transparent matrix, which folds over the right hand page of the book, enable the student to carry out experiments in conjunction with the theoretical work.

Purpose built machines of this type will probably prove useful, but it must be stressed again that a machine does not form an essential part of programmed instruction, nor can it be expected that a single type of machine will be useful in all circumstances.[13]

Optimum integration of machines into the training system is still under investigation. The new training centre at the Steel, Peech and Tozer works[14] maintains a ratio of instructors to apprentices of about one to ten, while making use of teaching machines, feedback classrooms[7,8] and electronic simulators. Pupils in a large Moscow secondary school learn history, languages, geography and chemistry only in specialised teaching laboratories.[14]

EFFECTIVE AND EFFICIENT TRAINING

Programmed instruction is an *effective* method of instruction because it requires a precise specification of the training goal and because validation of the programme ensures that this goal is in fact reached. Programmed instruction can be—and has in most cases proved to be—an *efficient* method in terms of costs and investment. This depends in part on the way in which the instruction is administered and integrated into the general organisational set up, but further research is required to determine the most efficient forms of programme construction and presentation.

A truly effective and efficient method of learning should, however, achieve *more* than is generally specified by means of the immediate training objective and criterion test of acceptable task performance. It should *fit the learner for further stages* in his education, training, vocational or professional work. The learner should be able to *transfer* and *enlarge* his knowledge and skills in new situations.

The training of an engineering apprentice should not only be concerned with his present proficiency, but also with *maximum adaptability* and *transference of skills* to new processes.

The science student should not be presented with out-of-date ideas and oversimplifications because these appear to increase proficiency as measured by examination success. The student must be introduced not only to modern scientific information, but also to the methodology of scientific thought and information.[15]

More must be done than just teach the student science, he must be equipped with the intellectual and manipulative skills which will enable him to think and act as a scientist. This essentially implies a *more throughtful and wider view of specification of instructional objectives and programme content.* It is sometimes assumed that 'learning *by* discovery' is necessarily the best method of achieving this end. This, however, neglects the fact that *learning by discovery*, sometimes in the form of guided trial and error learning, is not always the best method of *learning to discover.*

Any instructional process is designed to lead the learner to performance of a novel task by providing him with skills and concepts additional to those he possessed before the instruction. The learner's behaviour is changed from the initial to terminal behaviour (see Fig. 12.1).

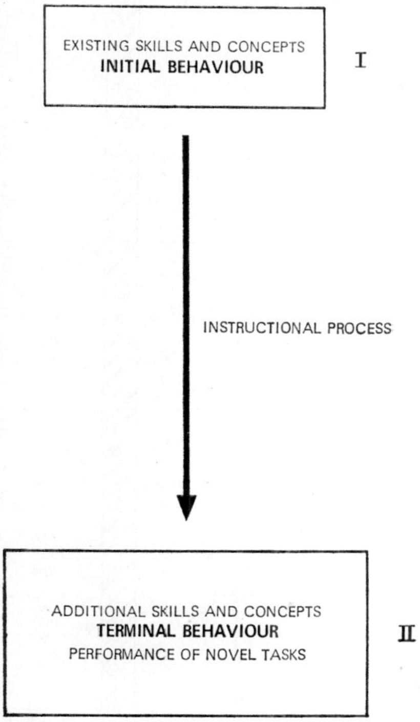

Fig. 12.1

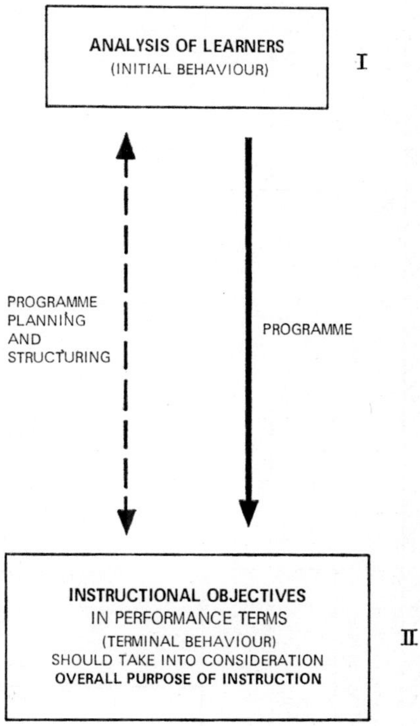

Fig. 12.2

Programmed instruction requires a detailed consideration of all relevant factors. It is particularly essential to consider the overall purpose of the training (see Fig. 12.2). It may be useful to note that although the programme leads the learner *forward* to the new task, it is often preferable to plan and structure the programme *backwards* from the objectives. This is discussed further under 'Programme Construction'.

Further expansion of Fig. 12.2 gives Fig. 12.3; and this indicates that the usual 'terminal behaviour' specification devised for an instructional

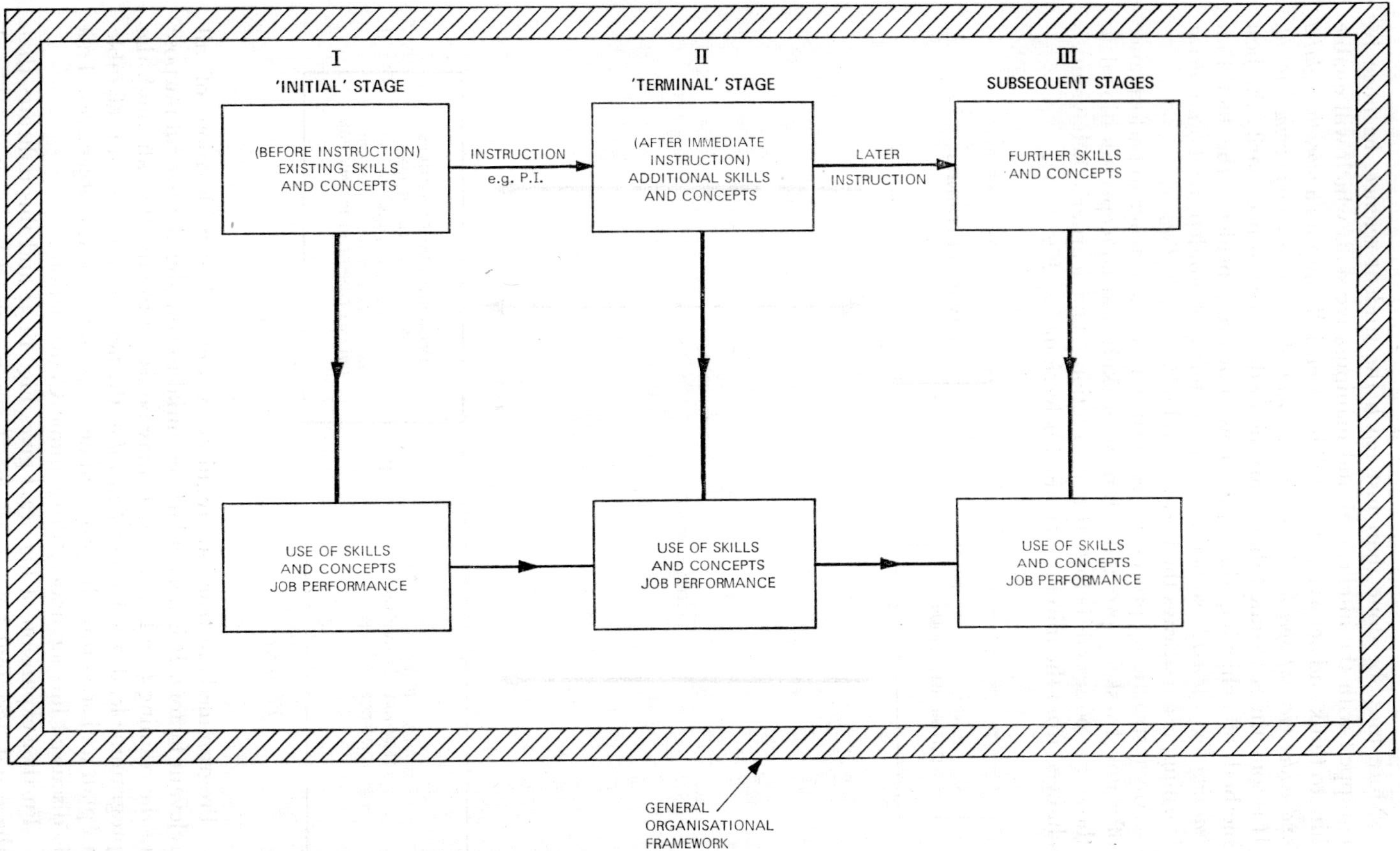
I
'INITIAL' STAGE
(BEFORE INSTRUCTION) EXISTING SKILLS AND CONCEPTS
INSTRUCTION e.g. P.I.
II
'TERMINAL' STAGE
(AFTER IMMEDIATE INSTRUCTION) ADDITIONAL SKILLS AND CONCEPTS
LATER INSTRUCTION
III
SUBSEQUENT STAGES
FURTHER SKILLS AND CONCEPTS
USE OF SKILLS AND CONCEPTS JOB PERFORMANCE
USE OF SKILLS AND CONCEPTS JOB PERFORMANCE
USE OF SKILLS AND CONCEPTS JOB PERFORMANCE
GENERAL ORGANISATIONAL FRAMEWORK

Fig. 12.3

programme should not be considered terminal as such, but should take into consideration instruction at later stages and transfer to different tasks.

The second section of this chapter discusses the *Concept Approach* to programming.

THE CONCEPT APPROACH TO LEARNING

MEANINGFUL LEARNING

The purpose of instruction—by a programme or other means—is generally to enable the learner to perform a novel task in an efficient manner.

Meaningful learning, in contrast to rote-learning, *involves concept formation*, plus the acquisition of relevant skills.

Rote-learning eventually enables the learner to repeat a sentence, an operation, etc., automatically and correctly, but without necessarily appreciating the meaning and significance of the task. Learning which involves concept formation is likely to lead to a far more efficient build-up of mental and manipulative skills.

A chemical process worker may for instance be trained to consider that efficient working of a filter press involves carrying out a number of operations in a predetermined sequence. He is trained to perform the task in the particular operational situation,[16,17] and is not instructed in the underlying concepts, for instance *why* the job is performed in the given way. An operative, who knows that the press must yield a well-washed, compact filter cake in the shortest possible time with minimum loss of material will be able to *transfer* to different types of presses without much additional training.

CONCEPTUAL CONTENT OF USEFUL TRAINING

All levels of meaningful—useful—learning involve some concept formation. A child who is familiar with the meaning of the term 'dog' can *associate* certain essential qualities with this term—even though his own pet dog may be of a different breed from his neighbour's dog. The child can also *discriminate* between dogs and other animals. The zoologist, however, has a deeper appreciation of the concept of 'dog'.

The degree and depth of concept formation, and particularly the extent to which different concepts must be related, vary with the tasks which must be performed when using the concepts.

For instance, it would be expected that the conceptual content of useful training should increase within each of the series given below:

1. Copy typist———→ Secretary, Clerk————→ Manager.
2. Semi-skilled worker→ Technician——→ Technologist→ Pure scientist.[18]

The conceptual content of lower levels of education and training may be small, it is, nevertheless, an essential part in most cases. A copy typist, for instance, must appreciate concepts such as neatness, accuracy and so on.

At the higher levels, correlation of concepts—chaining and strategies—is involved in processes such as judgment and decision making. There is presumably an upper limit to the intricacy of conceptual structure which can be imparted directly to the learner. It is doubtful whether we can teach 'leadership', by programme or any other means, and expect to be able to *measure* the effectiveness of the instruction. A subject or task which is too vague to be defined cannot be taught with guaranteed success. The rigorous analytical approach required in programmed instruction may, however, help to break down the subject into sub-topics and skills which can be transmitted effectively. It may be impossible to teach an individual how to be a good manager, but we can at least make him aware of what good management entails—according to current theory—planning, co-ordination, motivating and so on.

Application of behavioural analysis to concept formation and decision making processes, such as medical diagnosis, trouble-shooting procedures, has met with some success. Some decision making processes are said to come within the domain of instructional technology.[18]

DEFINITION OF CONCEPT

Concept formation is an important part of meaningful learning at almost any level. It involves a process of *abstraction*. A number of definitions of concept have been formulated. For example:

1. A concept may be defined as a class of instances all having a particular property. Concepts may be formed, therefore, by having experience of a range of examples that belong to the concept, and examples which fall outside that class.[19]
2. Generalisation *within* classes and *discrimination* between classes, this is the essence of concepts.[20]

The first few frames of a programme issued by Pfizer & Co. for medical practitioners deal with the concepts of *allergy* and *immunity* by stressing similarities and differences between them. A child can be taught to recognise triangles irrespective of colour and size.

USING A CONCEPT

Concept formation can be demonstrated only by appropriate behaviour, i.e., by *using the concept in a given situation.* Individuals at different levels of proficiency cannot be expected to use the concept in the same way. *Are they in fact using the same concept?* Take the concept of '*boat*' as an example. It can be said that the concept of boat is the general idea formed by a process of abstraction and the selection of the essential qualities in which all boats are alike. Different classes of individuals will, however, detect *different* essential qualities in which all boats are alike.

A child playing with a toy boat, primitive man fashioning a boat from a tree trunk and a marine engineer cannot possibly associate the same essential characteristics with the concept of boat. *Concepts have meaning only if they are related to a particular set of circumstances and, more precisely, to behaviour by a given individual, or group of individuals, in a particular situation.*

The concept of 'using a pencil' is analysed in Chapter 11, p. 121. This analysis can be carried further by relating to the individual who is using the concept. A schoolboy, an artist, a draughtsman do not have identical conceptual structures for 'using a pencil'. Some years ago, while engaged on research and development work on ball-point pens, 'using a ball-point pen', conveyed to the author ideas such as: correct gap size between ball and housing, mechanical wear in the system, chemical composition of the ink and the effect of this on the viscosity at different temperatures. The author is no longer actively engaged in this, and generally thinks of 'using a ball-point pen' in much the same way as any other consumer.

Instructors are concerned with more than just teaching a concept. They should be clear on *how the concept will be used by the learner.* Instruction in 'properties of water' may well be different according to whether we are training engineers, biologists, power station operators or plumbers. Some underlying common principles may be relevant, but it must be kept in mind how these will eventually be used. The economist does not look at a tax return form in the same way as an inspector of taxes.

In law, a patentable invention must not be an *obvious* development of known articles and processes. It can be asked 'But obvious to whom?' The legal answer is: 'The man skilled in the art to which the invention relates'. Like other concepts, that of *obviousness* acquires meaning only when used by a definite type of individual in given circumstances.

Programmed learning demands precise instructional objectives, stated in behavioural terms, and is, therefore, well suited for teaching *concepts of use* in given situations.

OPERATIONAL AND CONCEPTUAL DEFINITIONS

Confusion about underlying concepts is sometimes due to lack of appreciation of what 'conceptual definition' of a term means. Advertisements, for instance, may state: 'Latest concept in luxury living: large lounge, four bedrooms, hot air central heating, patio . . .' and so on. These do *not* constitute a conceptual definition of living luxuriously. They are merely —in certain circumstances—the outward *operational* manifestations of this concept. A *conceptual* definition of luxury living might include: comfort, entertaining, relaxation, privacy, etc. . . .

This point was well emphasised in a recent report on the work of an architect.[21]

> 'Peter Aldington believes strongly that adequate research on how people want to live is in fact more vital than the design; he says that with an adequate brief the house designs itself.'

The design of the house is the *operational* outcome of the *concept* of '*how people* want to live'.

The pattern of organisation of a particular business is the operational outcome of the concept of 'running an efficient business'. The operational aspect of 'failure of a machine' might include: gets hot, ceases to work, etc.; the conceptual aspect would include the underlying causes of failure.

The recent controversy about the labelling of wines can perhaps be analysed by contrasting operational and conceptual aspects. The wine trade emphasises operational aspects, i.e., that the label implies a given taste, bouquet, colour. The label on the bottle, however, makes a much wider appeal to the purchaser, because, in addition, it conveys to him the conceptual association between certain French wines and gracious living.

Both operational and conceptual definitions must be related to particular situations. Consider for instance possible definitions of '*good tea*'. Examples of operational definitions are:

1. Made according to instructions (girl serving in a canteen).
2. Minimum content of certain phenolic compounds (analytical chemist).

The conceptual definition of good tea could include: satisfied customers plus low cost (restaurant manager).

Training in the production of 'good tea' would involve a detailed preliminary analysis of the concept to be acquired by the learner and particularly of the way in which the learner will *use* the concept. It is not always easy to draw a clear distinction between operational and conceptual definitions. This may be associated with difficulties in formulating any kind of conceptual definition in a given instance. There is no single, conceptual definition of a 'good manager', but a knowledge of some of the newer techniques of management may possibly be an operational manifestation of a 'better manager'.

CONCEPT FORMATION. ASSIMILATION AND ACCOMMODATION

It has been seen that instructional objectives must be formulated with reference to the use of concepts within the immediate organisational system. More than this, however, is required for training of maximum usefulness.

Subsequent stages in training and education, and *future* tasks should also be considered. A barrier to further learning must not be created; present training should be of assistance in later training.

Experimental evidence has shown on the one hand that learning a second task can interfere with efficiency of performance of the first task,[22] and on the other hand that consecutive learning tasks can reinforce one another.[23] These two types of results have been integrated into a theory which distinguishes between two different processes for concept formation.[24]

Assimilation occurs in circumstances where a learner can fit new concepts easily into an existing conceptual structure. What he has learnt before *helps* him to assimilate the new concept.

Accommodation becomes necessary when new learning does *not* fit into existing concept categories, which then have to be reorganised. In practice, this may lead instead to *distortion* of the new concept. The learner distorts the new concept to make it fit what he already knows. This is a common trouble in post 'A' level science teaching. Students who have been presented with oversimplified and frequently out-of-date concepts become bewildered and disorganised when they are told to 'forget what you have learnt'. Their earlier training is *worse* than useless, it *interferes* seriously with later work. This represents a *wasteful* and *damaging* learning process. The world wide movement towards reform of science curricula should help to prevent this in the future. Examples are the Nuffield Science Teaching Project in this country, the Chem Study Project in the United States, the Unesco Pilot Project for Chemistry Teaching in Asia.

An efficient learning process should aim at creating a conceptual structure into which later learning can be fitted as far as possible by simple assimilation.

Programmed learning, with its carefully analysed objectives and content, its tested efficiency, can hope to achieve this aim with a greater degree of success than an instructor who, in the normal classroom situation, lacks the time for an analysis in depth of the instructional process and generally does not subject the results to the same rigorous performance testing.

THE CONCEPT APPROACH

The concept approach to learning takes careful account of the conceptual structure acquired by the learner and ensures that these concepts can be usefully applied in subsequent training and job performance tasks. This is possible only in cases where the *conceptual structure in the instructor's—or programmer's—mind is sufficiently widely based.* A teacher whose own knowledge and experience is limited to the immediate task and little else simply cannot adopt a sufficiently widely based view of the training. Programmed instruction can overcome this by bringing to the learner the benefit of instructors who do have this wide experience and who, in addition, have been able to spend sufficient time on the construction of an instructional medium—the programme—for transmitting these concepts efficiently.

The learner, too, must be aware of the overall purpose of the immediate instruction, he cannot, however, hope to have the same depth of insight. The lecturer and student cannot, for example, attach identical meaning to terms such as 'electron' or 'laser'. The lecturer—or programmer—must decide which *particular aspects* of a given concept are most usefully transmitted to the learner at a given stage of instruction.

The concept approach is summarised in Fig. 12.4.

PROGRAMME CONSTRUCTION

One method of programme construction involves the analysis of the subject matter by breaking it down into *rules*. The relationship between rules is then investigated by means of a *matrix*; *concept areas* are identified;

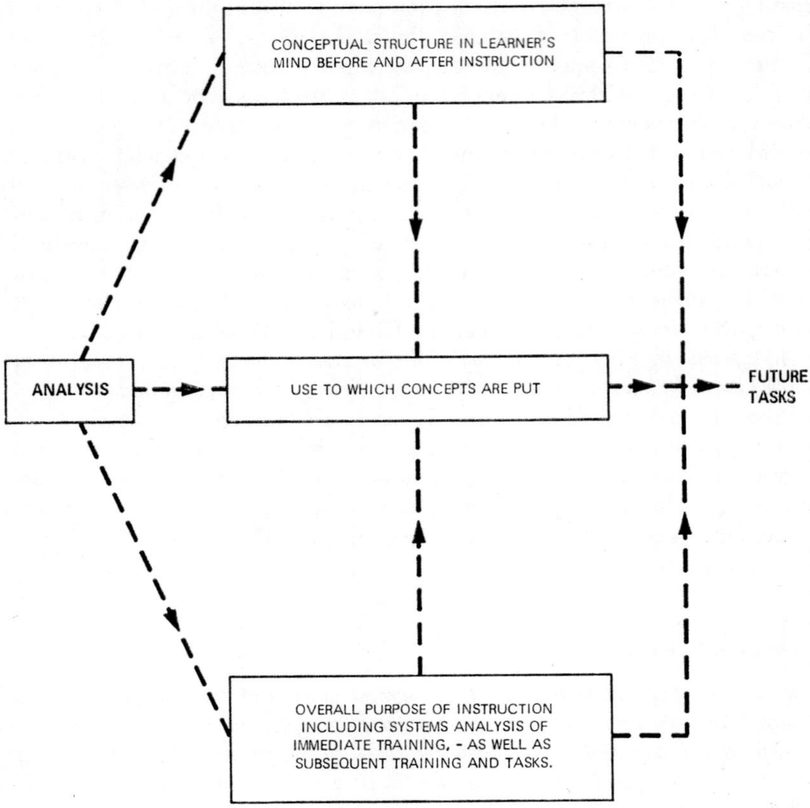

Fig. 12.4

a *flow chart* is constructed.[25] Recommendations include 'proceed from the known to the unknown; proceed from the simple to the abstract'.[26]

This method puts too much emphasis on the structure of the subject matter as something divorced from the learner. The behavioural aspect, how the learner will use the concepts, receives relatively little attention. Moreover, the rules which are formulated often lack precision,[27] relationships between rules is open to more than one interpretation, possible interference between concepts is disregarded.[28]

The author uses a different method. This involves essentially a system of analysis of:

(*a*) Instructional objectives
within
(*b*) the framework of the purpose of the immediate instruction
related to
(*c*) the general organisational situation
and
(*d*) future training tasks.

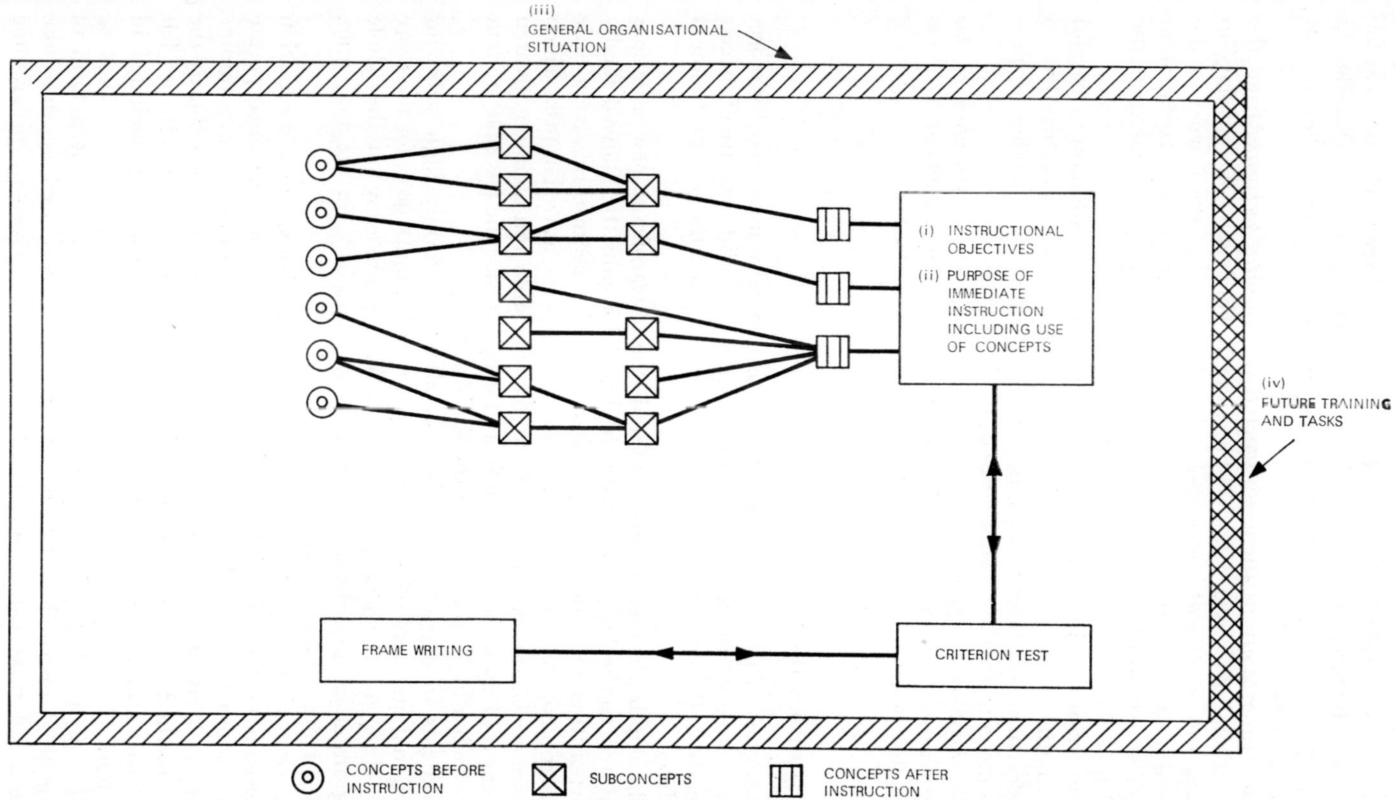
(iii)
GENERAL ORGANISATIONAL SITUATION
(i) INSTRUCTIONAL OBJECTIVES
(ii) PURPOSE OF IMMEDIATE INSTRUCTION INCLUDING USE OF CONCEPTS
(iv)
FUTURE TRAINING AND TASKS
FRAME WRITING
CRITERION TEST
CONCEPTS BEFORE INSTRUCTION
SUBCONCEPTS
CONCEPTS AFTER INSTRUCTION

Fig. 12.5

'What the learner is likely to be doing two or three years later—as far as this can be foreseen'—is considered, as well as the usual 'What should the learner be able to do after completing the programme that he was not able to do before'.

Each programme is then constructed by working *backwards* from the instructional objectives, i.e., deciding in turn which sub-concepts have to be built up, over and above those already known to the learner at the beginning of the instruction. This method is essentially an adaptation of Gagné's method of sequencing by means of a progressive analysis of the final and then interim objectives.[29]

The writing of individual frames has involved, in addition, a careful examination of terminology, phrasing, analogies, etc., to ensure that no interference with later work was being built up.[30] This method of programme construction is summarised in Fig. 12.5.

The construction of a programme requires continuous consideration of the conceptual structure which is evolving in the learner's mind, and also of the use to which these concepts will be put in subsequent training and tasks.

THE CRITERION TEST

The success of programmed instruction is generally measured by scores in the final criterion test. Whether or not this is a valid measure of success depends on the exact nature of the criterion test and the interpretation of results.

The depth of concept formation and ease of transfer to new situations is related to the learner's general ability. This must be allowed for by incorporating into the criterion test tasks which require relatively little correlation of concepts and transfer to novel aspects, as well as tasks which require *maximum* correlation and transfer. The able and highly motivated student is likely to attain a high score for both tasks, the less able student will reach a high score only in the simpler task.

Students in one case[31] reached an almost uniformly high score for individual questions of the criterion test: average mark 90% for the more able, 83% for the less able students. Writing an essay on the whole programme led to different results: average mark 62% for the more able, 35% for the less able student.

It seems that programme validation figures which show no correlation between test score and ability are obtained because of failure to incorporate into the test tasks which require use in depth of the whole programme material. Writing an essay on the whole programme is not necessarily the best method of assessing depth of concept formation. It is felt that *some* task requiring maximum correlation should be included in the test.

This approach to programmed instruction can represent a form of 'branching' instruction. Learners who do not score well on the more generalised tasks can be given additional instruction—by programme or other means. The remaining students will not require this.

CONCEPTS AND FACTS

Programmed instruction has been said to be most suitable 'in subjects where considerable factual information has to be assimilated'.[32] In another instance, 'the more abstract conceptual topics are better by far'.[33]

This apparent conflict of views may possibly be due to confusion about two points:

1. *Learning facts* is equated to *rote-learning*.
2. The *conceptual* content of *facts* is ignored.

'Polythene is made by the polymerisation of ethylene' is a factual statement, which, however, has little meaning unless the concept of polymerisation is understood.

The example below examines further the relation between concepts and facts. Supposing for instance that a programme criterion test asks: '*Columbus discovered America in 1492. True or false?*' If the instructional objective of the programme included simple rote-learning of 'Columbus discovered America in 1492', then the above question may *appear* to test this.

But what is really being tested: Who discovered America? When? What did Columbus discover? What is meant by 'discovered'? A meaningful and precise instructional objective, which included consideration of concept formation, should have sorted all this out.

The question can then be rephrased—incidentally, avoiding the appearance of dogma. '*Columbus discovered America in 1492. Do you agree with this statement?*'

The learner who does *not* agree can then be asked the reason for his decision: It was not Columbus, the date was different, the learner wishes to have first a precise definition of the meaning of 'discovery of America', for instance whether the discovery must lead to significant political and economic consequences.

The learner who agrees with the above statement could be asked to read and comment on an article entitled 'Yet Another Discoverer of America'.[34]

This article discusses rival claims that Leif Ericsson, or possibly some Chinese or Japanese, discovered America. It is particularly concerned with the claim that a Welshman, *Madoc*, got to America three hundred years before Columbus.

To quote from the article: 'Discovering America seems to be on the verge of turning into an international sport in which the rules are none too clear and the prospect of successful appeals for foul play all too frequent . . . There is another point about Madoc, though. *A discovery can be said to be a discovery only when the explorer comes back to tell his country about it.* Madoc did not.'

Before deciding whether or not a programme is a suitable medium for teaching 'the discovery of America', an instructor must sort out his own concept of *discovery*.

'$1 + 1 = 2$. True or false?' can be a test question in an elementary arithmetic programme. '1 + 1 = 2. Do you agree with this statement?' could serve as an introduction to different number systems. In the binary system, 1 + 1 = 10. On the whole, a programme is 'what you put into it'.

CONCLUSION

Recent developments strongly indicate that programmed instruction will play a fundamental and integral part in the future technology of education and training.

The rapidly increasing pace of technological and economic changes requires training which leads to maximum adaptability to new situations. The concept approach to learning ensures that this is achieved. It does so by its analysis of the conceptual structure which is to be formulated in the learner's mind, of the use to which these concepts will be put, and above all, of the overall purpose of the particular training. Concepts can be taught by programme, providing this overall purpose is kept in view throughout, and providing a sufficiently imaginative approach to programme construction is adopted. The need for high standards of programming is crucial.

Insistence on properly and precisely formulated instructional aims, and defined criteria of instructional effectiveness, checked by means of validation, will then become the most valuable and enduring features of programmed instruction.

REFERENCES

1. LYSAUGHT, J. P., and WILLIAMS, C. W., *A Guide to Programmed Instruction*, J. Wiley, 3 (1963)
2. LUMSDAINE, A. A., and GLASER, R., *Teaching Machines and Programmed Learning, a Source Book*, National Education Association of the U.S.A. (1960)
3. MINISTRY OF LABOUR, *Visual Education Yearbook*, H.M.S.O., 12 (1966)
4. MINISTRY OF LABOUR, *Memorandum No. 3, The Use of Programmed Instruction in Industrial Training*, H.M.S.O., 4 (Feb. 1966)
5. GLASER, R., *Teaching Machines and Programmed Learning, Vol. 2*, National Education Association of the U.S.A., 685 (1965)
6. GLASER, R., *Teaching Machines and Programmed Learning, Vol. 2*, National Education Association of the U.S.A., 66 (1965)
7. HOLLING, K., *Progrd Lrng*, 17 (May 1964)
8. MOORE, D. L., *Progrd Lrng educ. Technol.*, 37 (Feb. 1967)
9. *Programmed Instruction in the British Armed Forces*, H.M.S.O., 3 (1966)
10. COLDSTREAM, P., *Financial Times*, 'Programming Skills for Manager and Craftsman,' 21st Jan. (1965)
11. *Programmed Instruction in the British Armed Forces*, H.M.S.O., 23 (1966)
12. *Programmed Instruction in the British Armed Forces*, H.M.S.O., 8 (1966)
13. HERIOTT, J. D., *Industr. Train. Int.*, 226 (Aug. 1966)
14. *The Times*, 6th Dec. (1966)
15. COULSON, E. H., and NYHOLM, R. S., *Educ. in Chem.*, 232 (Sept. 1966)
16. GANE, C. P., HORABIN, I. S., and LEWIS, B. N., *Industr. Train. Int.*, 160 (July 1966)
17. U.S.I. GREAT BRITAIN LTD., *Tutor Age*, 33 (Autumn 1966)
18. GLASER, R., *Teaching Machines and Programmed Learning, Vol. 2*, National Education Association of the U.S.A., 447 (1965)
19. LEITH, G. O. M., *Visual Educ.*, 9 (Mar. 1966)
20. KELLER, F. S., and SCHOENFIELD, W. N., *Principles of Psychology*, Appleton Century Crofts (1950)

21. *The Times*, 14th Sept. (1966)
22. ENTWISTLE, D. R., and HUGGINS, W. H., *J. educ. Psychol.*, **55**, 75 (1964)
23. AUSUBEL, D., ROBBINS, L. C., and BLAKE, E., *J. educ. Psychol.*, **48**, 334 (1957)
24. LEITH, G. O. M., and MCHUGH, G. A. R., *Notes on Programmed Learning*, 'The Place of Theory in Learning Consecutive Conceptual Tasks,' National Centre for Programmed Learning
25. THOMAS, C. A., DAVIES, I. K., OPENSHAW, D., and BIRD, J. B., *Programmed Learning in Perspective*, Lamson Technical Products Ltd. (1963)
26. THOMAS, C. A., DAVIES, I. K., OPENSHAW, D., and BIRD, J. B., *Programmed Learning in Perspective*, Lamson Technical Products Ltd., 43 (1963)
27. THOMAS, C. A., DAVIES, I. K., OPENSHAW, D., and BIRD, J. B., *Programmed Learning in Perspective*, Lamson Technical Products Ltd., 76 (1963)
28. LEITH, G. O. M., *Tutor Age*, 23 (Autumn 1966)
29. GAGNE, R. M., *Psychol. Rev.*, 'The Acquisition of Knowledge' (July 1962)
30. SEDDON, G. M., *Visual Educ.*, 20 (May 1966); 15 (June 1966)
31. GLYNN, E., *New Educ.*, 21 (Aug. 1965)
32. BRYNMOR JONES REPORT, *Audio-Visual Aids in Higher Scientific Education*, H.M.S.O., 85 (1965)
33. *Harv. Busin. Rev.*, 118 (1964)
34. *The Times*, 6th Oct. (1966)

BIBLIOGRAPHY

General:

LUMSDAINE, A. A., and GLASER, R., *Teaching Machines and Programmed Learning, a Source Book*, National Education Association of the U.S.A. (1960)

GLASER, R., *Teaching Machines and Programmed Learning, Vol. 2*, National Education Association of the U.S.A. (1965)

MAGER, R. F., *Preparing Instructional Objectives*, Fearon Publishers (1962)

FRY, E., *Teaching Machines and Programmed Instruction*, McGraw-Hill (1963)

LYSAUGHT, J. P., and WILLIAMS, C. W., *A Guide to Programmed Instruction*, J. Wiley (1963)

MARKLE, S. M., *Good Frames and Bad*, J. Wiley (1964)

ROWNTREE, D., *Basically Branching—A Handbook for Programmers*, MacDonald (1965)

LEITH, G. O. M., *A Handbook of Programmed Learning*, University of Birmingham Education Department, 2nd Ed. (1966)

Surveys of Use of Programmed Instruction:

MURPHY, J. R., and GOLDBERG, I. A., *Harv. Busin. Rev.*, 115 (1964)

Programmed Learning Techniques and their Use in Apprentice Training, I.C.T. Engineering School, Croydon (Oct. 1965)

GLASER, R., *Teaching Machines and Programmed Learning, Vol. 2*, National Education Association of the U.S.A., Section 3, p. 647 (1965)

CENTRAL TRAINING COUNCIL, *Memorandum No. 3—The Use of Programmed Instruction in Industrial Training*, H.M.S.O. (Feb. 1966)

LEEDHAM, J., and UNWIN, D., *Programmed Training in Schools*, Longmans (1965)

LEITH, G. O. M., *Visual Educ. Yearb.*, 12 (July 1966)

Programmed Instruction in the British Armed Forces, H.M.S.O. (1966)

Publications which give information on available programmes:

Preliminary List of Commercially Available Programmes, National Centre for Programmed Learning, University of Birmingham

Register of Programmed Instruction, B.A.C.I.E., **2** (1965)

A Bibliography of Programs and Presentation Devices, C. H. Hendershot, 4114 Ridgewood Drive, Bay City, Michigan 48707 (1967)

DODD, W. E., and ENGLAND, A., *Programmed Instruction and Teaching Machines*, Stoke-on-Trent Central Library, 2nd Ed.

Programmes in Print, 1966, Association of Programmed Learning, 27 Torrington Sq., W.C.1

Teaching Machines and Programmed Learning, R. D. Gee, Herts, Hertfordshire County Council (1965)

13

Audio and Visual Aids

L. S. POWELL, B.SC., A.INST.P., A.M.O.I.P., A.M.I.E.E.

Head of the Department of Mathematics and Science, Garnett College, London

INTRODUCTION

The purpose of this chapter is to assist training officers in the assessment of the correct approach to the use of visual aids, to advise on the manner of their presentation and to help to decide which form of visual aid should be used to achieve the most effective result.

FACTORS INFLUENCING CHOICE

INTELLIGENCE

Of prime importance in determining which aids should be used is an appreciation of the intelligence level of the student. In general, it has been found that:

1. Individuals of high intelligence appear to prefer:
   Generalisations.
   Abstractions.
   Technical explanations.
2. Individuals of average intelligence appear to prefer:
   Concrete information.
   Multichannel communication, which includes photographs and complementary visual aids.
3. Individuals of lower intelligence appear to prefer:
   Line diagrams.
   Simple visual aids which satisfy the need of the individual to identify himself with the information being transmitted.

It is, therefore, important to remember that people of different intelligence react differently to the use of visual aids.

Aptitude and background must also be taken into account when choosing the most appropriate aid for making the most effective impact.

RETRAINING

When retraining older workers, their attitude to retraining will probably be one of resistance to the 'return to school' atmosphere. It is, therefore, important to create an atmosphere of lively participation. This can usually be achieved by the instructor refraining from the use of too much talk and chalk and by his concentration on simple visual aids.

Audio-visual aids, however, can also make secondary and subtle contributions to the efficiency of the learning to which they are directed. Well shown and properly handled, they are seen as evidence of the instructor's concern for his course members who, in their turn, become more ready to co-operate with him.

EXPLANATION OF WORDS

The instructor's choice of words is vital. Where the use of technical terms or phrases cannot be avoided, they can often be illustrated, either by a drawing, model or photograph. This visual 'explanation' will assist the message always provided that the aid itself is clear and appropriate and not in any sense a gimmick: the latter will weaken the confidence of the trainees in their instructor and distract from understanding.

THE POTENTIALITY OF THE CHOSEN AID

Illustrations are commonly used to show what things look like. A two minute film showing a particular machine, for example, may replace a much longer verbal description. Relationships expressed in graphical form can efficiently crystallise a welter of numerical details which, at best, should also be studied from a duplicated hand-out after their significance has been understood. Some design problems, work-study examples and ergonomic principles, can often be reduced to a series of superimposed diagrams. Such illustrations are efficient solutions to the technical problem of communication and commend themselves immediately to technical instructors.

Although the capacity of certain devices for attracting attention is quite evident, the irresponsible exploitation of this capacity brings whole presentations into disrepute. Individuals are prepared to give their attention to a striking illustration provided it is apt and useful, but they resent being treated like puppets.

Other aids may have the potential for focusing attention long enough for a series of related ideas to be fused together into a new concept. The need for this is greater with the training of individuals of lower intelligence who generally have a more limited attention span than their more able fellows. But this potential can be realised only if they are used with this end in mind. If individuals are to understand the action of a rectifier, say, they must attend for at least long enough to follow one complete cycle of events. The instructor who uses an aid as a focal centre for developing ideas should look at it himself as he uses it: if he looks at the trainees their eyes will move from the aid to him and the sequence will be strained, if not broken. If he is using a loop film, he will let them see it through without distracting them in any way. This need for the instructor to take second place to the aid and become its commentator is hard to learn but it may be critical to success in those fields where trainees must formulate concepts.

Perhaps the most usual function of an illustration is to show relationships of one kind or another; areas for sizes, lengths for speeds and so on. Where the relationships are of consequence, whilst the numerical data from which they stem are not, the latter should be stated in such a way as to give definition without confusion: the illustration itself should be uncluttered with this information.

On rare, but exciting occasions, an aid can evoke quite powerful responses by contributing to the learners' involvement in a problem. Typical of this type are the silhouettes of aircraft used in recognition training or the puzzle type of completion exercise. If these are to be most effective, they must be handled and presented as challenges.

Finally, some aids consolidate what has been done, or present as a general introduction, what is going to be done in a course. They bring together a wide range of material in a concise and acceptable form.

Summarising, aids can:

Invite co-operation
Attract attention
Supplement verbal information
Hold attention
Illustrate relationships
Challenge
Consolidate

But they will only do so effectively if the context in which they are presented is appropriate to the end.

TYPES OF AID

WRITING SURFACES

The ubiquitous chalkboard owes its popularity to its evident advantages. However well the instructor may be provided with pre-prepared visual information, moments continually recur when he knows that his instruction needs the kind of visual support that is given by illustration prepared on the spot in a form and at a pace determined by the unpredictable needs of the moment. It is then that he reaches for his chalk or whatever contemporary substitute he may have at hand. Very often the chalkboard contribution fits far more securely into lecturer–learner efforts at mutual understanding than the more clinically detached illustrations screened from projectors.

Chalkboards are cheap, they never go wrong (though they sometimes go grey) and they are always ready for use. The instructor who uses a chalkboard effectively and with familiarity is regarded as an experienced practitioner. For most situations, two adjacent boards are useful, the more centrally placed one for developing material which will be complete by the end of the period and the other for passing references, notes, calculations, technical words and so on, which will be erased immediately they have served their purpose. The arrangement and content of developed work on the main board should be planned in advance so that it fits the area available. Such a development can be used as a visual support to the final recapitulation of the training period. A typical layout is shown in Fig. 13.1.

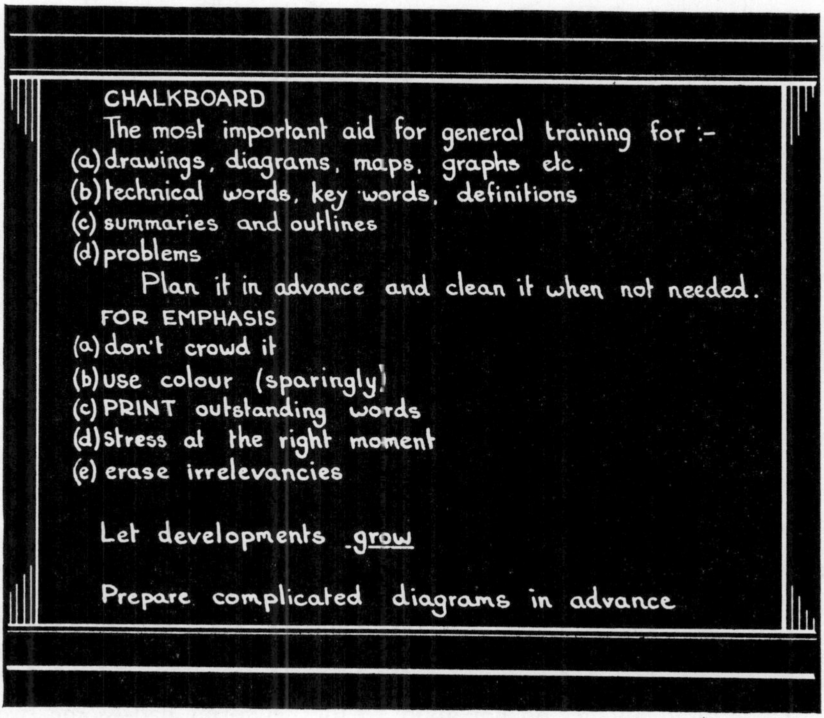

Fig. 13.1 Chalkboard

Where three dimensional illustrations are important, the chalkboard can be fitted with hinged 'wings' of perforated hardboard which can be swung forward and fixed to form the two sides of an open box. 'Lines' can be 'drawn' in space by means of elastic cords with small metal tags at their ends for anchoring them (Fig. 13.2).

One disadvantage of the chalkboard is the dust which spreads from even 'dustless' chalk. This problem is eliminated by writing on a glass or hard plastic writing surface with a water-based felt tipped marker. These surfaces must be cleaned with a soft damp cloth. Felt tipped markers are available in many useful colours and present a most pleasing appear-

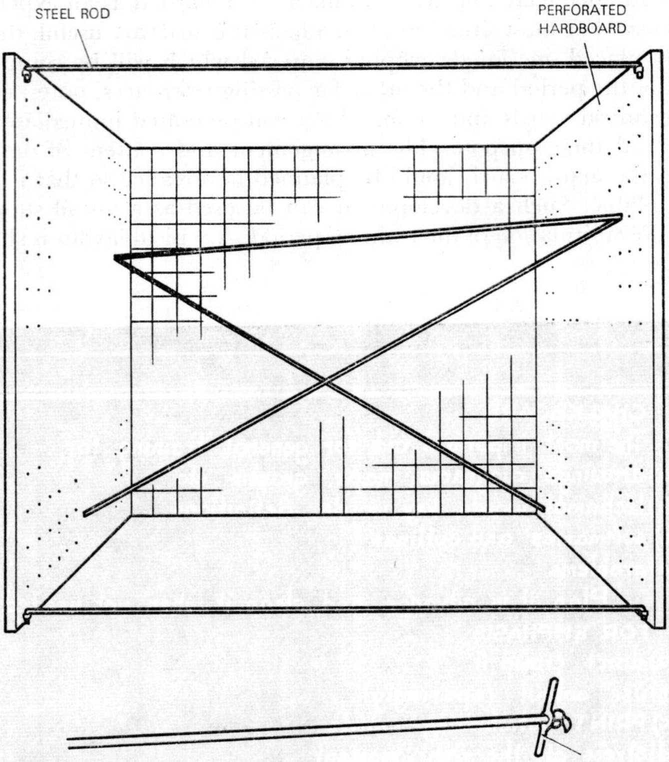

Fig. 13.2 Chalkboard with Three Dimensional Attachment

ance. The combination recommended for minimising eye fatigue is purple ink on a primrose yellow surface.

The chalkboard has two major disadvantages: it reminds trainees of school (which can deter those who associate school with failure) and it must be cleaned before further use.

These disadvantages do not apply to the newsprint pad—some two dozen sheets of white newsprint, fixed bookwise along their top edge by wire staples or wooden battens. The sheets should be about 40 in × 30 in for most purposes (Fig. 13.3). It may be used in exactly the same way as a chalkboard by marking on it either with a wax crayon or, easier still, with one of the many felt tipped pens now on the market. Work is not erased from the paper: either the finished top sheet is torn off and dropped into the wastepaper basket or it is folded over the back of the board so that it can be brought forward again for recapitulation.

One advantage peculiar to the newsprint pad lies in the fact that lightly drawn 2H pencil marks cannot be seen at a distance of more than a few feet. This enables difficult outlines to be drawn before the session begins or 'invisible' notes to be written on the pad.

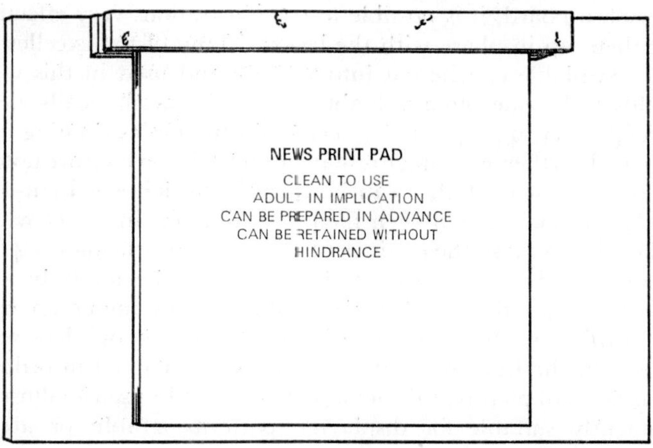

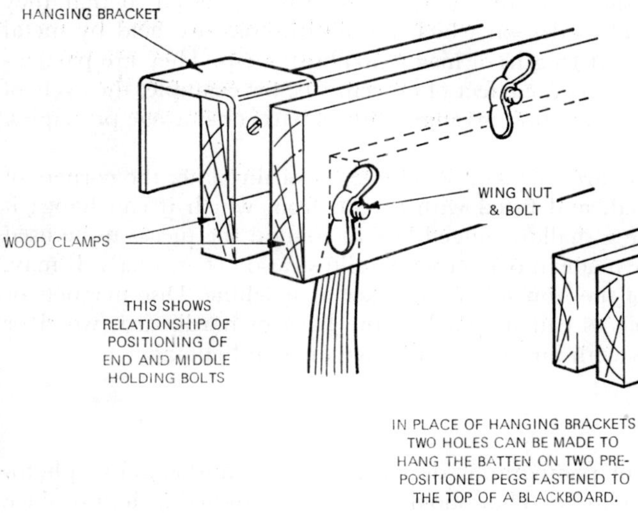

Fig. 13.3 Newsprint Pad

FELT BOARDS

A felt board consists of a piece of felt stretched tightly over a suitable board. Good quality felt, bought in lengths two yards wide, can be pinned over an old chalkboard. The surface should be brushed with a Boffin wire clothes brush to bring up the nap and give an attractive uniform and clean appearance to the cloth.

Paper pieces which are to be used as illustrations must be backed with lint, fluffy side outermost or with gridded flock paper. 'Copydex' is a suitable adhesive to use for this purpose.

With such a board, it is possible to use illustrations very effectively by building them up in phase with the lesson. Many of the excellent charts which are available can be cut into sections and used in this way. Not only is this technique often valuable, it can be aesthetically delightful and therefore very appropriate for certain forms of sales training meetings and the like. A further advantage is that large charts—which are notoriously difficult to store successfully—can be turned into feltboard cut-outs and stored in small packets and even filed in a filing cabinet; yet when they are displayed they make the equivalent of a chart measuring say 4 ft × 3 ft.

Diagrams can be drawn freehand or they can be made by drawing around the images thrown from book illustrations under an episcope. It may, in this case, be necessary to remove the episcope lens and hold it further from the instrument than the makers intended in order to get a sufficiently small picture. Of course, felt itself can be used for illustrations: it is eminently suitable for displaying patterns, graphs or areas, bar charts, etc.

MAGNETIC BOARDS

These consist of sheets of iron or tin-plate—often painted so that they will serve as chalkboards—on which the illustrations are held by metal disc magnets or short lengths of magnetic plastic strip. They are particularly useful for showing the effect of movements, for example, the cycle of a mechanism, flow diagrams, arrangements of parts or stowage principles, etc.

The magnetic board may consist of a sheet of tin-plate, the corners of which are rounded, and fitted with a cord from which it can hang; it may be made like a shallow hinged box to contain the pieces to be used on it, or it may be made into a permanent illustration (Fig. 13.4). It may, equally, consist of the iron side of a locker or machine. Disc magnets or plastic magnet pieces can be fixed to the back of hardboard, wood or cardboard shapes with an impact adhesive or with Sellotape.

PHOTOGRAPHS

The chief value of photographs derive from their authenticity: photographs seem never to lie. They should usually be studied at leisure since they carry a great deal of information and are generally less suitable than line drawings for direct teaching sessions. Of interest is the life-sized photograph which can be used to transport a working environment into a training room. Learning to operate dials and instruments fitted into such a photograph can give the trainee a sense of familiarity with the actual equipment which he would not get with the working parts alone. The same effect can be obtained with projected photographic slides or film on a back projection screen but in this case the area behind the screen is wasted. The latter technique is valuable for training in shooting, driving, navigation and so on, where responses to moving signals need practice.

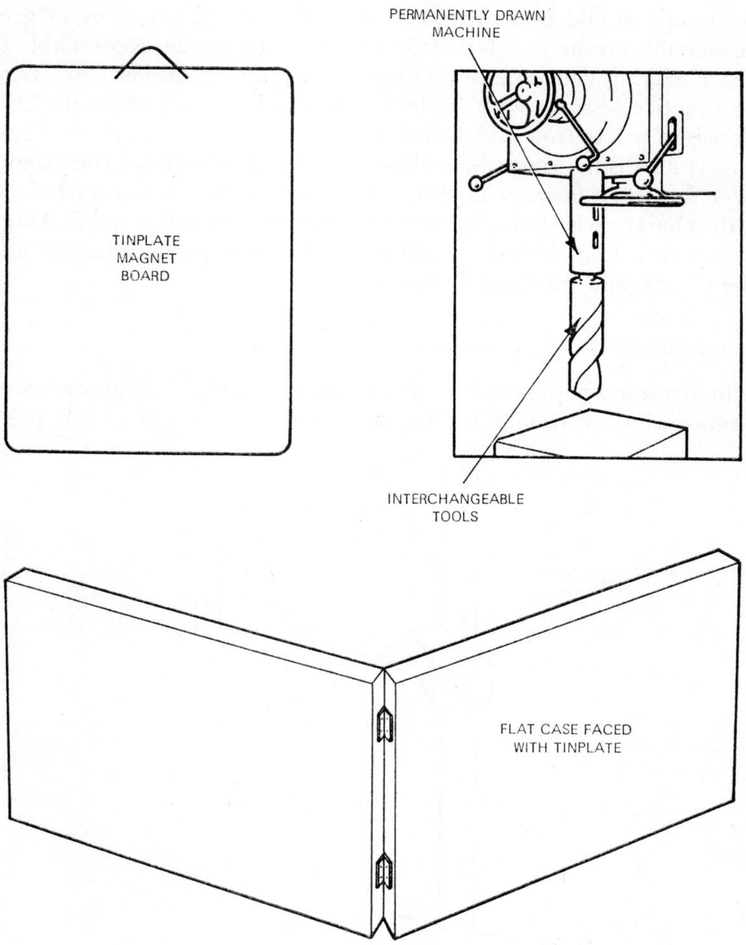

Fig. 13.4 Magnetic Board

Photographs can be important records of the development of projects which last over a period of weeks: they can be brought together at the end for a final recapitulation. Polaroid photographs which are 'instantly' developed are invaluable for 'freezing' movements during a demonstration or practical training session.

CHARTS AND DIAGRAMS

Charts are usually considered under two headings: those which are used by the instructor during a training session, and those which are for permanent display. Training charts should carry the absolute minimum of information since this will be supplemented by the instructor himself. Focal centres should be brightly coloured in, say, red or yellow whilst backgrounds should be in white or neutral tones of, say, grey or green. Professionally produced charts take time and are costly. Reasonable skill at chart making can usually be acquired by the enthusiast after two or three practice sessions with sable brushes and poster paints, under the guidance of a lettering and layout expert.

Charts for display must be sufficiently striking to attract the attention of those for whom they are intended. One method is to paper a whole wall with the chart or with lining paper on to which the chart is painted. Another is to present it in a three-dimensional display and, perhaps best of all, to fit the charts on a movable frame.

THE OVERHEAD PROJECTOR

This increasingly popular piece of equipment consists of a light box with a large glass window (usually 10 in square) in its lid (Fig. 13.5). Light passing

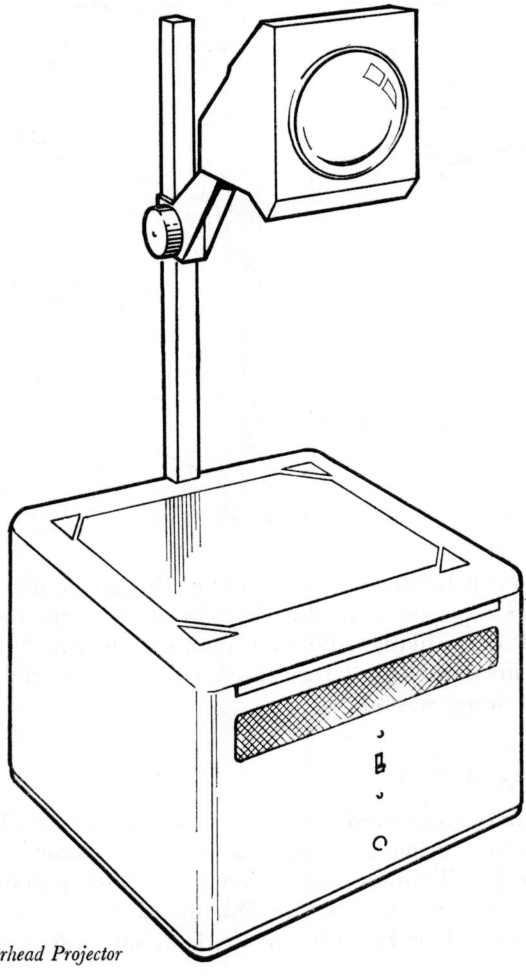

Fig. 13.5 Overhead Projector

upwards through this window can be focused on a screen, where it makes a picture of some 5 ft square. Room lights need not be dimmed.

It can be used as chalkboard substitute and, indeed, for this purpose most instruments are equipped with a roll of acetate film. This is 10 in wide and usually 50 ft long; it unrolls across the top of the window, on which it lies flat, to another roller on the opposite side of the box. As the instructor writes on the film, so the image of his writing appears on the screen. If he turns one of the rollers, the image moves across the screen and more writing surface is brought into position above the window.

Transparent squares of acetate film mounted in cardboard frames are to the overhead projector what the lantern slide is to the slide projector. The basic difference between these two forms of transparency stems from their considerable difference in size, and the consequent greater ease with which those for the overhead projector can be made.

Being so large, slides can be drawn or traced by hand either with suitable pencils, or with a pen and transparent coloured inks or opaque black ink. They may be fixed into mounts with adhesive masking tape. A wide range of slide-making materials is available including sheets of adhesive letters, adhesive shading and coloured foil. Transparencies which are the same size as the original from which they are made can also be prepared by photocopying methods. Dyeline photocopying produces images from black originals in any chosen colour and these are particularly suitable for presentation as a series of overlays.

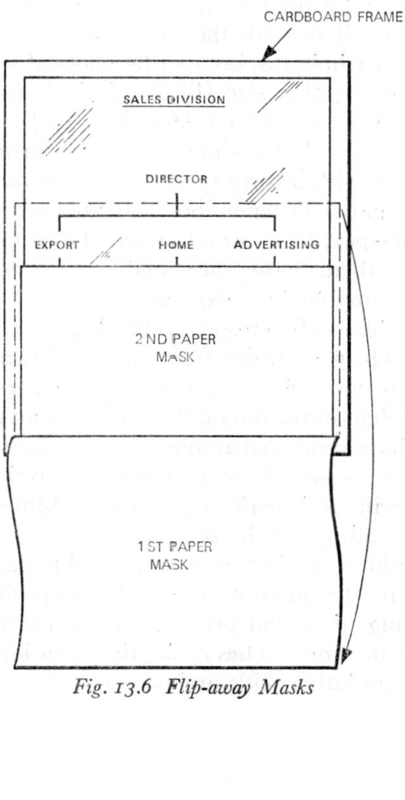

Fig. 13.6 Flip-away Masks

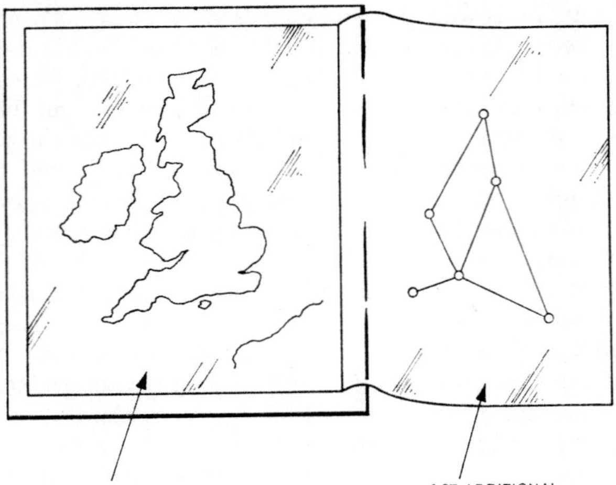

Fig. 13.7 Flip-over Transparencies

The quickest method of producing transparencies from sheet copy of suitable size is thermal copying: a dry copy, ready for projection, is made in about four seconds. Indeed, the fully equipped training department should possess materials for preparing hand-made transparencies, a flat-bed copier of the diffusion-transfer type and a thermal copier.

Transparencies can be used in the same way as training charts, or they can be made to fit in with the development of a training session. Thus, for example, a whole display can be exposed in stages by hinging back paper masks of suitable size (Fig. 13.6). Alternatively, a diagram can be built up from a series of overlays (Fig. 13.7). Parts of a diagram can be made to move with a sliding motion by fitting suitable acetate guides to the slide, or with a turning motion by pinning the moving part to the slide with a paper fastener. Sophisticated working diagrams can be built up from perspex, balsa wood or metal and more than repay the time taken to make them if they are used instead of separate diagrams for producing an animated film sequence.

Images of actual things often appear remarkably realistic on the screen: leaf forms, small tools, gear trains from small instruments, toy cars and people, beakers of water and so on. Consequently, it is often possible to carry out an actual demonstration on the surface of an overhead projector. AVO Instruments have produced ammeters, voltmeters and galvanometers which stand on the projector and actually record readings whilst a demonstration experiment is performed, thus enabling a very large group of trainees to take readings at the same time.

Most of the overhead projectors now available can be used on their sides provided the air inlets are not covered. In this position, and projecting images on the ceiling, they will project demonstrations with equipment occupying some six or more inches of depth—titration of liquids, puppet shows, certain manipulative skills and so on.

EPISCOPES

An episcope projects light reflected from the opaque surface of a picture either in sheet form or in a book. No episcope gives a clear image in a well-lit room and most of them are suitable only in a completely darkened room. A large training department might find one useful for the preparation of charts and for occasional use as a projector in the training room, but, in general, it does not earn its keep.

DIASCOPES

Standard slide projectors are obsolescent and find a place only in specialised settings such as the lecture theatres of learned societies. Two-inch square transparencies are probably the most valuable slides, particularly if they are projected from a *reliable* slide projector with automatic slide changing facility, remote control and a quartz-iodine light source. Such a projector can be used in a not-too-brightly-lit room, it is more versatile than a film-strip projector in that slide sequences can be modified, and it can be coupled, through a synchroniser to a tape recorder so that the slides are changed by inaudible impulses which the instructor has put on the tape.

CINE FILM

16 mm training films with optical sound tracks are available in abundance and many of them are excellent. But research has shown that there is a tendency on the part of instructors to place undue trust in the learner's capacity for translating what he sees on the screen into understanding and subsequent action. It has also shown that polished colour films produce no better learning of subject matter than the same films crudely produced in black and white; that questions spliced into the film can improve learning; and that prefilm tests for focusing or intensifying attention give significant improvement in learning. Indeed, the 16 mm film is an expensive training device which, by its smooth sophistication, can seduce both learners and instructors to a sense of false security. But, provided instructors guard against interpreting, 'I see', for 'I understand', and provided films are used as coaching devices or for attitude orientation, or provided that they are integrated into a course containing adequate tests and assessments, they can contribute very positively to a stimulating learning climate. They are particularly useful for presenting case studies to middle management groups, and for orientation purposes on induction courses.

Generally, short films are more useful training aids than long ones. They usually make considerably more impact if they are shown twice, separated by a discussion session during which attention is directed towards significant features which make some clear demand upon ability to observe and interpret. The instructor should always preview the films he intends to show and make relevant notes as he does so. From them, he can prepare an appropriate setting for the film and, most useful in many circumstances, can set 'before and after' tests on the learning objectives.

Film projectors suitable for magnetic sound as well as for optical sound make it possible to record and play back commentaries on magnetically striped film. This may be useful where the optical sound is in a foreign language or where its content is unsuitable. It is also necessary where films are to be made by the training department.

16 mm film projectors are precision made machines. They deserve skilled care and, ideally, operators should be trained in their proper use. A projection room adds considerably to the 'finish' of a presentation but contributes little or nothing to the teaching power of the film. Sales training departments and senior staff colleges should include small and fully equipped cinemas.

Where films are to be made in the department, the relative economy of 8 mm equipment and stock is no longer much offset by loss of quality. Super-8 film gives considerably improved picture quality and carries

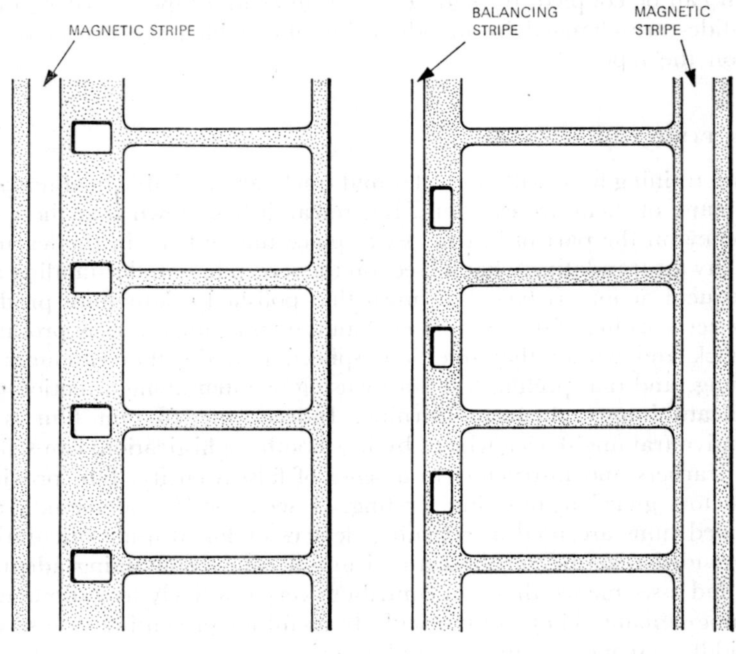

Fig. 13.8 Striping on Standard-8 and Super-8 film stock

more space for sound tracks (Fig. 13.8). A magnetic sound track can be affixed to the film at low cost by most photographic dealers, or a film striper costing a little over £10 can be bought for this purpose. Super-8 film can accommodate optical sound without any problem of interruption from the sprocket holes.

By far the most useful training film is the cassette-loaded 8 mm loop film. This consists of up to 50 ft (4 min) of standard 8 mm film which is sealed into a plastic cassette. Projectors for these loop films are available in one of two forms, the smaller which measures about 9 in x 5 in x 6 in and projects on to a small separate screen; and the larger which consists of a projector contained in a back-projection box, the screen of which measures about 18 in across (Fig. 13.9). The latter weighs about 34 lb and can be used in a normally lit room.

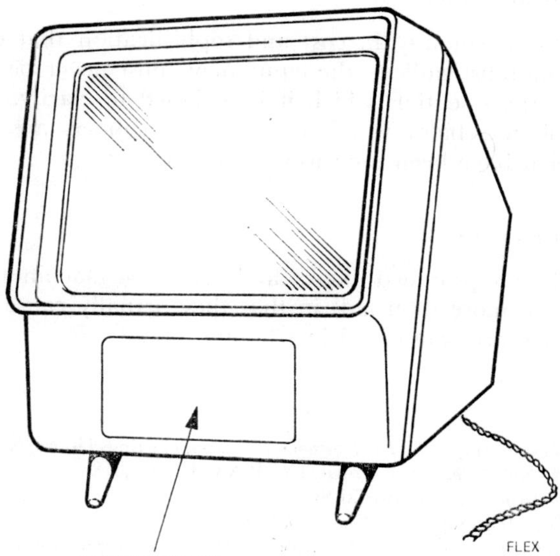

Fig. 13.9 Film-loop Projector

The projector is loaded by pushing the cassette into a slot; the instrument is then switched on and the film shows over and over again until it is switched off. For further details see, *A Guide to the 8 mm Loop Film*, by Powell and Powell, published by B.A.C.I.E.

CLOSED CIRCUIT TELEVISION

This device can be valuable for fairly obvious reasons: it can relay to many groups in many places. The camera can look at microscopic details or 'hidden' parts which everyone can then view under guidance and it can 'look in' on interviews and case studies of various kinds without interfering with them. Most important is the fact and impression of immediacy which it offers, and it is this quality which, in the end, should be evaluated in justifying its cost. Two cameras are many times more versatile than one and cameras with direct vision view-finders are the simplest to operate. Typical of the special use of this equipment would be in viewing simultaneously on one screen the two sides in a debate as a case study for analysing the psychological factors which influence decision making: this would be discussed with the viewing group whilst it proceeded.

The value of C.C.T.V. can be enhanced by a video-tape recorder. This can be used for immediate playback in training such people as sales staff, who derive enormous value from 'seeing themselves as others see them'. It can also be used to *store* recorded events which are deemed to be of appropriate merit.

MODELS AND SIMULATORS

These vary so enormously in cost and sophistication that one general comment alone must suffice—the equipment must never be so complex as to obscure the essentials which it is designed to clarify. Cardboard, toys, soap, plaster, balsa wood, elastic sheet rubber, are among the materials which have been used to good effect.

TEACHING DEVICES

These include programmed books and teaching machines of various kinds. They are more than aids in that they actually cause learning to take place: they are considered in Chapter 12.

BIBLIOGRAPHY

COSTELLO, L., and GORDON, G. N., *Teaching with Television*, Hastings House, New York (1961)
POWELL, L. S., *A Guide to the Use of Visual Aids*, B.A.C.I.E. (1961)
Using Sound Film Strips, The Industrial Society (1962)
Audio Visual Aids in Higher Scientific Education, H.M.S.O. (1963)
BROWN, J. W., and THORNTON, J. W., *New Media in Higher Education*, Association for Higher Education of the National Education Association, Washington (1963)
POWELL, L. S., *A Guide to the Overhead Projector*, B.A.C.I.E. (1964)
WITTICH, W. A., and SCHULLER, C. S., *Audio Visual Aids—Their Nature and Use*, Harper and Row, New York, 4th Ed. (1964)
ATKINSON, N. J., *Modern Teaching Aids*, McLaren (1966)
TAYLOR, E. A., *A Manual of Visual Presentation in Education and Training*, Pergamon (1966)
POWELL, G. H., and POWELL, L. S., *A Guide to the 8-mm Loop Film*. B.A.C.I.E. (1967)

14

The Project Method

J. IMRIE, M.B.E., B.A., A.M.B.I.M.

Visiting Lecturer in Management Studies, Portsmouth College of Technology

INTRODUCTION

In the context of this chapter, the project can be defined as an individual or group exercise which involves investigating a particular problem or theme and then presenting a written report on the results of the investigations. In this account emphasis is on project work with adult students which for many years has been a prominent feature of management studies courses in some technical colleges. In industry itself, the use of the project method for training purposes involves exactly the same principles. It has—as yet—not been widely used except in firms with advanced training programmes. Since the 1964 Industrial Training Act, however, there have been plenty of indications that as training specialists are increasingly employed, project method is likely to be widely used by them. For example, current technical college courses for the training of training officers regularly use the project method in a variety of ways both for individuals and for groups. It is likely that in the future the project method will be increasingly used by industry itself not only to explore and develop potential in individuals but also to probe new areas of knowledge and progress. (E.g., see the technical graduate group projects in Annexe C, p. 29 of the 'Bosworth' report on Education and Training Requirements for Electrical and Mechanical Manufacturing Industries, H.M.S.O. (1966).)

The purpose of including projects in certain college training courses is:

1. To give students practice in original investigation and/or research.
2. To develop their ability to plan the kind of investigation that is needed to deal with a particular business situation; to find, select and arrange the material needed for an investigation to the depth required by the nature of the project; to analyse information that has been obtained, to draw appropriate conclusions from it and to make any necessary recommendations that would seem logically to suggest themselves; and finally, to present information in a concise and logical manner.

The primary purpose is thus to provide a practical application of the basic ideas studied during the course and to associate the subjects studied with typical technical or business problems that the individual may encounter in his own work situation. A secondary, but still important, purpose is to give some training in preparing and writing the kind of report that is required, or may be required, arising out of his present or likely future responsibilities.

As the project is meant to be a practical exercise in analytical techniques, its value increases with the level of student being trained, with possibly a maximum application in all forms of executive development and management training. To be fully effective here, it is a form of training which requires a degree of maturity on the part of the student, and a reasonable level of both intellectual ability and of practical experience in industry or commerce. In executive development and management training courses, one of the problems is to show the connection between basic theories and techniques of management which are being introduced to the students, and management in practice, of which they may have some experience and for which they are being trained. Use of the project method as an integral part of such training courses helps to build on a student's background and experience and to involve him in the application of both old and new knowledge in a practical way.

It is for reasons such as these that in recent years the project has played an increasing part both in residential courses of the executive development type (in the form of 'syndicate' studies on set topics) and in courses held in colleges of technology for the Diploma in Management Studies. In the latter, it is increasingly common for the project to be a fundamental part of the course, and for its assessment to be an important part of the 'Award' of the Diploma, carrying up to 25% of the total marks.

CHOICE OF PROJECTS

The value of introducing project work into training courses will depend to a large extent upon the suitability of the projects carried out by particular students. This in turn will be closely related to the type of course, its duration, the time that can be allocated within a course to project work and the background of the students.

In comparatively short residential courses for supervisors or those of the executive development type, one of the major objectives may be to promote corporate activity, with each student involved to the maximum extent consonant with his background and experience. Equally, the aim is to extend the student intellectually, by associating whatever he does with the main theme or themes of the course. Finally, in courses of this nature, only a limited time is available for work other than the main lecture and discussion periods.

In these circumstances, there is probably little or no scope for the individual project. The project would probably be a group exercise, of the 'syndicate' type, where small groups of students will collectively study a set topic and subsequently present a joint report of their findings for critical analysis by the directing staff of the course. The theme of the project must be carefully tailored to the level of student and to the time available for discussion and presentation; there is no value in encouraging superficial study of a subject that is beyond the capacity of the students to explore in reasonable depth.

In long-term training courses, however, the approach to the choice of project could be different. Take, for example, a course such as a two to three year course on a day-release or block-release basis for the Diploma in Management Studies.

Here the main factors determining choice of project could be:

(*a*) The project will be done in the student's own time, to supplement course attendance periods. More time is therefore available; the probable time, desirable and available, can be determined by directing staff within reasonable minimum/maximum limits. The element of flexibility in the time available may enable a variety of projects to be undertaken.

(*b*) The general aim of the course broadly stated would be the study of management techniques and principles, together with the development, through practice, of skill in dealing with management problems and situations. Thus the method of approach must cover the provision of a body of basic knowledge as well as emphasising a practical and analytical approach to management problems.

(*c*) While the level of basic qualifications of course members is fairly uniform, students normally come from firms differing considerably in type and size, and will differ to some extent in maturity and experience.

In these circumstances, a typical project scheme might be any one or a suitable combination of the following types:

Type 1 Individual projects dealing with a suitable managerial situation, associated with the student's past or present experience and the firm with which he is employed. It is, of course, stressed that more than mere descriptive treatment is required; that the student is expected to make a critical analysis of the problem as described, with any appropriate findings and recommendations.

Type 2 Individual projects on subjects suggested by directing staff of the course, involving some original study—such subjects to be closely associated with those being studied while the project is being prepared.

Type 3 Attaching students to some firm, or a department of a firm, for a limited period, to investigate and report on a situation or problem that has been agreed with the firm or department.

Type 4 Throughout the course, each student to act as chairman of a small group of the students, and to submit a report on subjects set by the directing staff. Each student to have to submit at least one project report.

It might be useful to illustrate the above by reference to a particular course for the Diploma in Management Studies. The course is divided into two stages and students must qualify in the first before proceeding to the second, and final, stage. The first stage, building on normal previous experience (mainly of a specialist junior to middle executive grade) deals chiefly with management tools and techniques, to ensure a reasonably uniform background knowledge of them in preparation for the study of management in more depth in the final stage of the course.

Against this background, projects are required in both stages of the course. In the first stage, *Type 1* above has proved to be most suitable for the purpose. The student is invited to submit a project on a theme associated with his present or recent work interests. The presumption is that the material required will be readily available and thus no impossible task of investigation or research is imposed on him; that he can without undue difficulty apply the new knowledge he is acquiring and the 'management approach' to which he is being introduced to a practical situation with which he is familiar, and of which he recognises the importance. This kind of project should therefore be within the student's competence and yet be a means of consolidation of knowledge, and of training in the analytical approach which is so much a factor of success in the final stage.

Examples of projects of this character done successfully in recent courses are:

1. A method of production planning and control for batch-process manufacture. (A factual description and critical analysis of a new method initiated by the author in his own firm.)
2. The role of a pre-production department in a semi-conductor firm. (An organisation and co-ordination problem—author a production supervisor in the firm.)
3. Shift maintenance in a generating station. (General problems—author a shift charge engineer.)
4. The economics of employing female labour in a technical section. (Primarily a cost-factor study, with related problems of training and labour turnover.)
5. The management of quality in an electronics firm. (Author—a chief inspector, largely responsible for quality control.)
6. The organisation of product research and development. (In a pharmaceutical firm.)
7. Stores organisation.
8. A statistical study of the value of intelligence tests in selection procedures. (Applied to a particular large organisation.)
9. A pilot scheme of work study in a building and civil engineering organisation.
10. The selection and training of chargehands. (In a particular firm.)
11. The stocking of tools in tool-stores. (An example of a group project, done by four students in the same firm, dealing with a problem suggested by the firm as one requiring study at the moment.)

Projects such as these are limited in scope. They tend to deal with a comparatively narrow area of activity in a field of specialised management and with broad issues. In the final stage of the course, therefore, where the emphasis moves rapidly to an evaluation of principles, and where the integration of the various aspects of management becomes important, the type of project is changed. Normally the project given might be such as is indicated in *Type 2* or *Type 3* above.

In the first of these, the student would be given a choice of subjects suggested by the directing staff. The main objectives might be to encourage students to carry out some original study, to select and to use correctly material already acquired during the course or obtained from suitable published sources, and to promote correlation with previous studies. In these circumstances, suggested subjects would cover a wider field than those mentioned above, but would still be closely and practically related to the course content.

Again, this might be illustrated by example of projects actually given to Diploma students:

1. Some problems of growth in small and medium-sized manufacturing firms.
2. Financial policy in a medium-sized manufacturing firm; its determination and implementation in varying conditions of trade.
3. The maintenance of vitality in large-scale business enterprises.
4. A critical analysis of the specialised and managerial responsibilities of the student's position in his firm.

These projects are moving towards those of the university thesis but, provided the student is given reasonable guidance by tutors, they can provide a useful intellectual discipline, a means of widening horizons and encouraging an outward-looking approach and of appreciating the interrelationship between the various subjects studied during the course.

A project which provides these and at the same time increases practical association of basic ideas with existing business situations is that outlined in *Type 3* above. There the project is associated with existing practice in one or more firms. An example may illustrate the approach. The theme of the project was 'A comparative study of organisation and industrial relations in two firms'. The students made a number of visits to two selected firms in different industries and of a different size. They were given an introduction to the two firms as whole companies (company structure; objectives and policy; basic problems) and were then required to make a study, collectively and individually, of the existing organisation structures and of industrial relations and personnel policy and practice in the two firms, comparing and contrasting them. Information was obtained from talks given by, and discussed with, leading company executives and by personal observation, and consolidated by group discussions by students under guidance. Finally, students were required to submit individual reports on their findings.

It is also possible and often valuable to select a project which combines the general aspect of individual or group investigations and the typical 'business game'. An example is given at the end of the chapter.

SETTING PROJECTS AND CONTROL OF PROJECT WORK

Although the essence of project work is that it should represent the results of work done independently by individuals or small groups, it is important that the work is adequately directed by the training staff. The right amount of guidance has to be given to ensure that the project develops along the right lines in accordance with the 'object of the exercise'. At the same time, the purpose is to stimulate original thought and individual effort, and any excess of guidance or advice by directing staff will tend to nullify this aim.

It is desirable that some guidance be given at the stage when the project is initiated. If, for example, students are asked to investigate and report on a technical or managerial situation closely related to their own recent experience (*Type 1* above) they may be invited to suggest possible topics, after the general nature of the project work has been explained. But the suggested topics should be discussed with the directing staff, to ensure ultimate choice of one which is really relevant and appropriate, and which is reasonably within the competence of the student, from the point of view of time available for the work and of his probable ability to make an investigation of the situation to the depth considered desirable for the particular course.

Again, if the project has a more academic content (such as *Type 2* above), it is reasonable that the student be given some guidance as to source material. In either case, it is desirable that what may be involved in the project should be briefly discussed with the student, so that when he begins to prepare his material he will be proceeding along the right lines.

It is also beneficial for the student to be asked to prepare an outline scheme, with, for example, main headings and sub-headings, for further discussion with directing staff. At this stage, further clarification of the intended scope and method of approach can be made. Then the student is on his own, except that provision should always be made to seek further guidance if he runs into difficulties.

It is inherent in all this that a tutor should be allocated to a group of students, so that guidance along the lines indicated is always available.

ASSESSMENT OF PROJECTS

Completed projects should be carefully assessed by tutors and reports on them prepared, not only as an indication of relative merit, but also for the benefit of the students. It should be remembered that the project method should be used in suitable training courses as a means of providing for individual study in depth, as a test of ability, and also as a teaching medium. Comments on the general approach and detailed comments on matters of fact and how the project has been presented can be given to the student and afterwards discussed with him. A thorough two-way discussion on these topics can be very useful in the correction of individual weaknesses.

Teaching value can be increased in suitable instances by using projects as case-study material for subsequent study and discussion by members of the course. This is particularly so where an individual investigation has been made of a 'real-life situation'. The material itself is often valuable because of its association with current and practical problems. If it is presented on the course by the author of the project, his first hand knowledge and explanations increase its value. The author also benefits from having to amplify, explain and frequently to justify his facts and findings to a critical if still friendly, audience.

CONCLUSION

The project method can clearly be used in various types of college training courses. It has an obvious place in 'technical' courses, where the main emphasis is on the teaching of a skill or technique and where practical application of basic knowledge is fundamental. Equally, it has a place in courses where emphasis is on the development of the student's ability to analyse critically and express himself clearly and concisely. Project work undoubtedly requires a certain intellectual discipline and maturity of approach.

These enhance the possibility of it becoming a potent and more widely used technique within industry. As the impetus of industrial training increases, there will be an increasing need for individual firms to use reliable methods for satisfying their own promotional requirements. More effective means than hitherto will have to be found for identifying talent among existing personnel as well as for assessing and measuring the potential of the newcomer. To this end, some firms are already using the project method as a form of rotational training. The pace of change itself opens up numerous areas for 'project' investigation in every firm. At the same time that new industrial training boards impose their levies they also offer refund awards towards the cost of high quality training. This is an environment in which many firms could encourage their training specialists either to study, understand and themselves initiate projects as an effective and profitable form of training or else they could be encouraged to use the technical colleges to advise them on how best to make the most of the project method in face of the training challenges which lie ahead.

TYPICAL PROJECT

1. The group will start with a private company which is already in existence.

Exercises

1.1 First task—to examine the following

(*a*) Existing policy.

(*b*) Existing methods of manufacture and sales.

(*c*) Existing organisation structure—nature of functions, responsibilities, etc.

(*d*) Existing financial situation (balance sheet, P. & L. a/c, etc.).

1.2 Second task:

To make a critical analysis of the above (indicating strengths and weaknesses; isolating problems, etc.). To make recommendations in detail on any action needed at the moment, with present scale of production.

2. Assuming that a decision has been made to expand the project.

Exercises

2.1 To carry out appropriate design and product research.
2.2 To carry out appropriate market research.
2.3 To prepare market forecast.
2.4 To review capital situation and capital requirements, e.g., space required; new plant; working capital aspects.
2.5 To decide how to finance.
2.6 To prepare sales, production, production cost budgets (based on real data of labour costs, material costs, etc.).
2.7 To decide on changes in organisation structure needed and personnel problems arising.
(All the above to involve the preparation and presentation of reports from individuals and collective discussion at 'Board' meetings.)

3. Assuming that one year has passed and using imaginary (but realistic) sales, etc., information.

Exercises

3.1 Analysis of results (sales, financial, labour, etc.).
3.2 Analysis of problems arising and recommendations thereon.

4. Similar exercise on further expansion (if time available).

5. In 3 above—occasional interpolation of particular problems.

6. At 'Board' meetings—introduction of miscellaneous short problems concerning, for example:
Management Development.
Community Obligations.

N.B.—This type of project can clearly be expanded or contracted to meet local requirements. Equally, of course, an entire project could be to determine the training needs of a firm or organisation. This could begin with guided exercises in analysis of the component tasks within particular jobs in order to discover, evaluate and document material for an actual training programme. Advanced work within firms might then involve close co-operation with training officers in investigating training needs more broadly. (See especially Chapter 7 on Job Analysis.)

15

The Case Study Method

R. H. LEANEY, C.ENG., F.I.PROD.E., F.I.W.S.P., A.M.B.I.M.
Senior Lecturer in Management Studies, Portsmouth College of Technology

INTRODUCTION

The purpose of this chapter is to describe the case study method and its variants and to relate them to the area of management and supervisory training.

The case study method was adapted from the teaching practice of the law schools and has been developed in the field of management training by the Harvard Graduate School of Business Administration. This development has extended over a period of nearly 50 years and the case study has become an accepted part of management and supervisory training.

The case study method is a loose and ill-defined concept but its basic philosophy is that real learning is only acquired through experience. It is assumed that this can be partly simulated in the study group, provided lively and uninhibited group discussion takes place.

Case studies are often regarded as another 'technique' to be incorporated in a programme of lectures, exercises and projects. In fact, the case study method evolves from a philosophy of education which runs counter to much of traditional teaching practice. The conventional lecture approach tends to create an essentially passive attitude on the part of students which is not broken by the more liberal discussion group technique. Students still regard the lecturer or teacher as the leader and controller of the learning process. Such a group rarely examines its own resources as a means for increasing capacity for learning. This conventional concept appears to have advantages in economy of time and effort but it is debatable whether or not the intellectual capacity of the student is increased.

In the case method, the teacher's role changes. He becomes much more a member of the group and moves from the focal point that he traditionally occupies. The discussion and exchange of ideas is across and around the group, not centred on the teacher. He emerges as a responsible member who participates to ensure that students contribute to the discussion, that major points are not overlooked and obviously irrelevant issues are not pursued. In this sense he remains the leader.

The basis of discussion is provided by the case. This is an account of a situation, real or fictitious, where problems presented are relevant to the area currently being studied by the group.

Students present and defend their thoughts on the problems presented by the case and test their solutions in open forum. They learn that facts are capable of being interpreted in a variety of ways, and that others see aspects of the problem they have overlooked. Out of this debate should emerge tangible possibilities on decisions, solutions and consequences. A well conducted case study session achieves a high level of student participation.

However, it is not so easy to assess the extent of increased learning with the case method, as with more conventional techniques.

The case cannot create in the study group the consequences and responsibilities that arise from decisions in the real situation. Nevertheless, the method provides a rich opportunity for the student to experiment in his relations with his colleagues and may well reveal weaknesses of which he was unaware. He can be exposed to complex and difficult problems, relevant to his actual work, enabling him to strengthen his skills of analysis and decision making. But above all, the case method can create a dynamic situation as a basis for learning and for exposing a student to a democratic group opinion.

REQUIREMENTS FOR EFFECTIVE RESULTS

THE CASE

There is now an abundance of case studies and selection can pose a problem. Ideally, a tutor or training officer will be most effective when dealing with cases that he himself has prepared. However, case study writing is not an easy task and the time for the research and investigation is rarely available. Catalogues of cases state the area in which the case is set, i.e., finance, production, etc., and provide a brief description of the problems presented.

The initial selection may not prove satisfactory in use and some experimenting may be needed before a satisfactory programme can be developed. For example, it is not unusual to find that a particular case is a success with one group but fails with another. Abandonment of a case which has been carefully selected is rarely justified on the grounds of only one or two poor sessions.

Case studies are often criticised on the grounds that they do not present all the facts of the situation. Apart from the obvious reply that there are practical limits imposed by economics and time, there is a more fundamental reason. In real life, easy access to all the facts is rarely, if ever, possible, and even when it is, many of them are ignored. The real situation is often dealt with against the pressures of time, which intensifies the problem. In defence it can be argued that a well prepared case will present sufficient facts for its purpose, even though the writer may have deliberately excluded certain information. A further criticism, often made by students, is the absence of any information on what actually happened, the 'answer to the problem'. Harvard has always replied that the actual solution adopted may not represent good practice and that, in any event, a range of differing but equally satisfactory decisions could be reached. It is a prime purpose of the case that the students seek their own answer and attempt to justify it to their colleagues. Out of the debate will often come a concensus of the group on decisions and actions.

THE CASE TUTOR

It must be emphasised that there is no set method for using case studies. The purpose of the course, the type of student and their attitudes, the experience and knowledge of the tutor, the case selected are all variables which defy the attempt to define a routine procedure. The changed role of the tutor has been previously referred to. Sufficient time for case preparation, selection of suitable accommodation, comfortable and free from disturbance are obvious arrangements that the tutor must make. His own preparation must include a very thorough study of the case, to obtain the knowledge necessary during the session. A tutor will invariably form his own conclusions about the solution but it would be wise to be prepared to find that students present an aspect that he has overlooked, which forces a modification of his own viewpoint; a somewhat demoralising experience initially! The tutor introduces the case and may find it necessary to initiate discussion but he should avoid this discussion being centred on him. Given a good case with well prepared and alert students, a tutor may well find his role to be a completely passive one. This raises the question of the duration of a session. Within practical limits, the case should last as long as students retain interest in it.

Working to the timetable can be disastrous and the tutor should always be guided by the behaviour of the group. He may discreetly intervene to divert side tracking discussion back to principal issues; and he will inevitably encounter the monopolist who may justify more forceful stopping tactics, but so long as discussion is lively and well distributed over the group and decisions are emerging, his role should be an unobtrusive one.

He may conclude the session by summarising the main points discussed referring to other aspects that escaped the student and may wish to state his own viewpoint; but if he does so he must expect to defend it, possibly against the full group's onslaught.

It is quite possible that the task of concluding may be ably performed by members of the group; and the tutor's final task may simply be to close the session.

THE STUDENTS

It is not possible to be dogmatic about the composition of a student group. Background and experience of individuals will usually be similar on an industrial training programme. However, there are advantages to be derived from the differing viewpoints of the mixed group, provided that relative 'rank' does not inhibit discussion. Students must be allowed adequate and proper time to study a case before the session. Programmers who assume that the late night hours are satisfactory are being unfair to students. Basically students should be intelligent, reasonably articulate and able to participate in democratic discussion.

VARIATIONS OF THE BASIC CASE METHOD

Over the years, variations of the traditional approach have emerged which emphasise the flexibility of the method.

ROLE PLAYING

One inherent danger in the conventional case session is the tendency for students to take the viewpoint of the detached observer. This can create a somewhat clinical attitude. Role playing requires that members of the group assume the identity of the character in the case and their contribution is made from this standpoint. They are required personally to associate themselves with the situation presented in the case, and provided it is not treated frivolously, the student can acquire a much more intimate insight into the problems presented.

Role playing can be particularly effective as a means of attempting to explore situations where significant disagreement has arisen amongst the group.

'IN-TRAY' CASE METHOD

This is a role playing technique with a major difference in emphasis, since the student initially works on the case as an individual. He is required to assume a managerial role and the case material consists of the typical contents of a manager's 'in-tray'. The student is required to attend to the contents within a stipulated time, say one hour. The follow-on session is a discussion of the problems presented by the in-tray material.

Each student has the opportunity of comparing his decisions with those of his colleagues. This variation is primarily an exercise in limited-time decision making, and with carefully selected documents can provide valuable experience, particularly as a testing ground for assumptions and ability to establish priorities.

INCIDENT PROCESS METHOD

This variation was developed by Paul Pigors of Massachusetts Institute of Technology and attempts to create greater realism by requiring students to obtain the facts of a situation of which they have no previous knowledge by direct questioning. The case is initiated by the announcement, in the most limited of terms, of an incident. By questions to the tutor, the group seeks to determine the facts of the incident and its environment. Unlike the conventional case, students have the opportunity of obtaining as much information as they need. When they feel they have sufficient facts, they attempt to define the problems and then proceed, as in the conventional method, to decide what should be done. The incident process method is probably a more realistic one than the text case but it suffers the limitation of being confined to issues that do not require a large amount of supporting data which demands detailed study by the student.

THE SYNDICATE METHOD

The syndicate concept has been developed particularly by the Staff College at Henley. The student group is directed into sub-groups or syndicates and the treatment of the case is more formal than in the Harvard method. Each syndicate considers the case in a group, unlike the case method, conclusions reached are formally presented by an elected chairman to the total group. The total group may then comment on and criticise each syndicate's proposals, whilst the syndicate chairman acts as the principal exponent for his colleagues' decisions. It is quite usual for individual members of a syndicate to enter the discussion following their chairman's presentation and support him where necessary. Whereas the incident process method approaches practical conditions of problem identification and decision making, the syndicate method emphasises the equally real problem of formally presenting proposals and decisions to a critical audience and defending them in open discussion.

FILMS, FILM STRIPS WITH RECORDINGS

These visual and audio aids are designed to give a greater degree of comprehension of the case. Generally they are only suitable for presenting relatively simple issues but they are most useful for introducing students to the case method. Films have the particular advantage of presenting a situation as it actually occurred and students learn to face the difficulties of observing and listening objectively. Play back is not normal practice since this facility cannot be granted in real life. There are certain outstanding 'case study' exceptions to the short film; for example, the full length film (90 minutes) '*Twelve Angry Men*' provides excellent material for intensive analysis of the subject of attitude changing within a group.

CONCLUSION

This chapter can only provide the introduction to the case method and survey its principal variants. A brief bibliography is given and sources providing the largest selection of cases have been shown. The case method would seem to be of most use to the training officer engaged in programmes of management and supervisory training, yet there seems to be no reason why suitable case studies should not be used on courses for apprentices and trainees at any level. There is, therefore, abundant opportunity for experiment with case study methods in the field of industrial training.

As job and task analysis techniques are developed by the industrial training boards and by industry itself, opportunities will doubtless occur for documentation of a much wider range than hitherto of case material from British industry. Development and documentation of work from these sources might well justify grant awards by industrial training boards.

BIBLIOGRAPHY

MCNAIR, M. P., *The Case Method of the Harvard Business School*, McGraw-Hill (1954)

PIGORS, P., MYERS, C. A., and MALM, F. T., *Readings in Personnel Administration*, McGraw-Hill (1952)

HOWER, R. M., and GLOVER, J. D., *The Administrator: Cases on Human Relations in Business*, Richard D. Irwin Inc., 3rd Ed. (1957)

GAY, W., and CAMERON, D., *A Manager's Casebook*, Heineman (1967)

IVENS, M., and BROADWAY, *Case Studies in Human Relations, Productivity and Organisation*, Business Publications London (1966)

MARSH, A., *A Collection of Teaching Documents and Case Studies* (*Industrial Relations in Engineering*), Pergamon (1966)

Sources for case studies

Harvard University, Intercollegiate Case Clearing House, Soldiers Field, Boston 63, Massachusetts

Cranfield College of Aeronautics, Cranfield, Bedfordshire. Studies compiled by the Management Case Research Programme include: *Bibliography of British Case Studies in Business Administration* (1966); *Management Case Studies* (1965)

Newman Neame (Training) Ltd. Studies compiled include: *Human Relations Case Studies:* a catalogue of cases and supporting material (1966); *Case Studies for Management Training* (1953); *Case Study Writers' Circle*, a continuing series of booklets

16

Lecturing

J. SIMS

Senior Lecturer, Ministry of Labour Instructor Training College, Letchworth. Assistant Manager, Government Training Centre, Letchworth

INTRODUCTION

The purpose of a lecture is to develop a theme and to impart information. It can be the speediest but not necessarily the most effective method of teaching. It enables the maximum amount of information to be offered in the minimum time. A 'classroom lesson' situation, although more time-consuming and more difficult for the teacher to master, is usually far more effective for imparting smaller amounts of information in the same time since at all ages it involves student participation. During the lecture, the group or audience is obliged to listen. Participation can only be by taking notes of the facts and ideas presented and, where allowed, by questions to clarify specific points or to obtain the speaker's opinion at the conclusion of his lecture. In the instructor's lesson, however, the group will constantly be brought into direct participation by the use of discussion and the question and answer method described in Chapter 22.

LECTURING TECHNIQUES

A good lecture depends on good lecturing techniques and these are not easily acquired. Some people are fluent enough in everyday speech and because of this may be encouraged to believe that they would make good lecturers. In this, however, they are often mistaken. In lecturing, the volume and pitch of the voice, the speed of delivery, the effective use of pauses and emphasis, and the art of conveying meaning by facial expression must all be understood and mastered. It could be disastrous to ask someone to give a lecture without previous practice—not only for the audience, but also for the lecturer's self-confidence. A disastrous first attempt could inhibit and perhaps even destroy any lecturing potential the beginner might have: also some audiences would be slighted if addressed by an obviously inexperienced lecturer.

With lecturing, the old adage about practice making perfect can only be true if the methods being practised are correct, i.e., if both the initial approach and the method of instruction are correct. A person who diligently practises wrong methods can at best succeed only in doing the wrong things well! To deliver a lecture expertly requires hard work, forethought and planning.

It must obviously be the lecturer's aim to use to the utmost his hearers' capacity to listen, understand and retain the facts presented to them. He can best achieve these ends by ensuring that his lecture is pitched at an appropriate level and that the receptivity of the audience is kept working smoothly, and at the right speeds, throughout the lecture. To ensure this, the lecture must be effectively conceived as well as effectively delivered. Behind every successful lecture there must be perception, imagination, quality of thought, a logical framework and competent preparation.

THE FRAMEWORK

SYNOPSIS OF PREVIOUS LECTURES, IF ANY

If the lecture is one of a series, 2–3 minutes should be set aside before the lecture begins for a brief synopsis of the preceding session(s). This is one way of establishing immediate contact as well as continuity.

INTRODUCTION

If the talk is to last an hour, say, then 2–3 minutes should be allocated to the introduction. Careful thought should be given to this section of the lecture. The lecturer may draw from his personal experiences or he may quote historical facts connected with the subject, but he must not forget to make absolutely clear the object he hopes to achieve by giving the lecture. This is when he states his purpose and theme, and explains to his audience the way he intends to unfold his subject.

DEVELOPMENT BY STAGES

A lecture should be developed in definite stages. Each stage should constitute a clear exposition of one section of the whole subject. Ideally, each stage should not exceed 10 minutes in length. This will normally protect the audience from mental saturation. In a lecture lasting an hour, there could thus be four 10-minute stages, each followed by an intermediate summary and a question period. Some lectures, of course, would not easily divide into four equal parts; some might fall more naturally into three or five or six stages according to the number of key points. Normally a lecturer should never speak for more than 20 minutes without a break for questions. Many lecturers do not of course conform to this suggested pattern and as a consequence often lose the full attention of their listeners.

QUESTION PERIODS

Time allowed for questions obviously depends on the lecturer's assessment of audience reaction and interest and on his skill in seizing the right questions to provide himself with the opportunity to give powerful reinforcement to his subject. Generally, however, at intermediate stages, a 2–3 minute period should be the aim.

SUMMARIES

The intermediate summaries should be used to re-emphasise the most important aspects of the subject-matter covered up to that point, and should link each stage of the lecture into the next, thus ensuring effective continuity. The final summary should always be made *after* the last question-period has ended. It should recapitulate main points of each stage into a coherent pattern constituting a precis of the whole lecture. The introduction of new subject-matter at this point should be avoided as this would be tantamount to starting another lecture. The audience should leave the lecture-room with the theme and the key points of the lecture uppermost in their minds rather than thinking over the answer to the last question.

THE PREPARATION

After the framework has been decided, the more detailed work of preparation can proceed. It is hardly possible to give a good lecture without thorough preparation. The audience will expect to hear a well thought-out exposition not a series of haltingly delivered platitudes. Before he starts speaking, therefore, the lecturer needs to know exactly what information he wishes to convey to his hearers, the order in which to impart it and where the emphasis is to be placed. The following plan of action is suggested.

DECIDE ON THE SUBJECT

The subject or theme must be clearly formulated. With in-company courses the subject is usually taken from a programme, syllabus or curriculum of training, and should be attuned to any practical work known to be in progress at the time. It is at this stage that the overall theme and the timing of the lecture should be considered. These will determine how deeply it is intended to develop the subject. They will indicate too the likely number of stages required to break up the subject and give balance to its parts.

COLLECT THE INFORMATION

Text-books and background reading are usually invaluable as preliminary sources of material. Other useful information may be derived from professional papers and bulletins, from the press, from trade journals, and from manufacturers' publications. By drawing on his own knowledge and experience the lecturer should be able to sift his findings and present them in a manner which will make them readily understandable. Simple words make for clear meanings. All relevant material from text-books or other sources should be re-read very carefully to ensure where best to key it in. In this way the theme develops and a clear idea is retained of the general lay-out and the main and subsidiary points and conclusions. At this point a set of preliminary notes can be made. After the first reading, an attempt should be made to reconstruct the main structure from memory, arranging points noted in a logical order of importance. Anything that is not completely clear should be carefully noted in order that it may be given further study and if necessary be clarified later by further careful research. When the most important points have been isolated from the purely descriptive matter, the number of stages needed in the lecture will become apparent. The facts isolated should be separated into three categories:

1. 'Must know' Vital points which must be known if the object of the lecture is to be achieved.
2. 'Should know' Points which it is desirable, but not essential, to know.
3. 'Could know' Relatively unimportant points.

The facts learnt should be associated with facts already known and what has been read should be compared with material collected from other sources. Cross-checking of text-book references is advisable, and helps to consolidate new knowledge.

ARRANGE NOTES

The notes compiled from all sources should now be arranged in a logical sequence and a script outline should be prepared. Whilst it is not essential, a full script may be written out at this stage. This is time consuming but it certainly helps the lecturer to become much more familiar with his subject.

PLAN THE PRESENTATION

The finished script should indicate at which points in the lecture to use a blackboard or other visual aids. (See Chapter 13.) Every script should clearly show headings and sub-headings together with notes and points for special emphasis and any necessary sketches. Bold clear lettering is vital whether notes are lengthy or brief. All references to source material should be checked and cross-checked. The number of stages decided upon will indicate how many question periods to provide for. It cannot be over-emphasised that a clear plan of presentation will give the lecturer confidence in delivering his subject-matter, and will help to produce the best results.

GIVING THE LECTURE

Very few lecturers never use notes. Lecturers who are most effective have usually gone over their notes so many times and are so familiar with their layout that they are able to use them unobtrusively. However, no deliberate attempt should ever be made to conceal their use. Before starting to speak the lecturer should look around the audience to ensure that his audience is settled and ready to listen. In this way, he can establish audience contact from the start. He should begin by explaining crisply and clearly what he is going to talk about and what his procedure will be.

The main points on which he should concentrate may be summarised below.

SPEAKING EFFECTIVELY

The rate of speech needs to be slower than that used in everyday conversation as the speaker must be audible to those at the back of the audience. The volume and speed of speaking should be varied to add interest and emphasis and to avoid monotony. Clear, well articulated speech is essential at all times. The lecturer should aim too to communicate with every individual. He can create the impression that each member of the audience is receiving his attention by frequently glancing round and 'catching their eyes'. He should never speak with his back to his audience.

BEING UNDERSTOOD

The lecturer's exposition has to be clear and concise at all times. If abbreviations, trade jargon or technical terms are used, he must ensure that his audience is made aware of their meaning. Carefully chosen analogies and neatly turned phrases can often assist understanding. A positive approach to the subject is most effective; conversely negative instruction tends to confuse an audience and wastes valuable time. Sincerity and a sense of purpose always give strength. A lecturer should not only know his subject but, by his conviction, he should demonstrate that he knows it.

CREATING INTEREST

Enthusiasm and enjoyment should be evident in the speaker from the outset and he should aim to project these so that the audience too has an enjoyable experience. Liveliness, confident and frequent looking all round, well-timed pauses, ability to catch and respond to the humour of his audience, all help the good lecturer to create and rivet attention.

MANNERISMS

Mannerisms in speech or action should be avoided as they distract from the subject. A regularly mispronounced word, a word or phrase too frequently used are quickly noticed. Chalk tossing, 'washing the hands' with a stick of chalk, pencil waving, pointing the pencil, gesticulating without realising it, pulling the nose, gripping the chin, awkward posturing, tossing the mane and pacing backwards and forwards are among the most frequent mannerisms. Most of them can be avoided by clasping the hands behind the back or by firmly gripping the table or lectern with both hands.

ANSWERING QUESTIONS

An out-of-place question should not be dismissed summarily but should be noted and an answer promised later in the lecture. A questioner's inability to express himself precisely may sometimes give rise to what appears to be a stupid question. He should always be answered politely. A speaker should never allow himself to be provoked no matter how obtuse or provocative the questioner might be. The ability to give quick answers is expected from an expert and picking up the exact point of the question is equally important. Every question deserves a straight answer and if a speaker is unable to answer, he should on no account attempt to bluff as this is easily detected and will lower his standing. He should agree to find out the right answer and the questioner should be invited to see the lecturer at the end of the session. Questions may be classified into four groups:

1. Those which show that the lecturer has been understood but that some amplification is required.
2. Those which need not have been asked if the questioner had given the matter more thought.
3. Those which show that the lecturer has not been understood.
4. Those which side-track the issue (not to be confused with 1).

It is usually a good practice to repeat and sometimes to rephrase a question. This gives everyone a chance to hear it and it also enables the speaker to formulate his reply.

Type 1 questions are usually the commonest and give the speaker an opportunity not only to amplify points already made but if necessary to introduce new ones. Type 2 questions can usually be answered briefly, sometimes wittily but never unkindly. Type 3 questions may require fuller answers but here too replies should never be couched in terms which deprecate the questioner's inability to understand in the first place. Type 4 questions should be handled politely but need not necessarily be answered.

FINAL COMMENTS

As stated earlier the final summary should normally come after the last question. It should take the audience back to the beginning of the session to repeat the announced purpose, theme and range of the subject; and then with not more than a cryptic sentence or two on each stage, to emphasise main points, it should swiftly state any appropriate conclusions and end with a convincing message.

If the lecturer has conducted the session without a chairman he should not fail to thank participating members of the audience for their questions or for discussion contributions which they have made.

Finally, no matter how often a particular lecture is given the task should always be taken seriously and notes should always be gone over sometime in the immediately preceding 24 hours. If a lecturer does not do this he stands a good chance of regretting this omission. If he ever feels satisfied with the quality of his lecturing that is the moment when he should set to work to make improvements.

17

Discussion Leading and Group Training Trends

JOHN W. BARBER, M.A. (OXON.)

Senior Lecturer in Management Studies, Portsmouth College of Technology

INTRODUCTION

The purpose of this chapter is to stress the importance of discussion as a method of learning and to show the significance of skill in discussion group leading for both the line manager and the training officer.

The scope and range of some of the main forms of discussion leading will be considered and detailed techniques will be outlined for the conduct of effective discussion group leading within training.

DISCUSSION AND MOTIVATION

Discussion is a valuable means of stimulating and consolidating motivation in the individual through his own participation and through the effect and influence of other members of the group.

DISCUSSION INFLUENCES ATTITUDES

Lively group discussion does two things; it involves the individual and gives an appearance of free expression. This appearance may be deceptive since most people seek group approval for their views and ideas and almost subconsciously they are prone to modify them—sometimes temporarily—in face of the pressures and counter-pressures which group discussion produces. Individuals join in the discussion and also listen, perceive and seek approval or confirmation. Sometimes they try to convince others and sometimes they themselves may be convinced. To that extent group discussions may change opinions. In a truly co-operating group it may even change behaviour and attitudes.

In British industry new ideas about discussion are developing today. After 25 years of the 'human relations' approach to supervision, the idea is at last breaking through that succesful management and supervision at all levels should be based on participation through discussion between the supervised and the supervisors—not so much in terms of joint consultation with elected representatives—but more in terms of face-to-face discussion between individuals. Discussion on specific roles, responsibilities, work targets and measurable results is now being seen to be vital at each level between those in authority and those immediately below who will have to achieve particular work targets. This in essence is what the doctrine of management by objectives is about. Here quite obviously there will be increasing need for the skills of effective discussion.

TO PARTICIPATE IS TO LEARN

Meanwhile, and especially since the stimulus of the 1964 Act, industrial training is increasingly recognising that effective group discussion leading among adults will probably provide one of the most potent of all learning methods. Discussion group work is being seen to be more effective because it creates more arousal in the individual. At all levels it can stimulate active participation. As a good discussion continues it can very quickly arouse more interest and involvement and thereby create more positive motivation than the straight lecture achieves. Once motivation is generated through discussion this process in turn can become cumulative. If the same group meets regularly for a week or more on a residential course, then provided the discussion leader knows his job and provided every individual is free or comparatively free of all status anxieties, each finds himself caught up in a very dynamic form of activity. Almost imperceptibly attitudes tend to change. Teamwork develops. Each individual begins to know his group and within its pressures and counter-pressures begins to find his own particular role. He begins to find that he 'belongs'. The longer the same group meets in this way, the more pronounced these characteristics become. In this environment individuals may never be completely tuned in to one another but they are at least on approximate wave-lengths and are therefore disposed to learn from one another. The wise group discussion leader recognises that to participate is to learn and ensures that all are drawn in and that their contributions are developed, tested and explored. He does not so much try to teach the members of his group as to place them in a favourable position to learn. In terms of his own role in the adult learning environment he might recognise the aptness of the following lines:

> 'No man can reveal to you aught but that which already lies asleep in the dawning of your knowledge. If he is indeed wise he does not bid you enter the house of his wisdom, but rather leads you to the threshold of your own mind.'
>
> Kahlil Gibran AD926

TYPES OF DISCUSSION

In industry, interviewing is probably the most frequent and direct form of discussion between two or more individuals. It has been written about elsewhere in this book, and as a technique it can be shown to produce no perfect answers, especially to selection problems. In a different way, the same comment can be made on the tutorial situation except that this presents more opportunity for clearing away misunderstandings of facts and for seeing their significant relationships. An increasing need in the future will be for the training officer and the line manager to possess the knowledge and capacity to combine such special interviewing skills as those required for counselling, appraising and tutoring.

DISCUSSION IN LINE MANAGEMENT

An additional and even more important need is to discover and to consistently apply effective disciplines for formal and informal business meetings. Today these occasions demand specific decision making often within a permissive or semi-permissive framework. For example, there is the manager who regularly calls his staff together to plan and co-ordinate with them the work ahead. He leads discussions which are unique because of the work discussed, the interplay of personalities at a similar level, and the influence of his own qualities, authority and temperament. How valuable these occasions are depends very much upon his skill and sensitivity. The precise role he adopts is crucial. Most men who have authority assert it. The man who does so least autocratically, and who also leads a good team, is now seen to achieve best results. His discussion leading constitutes a genuine search and exploration of the potential within his group. He leads uninhibited discussions. Whenever he convenes such a meeting he usually makes clear beforehand what ground is to be covered or else he does this immediately his meeting starts. Alternatively, if the session is to be a brainstormer or a free exploration of common problems, then he makes this clear and explains why the session is being held, how long it will last, and what he hopes will come out of it. He does not seek to avoid conflict, nor does he try to shelve it by compromise. He aims by discussion to achieve penetrating analysis and practical solutions. He is not afraid, however, to delegate both responsibility and authority and he uses the discussion group to make this clear to everyone present. He does not see his role as a discussion leader involving the direct exercise of power but rather the exercising of an authority which derives from respect shown by him to the group and which is fed by ideas stimulated by the group's discussion.

He tries by his technical expertise and by knowledge of his group to analyse, with their help, the law and logic of each situation and in so doing to arrive at effective decisions on correct priorities.

DISCUSSION AND SELF-DISCIPLINE

These managers are nevertheless largely dependent on the co-operation and perceptiveness of every member of their group. This perceptiveness is a special skill which is also a form of discipline. It enables each individual to recognise that in group discussions there are limitations to his own freedom of expression. It can thus lead each to learn to distinguish between monopolising time and making effective contributions to the discussion. It enables the individual to participate by listening and by simultaneously observing behaviour of individuals and the reaction of the group as a whole; and of course, it is a skill which enables each to make his most effective contribution to the discussion at exactly the right intervals.

No man is ever a complete master of such a complex of social skills. Nor can even the most perceptive person in charge of such a discussion always apprise its undertones with complete accuracy. To a great extent he must rely on his own example. If it is good enough, it will stimulate each member of the group to greater perceptiveness and to impartial service. When the manager succeeds he provides for everyone a self-disciplined freedom of expression, which is combined with restraint and a sense of purpose and dedication to the job in hand.

DISCUSSION AND 'DOUBLE-TALK'

At worst, many opposite tendencies can prevail. Almost always these are related to deficiencies in the leader which are never more externalised than when he is leading a discussion with his own group. Rancour, double-talk and mistrust can then develop among everyone including the discussion leader himself. When this becomes prevalent, a kind of self-protecting sick conformity emerges. It carries with it a veneer of *bonhomie* and beneath this there is often cynicism, bitterness and frustration, side-by-side with blatant ambition and little real concern for the job except in so far as it may provide opportunities for advancement.

Every working group of equals or near equals has within it all of these possibilities and many worse ones. Part of every manager's job therefore is to detect and destroy them. If he is perceptive enough, most of them can be detected when he acts as a discussion leader of his group. If he comes newly promoted to such a situation and has high standards of integrity, impartiality, as well as perceptiveness, he can quickly make an astonishing difference. However, where these personal qualities are absent then the rot often intensifies. When this happens, all possibility of sincere co-operation within the group ceases. Output falls. No amount of group discussion will retrieve the position until the composition or the leadership of the group is altered.

DISCUSSION AND FORMAL TRAINING

The discussion group method has many applications in a formal training environment. Preceding chapters have already dealt with various forms of case method and project method. Each of these can involve a special kind of discussion group leading. With the usual form of case method, there can often be the appearance of an uncontrolled discussion. At its best, the leader by infrequent but subtle and penetrating interventions can often develop new approaches and new thinking. The group project also has many forms but the basic feature is that a group task is made clear and the group then becomes self directed and frequently appoints one or more discussion leaders of its own. Similarly, on many formal courses the full student group is often divided into smaller syndicates each of which reports back after it has analysed and discussed a given problem or topic. All of the Ministry of Labour's T.W.I. courses also use discussion method. Each of these approaches is a form of group exploration. All members participate and all learn from one another.

The previous chapter on lecturing techniques refers to the undoubted benefit of breaks for intermediate questions and discussion at intervals of about 20 minutes. As yet this practice is not widely accepted. It can probably operate with least inconvenience to the lecturer when his subject matter is distinctly factual or instructional, and therefore more easily divisible into self contained sections. However, where the theme of a lecture is mainly concerned with ideas or attitudes, even experienced lecturers often believe that intermediate discussion is an unnecessary and inconvenient interruption which may not only prolong the session but may interfere with the logical development of the theme. These assumptions ignore the fact that most groups are incapable of listening with profit for more than about 20 minutes. In industrial training, there are very few exceptions to the depressing evidence that listening to one voice for longer than this is not only exhausting but is of rapidly declining value to all except the very experienced and methodical note-taker. With small groups of a dozen or so, the most rewarding lecturers are those whose material is logically structured and who invite and develop questions and discussion at every important point. Perhaps every lecturer should normally incorporate intermediate discussion, and perhaps every lecturer should receive intensive training in the techniques of discussion group leading.

GROUP DISCUSSION LEADING TECHNIQUE

Most of the successful characteristics of discussion leading in line management are equally applicable when formal discussion group leading is the specific training method to be employed. As with formal lecturing, there are certain essential preliminaries.

THE PRELIMARIES

Personal comfort is important. Chairs should be reasonably comfortable. An open horseshoe grouping without tables is one good arrangement. Chairs which incorporate a swing desk top can be used. When tables are used, the aim should be to keep the open horseshoe shape. The point of this arrangement is to make all members of the group aware of the equalness of their role, and to ensure that all can easily see and be seen, hear and be heard. The leader's position in the gap at the points of the horseshoe should enable him if necessary to step into the arc of the group.

Good light and adequate ventilation are important. Ashtrays and note-paper should be to hand. If name cards are used, they should have the surname on both sides of the bottom half of each card and thus leave plenty of space for inserting first names or nicknames in large capitals. Name 'tents' are easily made by double folding ordinary foolscap or quarto sheets. They can be made more rigid if staples are inserted near the top edge.

If visual aids of any kind are to be used, they should be competently prepared and remain unobtrusive until required. Charting paper, chalkboard or an overhead projector meet most contingencies. The discussion group leader who fumbles for visual aids or cannot find chalk, chinagraph or duster cannot be defended. He simply has not bothered to pay his group the courtesy of preparing for them adequately.

THE ROLE OF THE LEADER

The crucial importance of the leader's role in line management discussions has already been emphasised. The role of the discussion group leader in formal training is equally important.

The starting point for success is respect for the individual and for the group. The leader must be dynamic and constantly alert. His interventions must sometimes be bold, but more often unobtrusive. Their effect must always be catalytic. He is there to stimulate creative thought and to draw it out from individuals. He can only do this well if he is completely sincere. Every good discussion leader recognises that on most subjects the collective interpretation of a group is often far more profound than his own. He therefore displays a genuine humility towards his task. He must perceive, guide and integrate the trend of group thinking. He cannot do this unless he combines all the time, enthusiasm, restraint, keen observation, good humour and the ability and courage to make quick and wise decisions, for example, on arresting one individual's contribution in order to allow expression to another.

If he is to fulfil his role, he must set certain limitations and avoid certain dangers. The session obviously has to be time-tabled. The subject to be discussed must be thoroughly studied. He must extract an appropriate amount of material to give to the group. This can present a danger point. If discussion is to be creative—a seeding ground for new ideas—then it would clearly be wrong and incongruous to have everything 'cut and dried' beforehand.

Any group can quickly detect the situation where prompts, interventions and the rephrasing and listing of responses from the group indicate foregone conclusions on the part of the group leader. When this happens, a group will feel that it is either being cheated or indoctrinated. The leader might even have all the anticipated responses listed on a concealed chart. When he has rephrased and listed the cued responses he has inveigled from the group, he may then proudly reveal the similarity between their conclusions and those given on the prepared chart. At that point, the group may justly feel that it has not so much been consulted as insulted.

How then should a good discussion be conducted?

OPENING THE DISCUSSION

The leader should state the objectives and time limit for the session in the opening sentences. He should explain the theme and briefly explore its possibilities. He should emphasise the value of relating discussion of the theme to each individual's experience. If certain areas of importance have to be covered, he should outline them clearly and if possible relate them to the group. At this point, to display or list not more than half a dozen provocative headings can be a useful stimulant for subsequent discussion. If the group is meeting for the first time, the leader's initial task is to create an informal friendly and lively atmosphere. At any stage, it is a mistake to talk down or to try to impress a group. In the early stages this can be disastrous. He must show sincere respect for the opinions of others. The leader should look around at everyone constantly. In this way each member of the group soon feels that the speaker is talking to them individually and is anxious to hear and listen to each opinion.

In opening the discussion the leader should not attempt to cover too much ground. Five to ten minutes should be ample time in which to begin to develop the theme, to give certain facts and to raise a number of questions. In the last minute he can concentrate on a selected area of importance; and he can end by stating two opposing views and then ask for opinions on them from the group.

HANDLING THE DISCUSSION

The leader must keep each member of the group under observation all the time. This can be done swiftly and without distracting any member who may be speaking. The technique is simply to split time and attention about equally between the speaking member and the rest of the group. When a lively discussion develops, the leader is constantly being asked questions. Almost always the correct action is to return the question to the rest of the group. Sometimes this is not easy to do. If the discussion is quickening this is usually the point where capacity to control and lead is most needed. Many kinds of action are possible. Perhaps it is enough to indicate two kinds which the skilled discussion leader may take and a third which the less experienced should guard against.

1. The leader when faced with a particularly challenging question may plunge into the discussion and deliberately take sides. He does so in order to identify himself with the group and in order to promote reaction. As speedily as possible, he must then take up the discussion from the other side. While it lasts, this chameleon-like behaviour can be carried on with great vigour, exhilaration and with complete sincerity. This dynamic procedure is usually termed in industry 'stirring-it-up' and that is the precise objective. The danger is that it may last too long and be done too often. The leader can apply this technique with greatest effect when he has a thorough knowledge of the subject, otherwise he may find himself in difficulty. His basic aim must always be to achieve all round participation. If he com-

mits himself others are much more likely to follow his example. He must at this time be perceiving as well as participating since only by doing so can he detect the precise moment to pause, give way and encourage others to resume.

2. The second type of action involves rapid decisions and a rapid return of the question to the group. The leader may pass the question directly back simply by repeating it and asking what others think. He may, however, decide to give implied approval or disapproval by adding straight or interrogative prefixes—'that's a fair (or good or sound or shrewd) question'—or 'is that a fair question?' and then repeating the question asks the whole group what they think. Occasionally he may direct the question to an individual. There are many possible variations of this technique. The question may be rephrased and given greater emphasis as it is repeated. If the leader believes that a new twist should be given then he swiftly adds an appropriate rider. He usually does this in order to guide the discussion in a new direction. For the same reason, he may acknowledge the question and then throw into the discussion a completely new question. Lastly, he can of course seize the question and make it the basis for an intermediate summary, and from this he can immediately proceed to a new area of discussion. His precise method and timing of the resumption of discussion by the group will depend on what he is saying and what he is seeing.

3. One of the commonest mistakes of the less experienced discussion group leader is the tendency to control the group too severely. This can have unfortunate consequences for him. Usually, however, most groups—especially at the beginning—are very tolerant. With an experienced leader, sub-groups may form and compete with surprising speed but with a less experienced leader they may do so very much more slowly. If the interventions and control of the leader have grown increasingly tedious before these sub-groups have formed, the group may suddenly resent the way they are being dominated and unite against the leader and isolate him. At that moment he loses control. Instead of leading the group he finds himself opposing it. Everything he says may be contested. This is a chilling experience and should never happen with a good leader. It can happen with a bad one who will then make the situation worse if he is foolish enough to take up a stiffened attitude which his injured dignity does not allow him to modify. This is an extreme example and illustrates the powerful pressure which a group can exert. The significant thing about it is that in such a situation every member of the group—except the leader—can usually see what is happening. Eventually someone assumes an unofficial leadership and in this way helps the leader to escape from his embarrassing predicament.

Each of these three kinds of action derives from questions addressed to the leader. These questions are clearly vital. When used by him with great skill, they can be a major factor in a stimulating and controlled discussion. Their complement is the type and form of questioning which he himself invents, and the way in which he uses this also to control and guide the discussion. This is partly conditioned by the kind of people to whom his questions are addressed.

Among the various characters who may be present in a discussion group are the over-talkative, the perfectionist, the aggressive type, the arguer, the grumbler, the shy person, the conformer, the rebel, the constructive critic and the carping critic, the intellectual, the cynic, the opportunist, the humorist, the hair splitter and the garrulous fool. Fortunately, the range is seldom as obvious and varied as that. Nor does any one group consist of people whose contributions are as distinctly separable. In action in a group, people are all much more part of one another. The essential distinctions which the discussion leader has to make are not therefore concerned so much with differences of personality (for which he must try to make due allowances) as with quick recognition and assessment of the quality and appositeness of contributions. If he can assess these accurately and at the same time can ensure that no one, including himself, is monopolising the discussion, then he can develop a group climate in which creative ideas, new knowledge and the changing of attitudes can be understood and accepted.

If the discussion leader is observing every move in the group and has very frequent eye-to-eye 'contact' with everyone, there is very little chance of awkwardness or loss of control. When he is not asking questions his activity will consist of intensive observation and listening. Occasionally, he may pull up short a contributor who threatens to make a speech. He can do this bluntly and yet tactfully. He can very boldly, but not rudely, interrupt by saying 'George—just a minute—you have made three very important points which I am sure we would all like to discuss before we go any further.' Strong control of this kind interspersed with apparent acquiescence is quickly respected by most groups. If the atmosphere is lively and if the discussion is evenly spread, all including the quieter members can eventually be drawn in.

Skilful initiating of questions by the discussion leader is one of the most effective means for coping with personalities and for guiding and controlling the discussion.

QUESTION TECHNIQUE

In general, it is much better to address questions to the entire group (overhead questions) rather than to a selected individual. There is an element of compulsion about direct questioning which most people dislike. They feel they are being dragged in rather than invited in. With good eye contact, the overhead question can be tantamount to a direct invitation, except that just as the leader can deliberately choose not to have his eye caught—so too can the member of the group do likewise if he is being silently signalled.

If a response is not made, the leader may assume that his question has been too broad and he may, therefore, narrow it down. Instead, for example, of asking again what kind of discipline industry needs today he asks what kind of discipline does *this* industry need today. Narrowing or broadening in this way can obviously be applied to almost any general question.

Once a particular area of importance has been opened up, he may if necessary give support and guidance by means of a battery of follow-up questions. His timing and selection of the appropriate question and his tone of voice whether lively, humorous or serious should always be encouraging. Very often he may enlarge on an earlier statement before phrasing the next follow-up question. In this way, he can also recapitulate to ensure that all are following the development of the subject.

Other main types of question include questions of fact. These can be used to best effect when the leader is conversant with the experience of individuals in the group. Here too, if he is maintaining good eye contact, he may prefer the overhead question. However, if the right atmosphere exists, most people who are familiar with the information required are pleased to be asked to supply it. This is the kind of situation where direct questioning may be welcomed, especially if a little forewarning is given such as the naming of the person before (not after) the question is directed to them. Here too, a series of direct questions to the same person may be the obvious and only appropriate way to proceed.

SUMMING UP

Towards the end of a discussion period, the leader may attempt a final pulling together of the subject. This is his last intermediate summary and in it he may indicate how far the discussion has gone in terms of the original objectives and what might be discussed before the session ends.

The decision to have intermediate summaries and the form and style they take will obviously depend on the type of discussion and on the approach of the leader. With instructional sessions, item by item summaries can often conveniently check and consolidate what has been learnt and can then be used to introduce the next item. With problem analysis sessions, the position though more fluid can sometimes be developed by means of intermediate assessments. On other occasions, the best thing is to have no deliberate interruption but rather to advance the discussion by means of the questioning techniques already described. With even more general discussion, the question whether to attempt any intermediate or final summaries often depends on how wide the discussion has been.

What the discussion leader does about a final summary will depend very much on the quality of the discussion. Theoretically, he ought to have some note of the main topics, viewpoints and conclusions. In practice, he is often so busy observing and controlling that there is no time to do this. One of the values of having had a displayed list of headings is that at the conclusion of a lively discussion, these will provide a framework on which the leader may comment, if only to emphasise how far short, or how much further, the actual discussion went.

In his final comments, the leader should always make a clear distinction between what the majority appeared to accept, what the main minority viewpoints are and what his own views are—always provided that he considers them apposite and worth giving. Finally, after explaining what follow-up there will be, he should thank everyone for their contributions and ensure that at that point the discussion does not start up again.

FUTURE TRENDS IN TRAINING BY DISCUSSION

The purpose of this final section is to outline three important areas in which future training of industrial groups by the discussion method could be intensively developed.

NEW OCCUPATION TRAINING

The first area is in new occupation training for unskilled workers or in retraining of older workers made redundant by new technological developments. In both of these situations, too little attention has been given in the past to the problem of dispelling the fears, doubts and inhibitions of the individual towards the training task which he faces. (See Chapter 27.)

These fears and doubts about his own capacity to succeed are intensified if he is faced with inept or inadequate training methods. To train for a skill by means of the group discussion method would undoubtedly be laborious and time consuming and ineffective if the training were in any sense haphazard or inadequately planned. With intensive job analysis translated into logically planned training sequences, it is possible to adopt a step by step approach in which each item is self-contained on much the same lines as the job breakdowns in the Job Instruction method of Training Within Industry (T.W.I.). It is then possible to do effective and streamlined training precisely because provision for discussion is a built-in component of each item of instruction. A three phase explain-show-do approach can then be so arranged for each item that discussion consolidates learning at each phase. Explain and discuss, show and discuss, and do and discuss—this is a method which works well because from the outset, it is based on participation and inspires and develops motivation. This is particularly true at the early stages when fear, inhibition and insecurity might otherwise prevail.

The lengthier preparation required to plan such training and to train instructors in discussion leading techniques is more than offset by the quick results which can be obtained. Skills which in the past may have taken years to acquire can be mastered in a month or six weeks; in sufficient depth for the requirements of specific jobs in a particular firm.

No force for learning is more powerful than the team spirit of a group of purposeful men who see a new future for themselves in attaining goals which have been clearly defined and which they have thoroughly discussed. Many trainers have yet to learn that informal discussion expertly conducted in the early stages of training is a valuable way of saving subsequent training time. High training standards for effective discussions create high morale and speedy learning.

SUPERVISORY TRAINING

The effective training of supervisors also remains one of the most important and difficult problems of industry. Recent research has indicated that former methods of teaching the skills, techniques and principles of supervision are useful but inadequate. They have been too general. To issue check lists of duties, responsibilities, desirable attitudes and company policies and procedures, and to give once-off courses on them are not enough. These are now being seen more as the framework in which the real problems of supervision arise.

In the future, increasing emphasis will be put on diagnosis, analysis and discussion of defined problems and of specific areas of difficulty associated with particular groups of supervisors. In this context, the role of the training officer as a catalyst and discussion leader can be an invaluable one. Bringing together, at first perhaps in a separate training environment, supervisors from adjacent sections and departments for the purposes of discussing their inter-related problems and of communicating their findings to senior management is likely to become the main substance of in-company training of supervisors in the future.

The first discussions must be planned with foresight within the existing framework of policies and procedures and in reference to actual problems of specific sections and departments. This thorough preparation for such training would, however, be completely futile and even harmful if the training were not followed up by appropriate management action and by continuing supervisory discussions at regular intervals. The dictum that training should be a process is particularly applicable to supervisory training discussions.

When middle managers themselves become experts in problem diagnosis and analysis and in conducting such training discussions, then this supervisory training process will achieve maximum impact and efficiency.

MANAGEMENT TRAINING

A third developing area is the training of middle and senior managers by discussion methods towards a clearer perception of their own personal image, role and behaviour within the hierarchy of their particular firm. In this context the 'T' group (T=Training) discussion method—sometimes called the group dynamics method—merits comment because of what may emerge from it.

It is a special form of discussion. The group is usually residential. Members are given complete liberty to play the truth game with one another. The group is leaderless except that somewhere in the background—not necessarily present at all sessions—is a psychologist who plays a 'sleeping clergyman' role in which he is often only roused to verbal activity at the request of the group.

The course members have no set agenda, though they may be told that they are present in order to learn how others see them. They often become totally frank with one another. This can be a wounding experience and can have a shattering effect on self confidence.

It can be revealing and rewarding for those able to appreciate it. Others who are hyper-sensitive or less mature may be upset and disturbed, whilst those who are fundamentally insecure should not be exposed to it.

In practice, most people are extremely sensitive about personal criticism of themselves or anything they do. Unless it is conveyed with great tact, it tends to be resented and rejected even when given in private. Given in a group, it offends their code of behaviour as well as their pride. It may touch to the quick in other ways too, for behind the appearance of what people are, they all may experience phases of uncertainty or insecurity. In the 'T' group, it appears to be comparatively easy by the pressure of group opinion to strip away the favourable illusions that individuals have about themselves. What is not so easily discernible is what takes the place of destroyed illusions when the course member returns to his usual environment.

It is true that for a time he may be more questioning, more observant and a better listener; or he may make a pretence of being these things because he thinks that is what the organisation expects of him. Perhaps it is also true that in terms of behaviour of individuals in an organisation, the 'T' group experience may tend to subdue, or blunt, dynamic behaviour and produce more conformity especially if the group is drawn entirely from that organisation.

More recent trends have been away from the completely unstructured 'T' group to much more normal forms of discussion group work in which self-led groups are given realisable aims such as examining the extent of systematic analysis and planning in their own approach to a problem solving situation. In this setting, self criticism may be added to group criticism of individual performance, but in reference to a given role and task as a member of a team. Again each group may have an adviser in the background. They may also appoint their own observers to report on performance of other members. To act as observer is in itself a useful form of training.

In some circles, the view has more recently been taken that the most effective way of influencing individual behaviour by group discussion is only by reinforcement and group approval of whatever is well done. This still leaves certain familiar questions unanswered, for example, by what standards do the approvers approve or omit to approve; but it may be that discussion group activities of this last kind could lead to much more penetrating insights into the difficult territory of personality appraisal by supervision at all levels.

Most of the criticism made of appraisal systems in the past has been due to their subjectiveness and emphasis on personality appraisal. Antagonisms develop when the individual assessed finds that because his personality traits are not approved by his superiors his work performance is given similar disapproval.

Just as the unstructured 'T' group's emphasis on personality factors is giving way to emphasis on co-operation in relation to specific tasks, so too appraisal of personality traits in the future should give way to appraisal by assessment of measured performance and results.

In a sense this brings the situation full circle to the impact on the individual group member when skilful re-inforcement is provided by a perceptive discussion group leader. In this context it has already been observed, for example on Training of Training Officer courses, that individuals then seek out, in private, from the discussion leader more information about themselves. They then seek rather than resent constructive criticism, especially in tutorials where the normal superior-subordinate line relationship does not exist.

This is where real progress on personality difficulties can begin. In future, the training officer may himself find that as an expert in the various forms of discussion group leading, individuals will ask him to perform a special kind of tutorial role in which he explains to them as their discussion leader what he has noted about their performance and their effect on others. In the future, if his discussion leading takes place at many levels, he may find that he also becomes a counsellor at each of these levels.

Discussion leading in all its forms can provide a successful, satisfying and powerful method of group learning, provided that it is recognised, as a skilled discipline which fails unless it is based on respect for the individual.

BIBLIOGRAPHY

KLEIN, J., *The Study of Groups*, Routledge and Kegan Paul (1956)
ARGYLE, M., *The Psychology of Interpersonal Behaviour*, Pelican (1967)

PART 4

TRAINING FOR SPECIFIC FUNCTIONS

18

Management Training

C. R. McLENDON, B.S., M.I.E.E.E., LL.D.

Vice-President and General Manager, General Precision Decca Systems Inc.

Col. A. L. WORK, B.SC., M.I.MECH.E., F.I.E.E., M.B.I.M.

Senior Lecturer, Department of Management Studies, Portsmouth College of Technology

INTRODUCTION—
THE CONTROVERSY IN MANAGEMENT TRAINING

Management training is probably the most controversial area for the training officer. There are those who say that there is no such thing as a body of knowledge which constitutes management studies, and that managers can be trained only through the experience of managing. Many resent the suggestion that they require training, and doubt that a training specialist can teach them anything about *their* job. A vast array of new books on management training is pouring out from the American publishing houses. Here in the U.K. a variety of management courses is being given the hard sell by management consultants, by commercial colleges and even by technical and other colleges financed by the public purse. Some of these courses are excellent, some are good, some are for specialists only, some are for beginners, others are for experienced managers, and some are just after your money. How is the harassed and badgered company training specialist to find his way through this morass? How is he to decide on the suitability of a particular work study course, whether 'T' groups are worthwhile, whether for example Henley or Ashridge meets most of the needs of his middle management, or whether he ought to run management training internally? This chapter aims to help the training officer to clear his mind, to strip management training down to its basic concepts, and to suggest a way through the maze of existing provisions.

AREAS OF MANAGEMENT SKILL AND KNOWLEDGE

It has been said that management is the art of getting things done through other people, but it is more than this. An initial look at the areas of skill

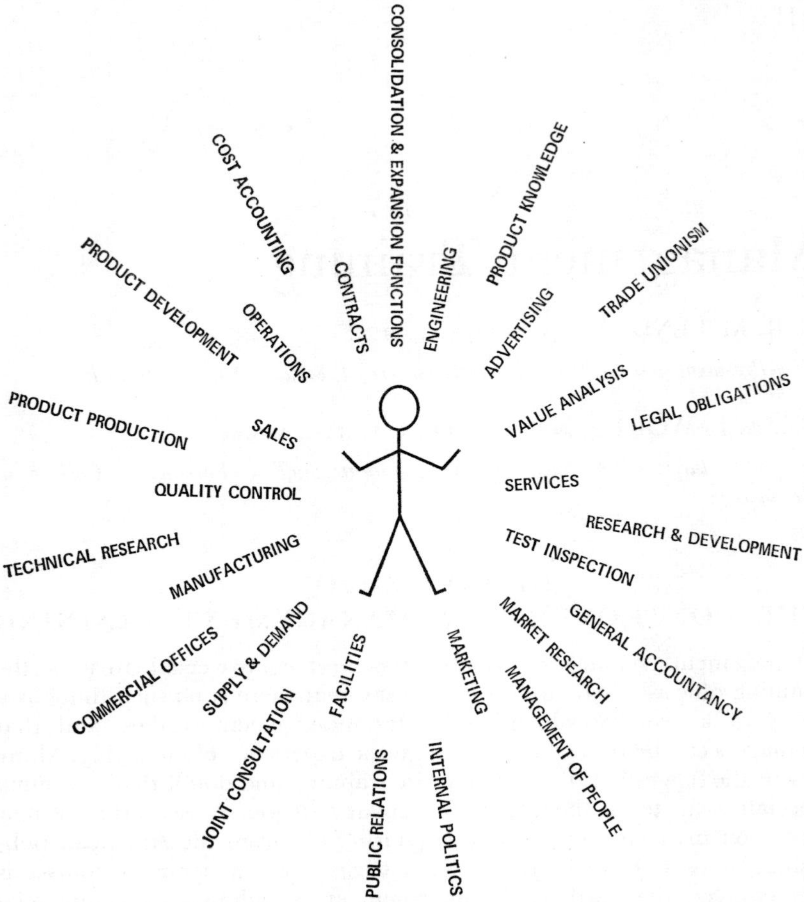

Fig. 18.1 Areas of Required Knowledge

and knowledge required conveys a complex picture. Managing people is involved in all of the areas of required knowledge shown in Fig. 18.1.

THE MANAGEMENT FUNCTION

In order to discuss constructively the problems associated with management training and development, an insight is necessary into the broad general nature of management and its tools. The manager plus the tools used to do the job can be defined as the 'management function'.

THE BASIC CONCEPTS OF MANAGEMENT

All managerial activities are a combination and manipulation of human and material resources in relation to time and cost. These activities can be duration (time) limited, resource limited or cost limited. The manager's tools, therefore, are combinations of these three basic elements within the limits they impose (see Fig. 18.2).

THE BASIC CONCEPTS OF MANAGEMENT
CAN THEREFORE BE LOOKED UPON AS:

DURATION LIMITED	-	T/RC LIMITED TIME UNLIMITED RESOURCES AND COST
RESOURCE LIMITED	-	R/TC LIMITED RESOURCES UNLIMITED TIME AND COST
COST LIMITED	-	C/TR LIMITED COST UNLIMITED TIME AND RESOURCES

Fig. 18.2 Managerial Activity

THE OBJECTIVES OF MANAGEMENT

Since the objective of management is to supply a product or a service in a given time within these (T.R.C.) limits, the expression 'unlimited cash and resources' has been used for convenience in Fig. 18.2. No management has unlimited cash and resources. There are, however, examples where time is so dominant that cash and men are not the limiting factors. There are also situations where the speed of performance—though important, cannot be increased by more cash, or by more men. An example is the speed of building an atomic submarine which may be limited by the sheer physical impossibility of getting more men working on it at any one time. Similarly, if the management job is to grow timber for a paper maker,

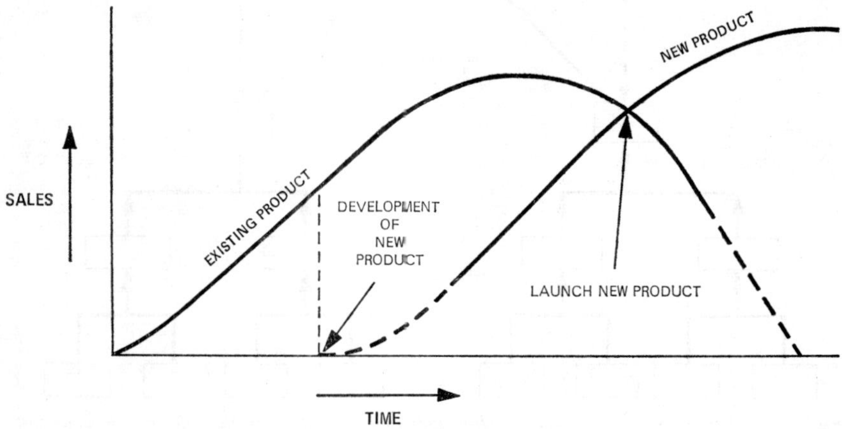

Fig. 18.3 Product Life

the time in which trees can grow is unlikely to be influenced by more money or more resources. Here the aim must be to grow as much suitable timber as is economically possible, on the smallest amount of land, in the most accessible place with the least cash and least number of people.

The above examples are extremes which emphasise that management is the science and art of using time, resources and cash in the manner which is most economical for the purpose of the enterprise. This, some will say, is a narrow view of management. If cash, resources and a viable product are available, a business exists. The life of any product is limited, and some business men will say that the art of management is to judge the development time, the crest of the demand, to forecast the death of the product and decide when to develop the next product (see Fig. 18.3).

THE ELEMENTS OF THE MANAGEMENT FUNCTION

From F. W. Taylor,[1] from Fayol,[2] Urwick[3] and others to Peter Drucker[4] and to Douglas MacGregor,[5] there have been various attempts to analyse

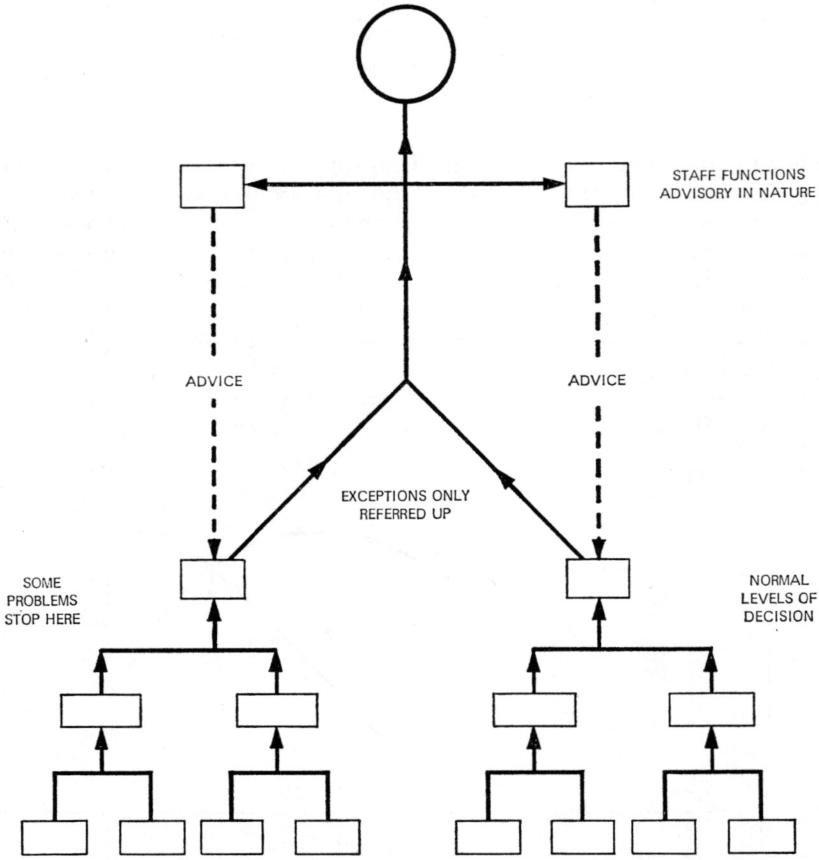

Fig. 18.4 Management by Exception. Information is fed upwards through line management. When circumstances arise which call for top decisions or functional advice, these pass downwards

management. It is enough here to mention just a few of the ideas and to provide references to which the training specialist may refer. Urwick breaks down management into six main functions—forecasting, planning, organisation, co-ordination, direction and control. All activities can be broken down to these six elements—including training.

THE ART OF MANAGEMENT

The art of managing is to balance activities so that the aims and objectives of the organisation are met.

THE EXCEPTION PRINCIPLE

Since management is so complex, one effective form of operating is through the principle of management by exception.

Management by exception is a process whereby routines of activity are established and management skills are applied to the exceptions to these routines (see Fig. 18.4).

MANAGEMENT BY DIRECTION

Management by direction is a method by which activities are directed to happen, and supervision is applied throughout the framework of activity. In general, nothing is started or stopped unless directed. Management

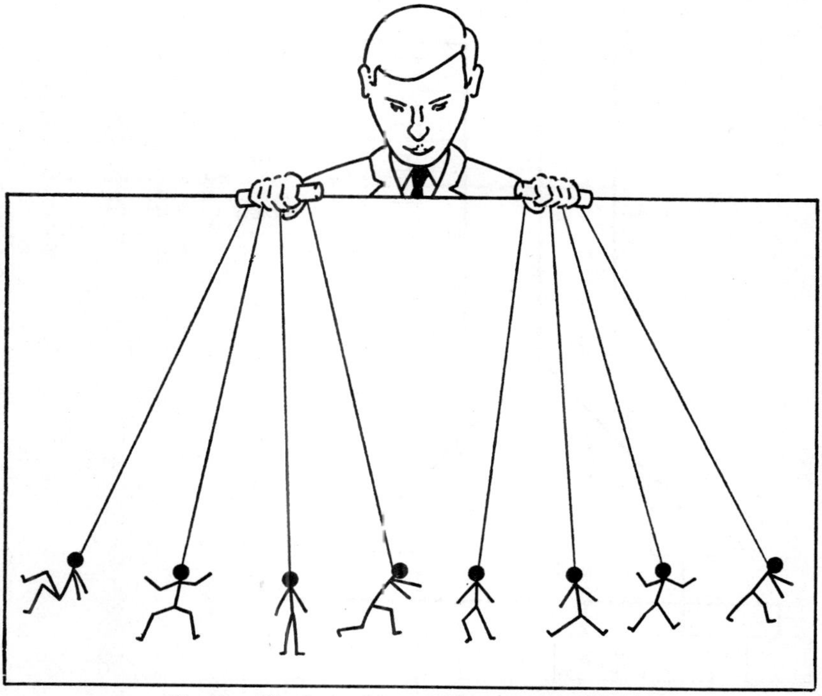

Fig. 18.5 The Puppet Master—Management by Direction

by direction is a time-consuming method when applied to large, complex organisations. It sometimes works well on the lower tiers, where a manager has only a few of the elements of an activity to cope with. Successful management by direction depends on the personality of the leader, the confidence of the rank and file in his decisions, and the willing acceptance of the authoritarian nature of the discipline within the organisation (see Fig. 18.5).

MANAGEMENT THROUGH FUNCTIONAL SPECIALISTS

The other extreme to management by direction is management through advisory staff. No decisions are made by the line manager without sounding out the advice of a large specialist staff. Although this large staff is sometimes justified for a large capital process plant such as a steel works or an oil company, the development of the computer to replace staff functions may reduce the need for this kind of organisation. When there are overlarge specialist staffs in manufacturing organisations the line manager can feel that his role is being usurped (see Figs. 18.6 and 18.7).

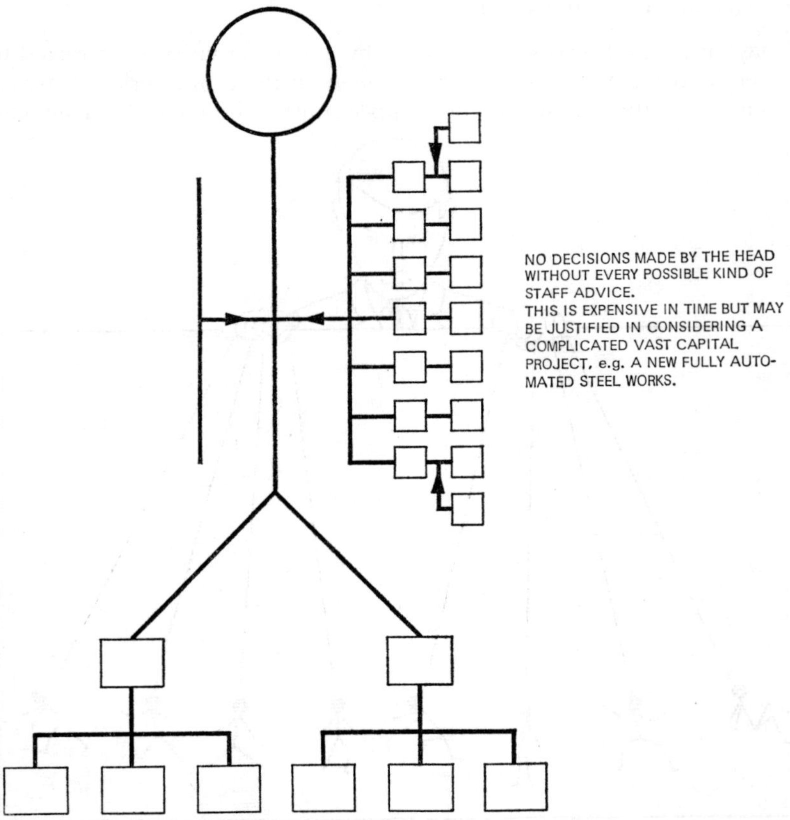

Fig. 18.6 Staff Management Orientation

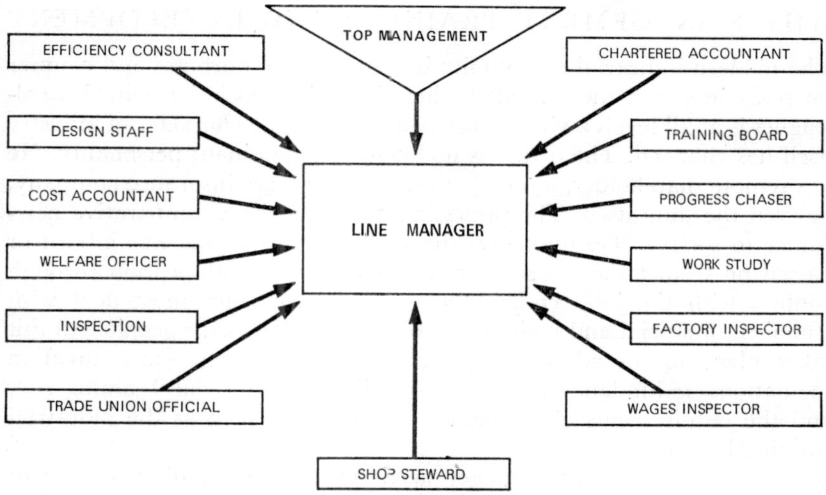

Fig. 18.7 Staff Directed Organisation—may submerge the line manager

POSITIVE AND NEGATIVE REACTIONS TO CHANGE

An individual's executive skill can often be judged by his reaction to proposed changes. Positive reaction is of fundamental importance. If the person comments 'Yes, we could do it if we had this or that'—this is an indication of positiveness. This positive approach requires an immediate grasp of a problem area and a mind which seeks a solution to the problem rather than looking for reasons why change cannot be implemented.

DYNAMIC MANAGEMENT

The management body, that is, the collective body of those who manage any enterprise or interest, can be either dynamic or passive or they can be partly one and partly the other. They can be either enlightened or unenlightened or again partly one and partly the other. Dynamic management is generally characterised by people with a positive approach who implement short- and long-term policies simultaneously. They make maximum use of management by exception and require that a large proportion of time on every level is spent on communications and training. Evidence of dynamic management is decision making delegated to the lowest possible tier, and profit and loss responsibility given to the lowest level in the management body. In this situation, distribution of responsibility is always accompanied by the required authority.

ENLIGHTENED MANAGEMENT

In appropriate types of organisation,[6,7] enlightened management policies result in a flexible organisation which can easily shift from product area to product area in order to meet the pace of modern industrial change. Enlightened management can only be achieved and maintained by continuous management training and development.

WHY MANAGEMENT TRAINING AND DEVELOPMENT

The modern industrial community is a complex of purposes and counter purposes in which the role of the individual in association with his colleagues is at all levels a vital component. Today, the character of industry itself has changed. The owner is no longer the dominant personality. He is a remote shareholder, a merchant bank, or a large insurance company, or even the state itself. The professional manager is a comparative newcomer to society. Traditionally, the skill of being a boss was inborn or passed on from father to son, or was developed by absorption through contact with the job. Today, the successful manager must deal with technical, financial and political problems as well as with people. All this takes place in an accelerating atmosphere of change—new taxation regulations, redundancy payments, S.E.T., the Industrial Training Act, and the requirements of semi-government bodies such as the industrial training boards.

Today, therefore, the manager needs to know something of current economics, law, accountancy, sociology, mathematics, scientific method and politics. On this knowledge, he bases a range of management techniques, such as budgetary control and standard costing, work study, operational research, labour relations and industrial psychology, and training methods. Thus, there is a need both for management education and for continued management training. This is the basis of all management development.

Carefully planned management development programmes are needed to prepare a cross section of younger managers, each capable in his own field, to learn gradually about the problems of a company as a whole and thus to enable each to move to a new level of decision making. Simulated situations and project studies can provide some practice for these men in the skills of analysing and correlating relevant information, but there is in the long run no substitute for experience on the job. A company has a certain limited capacity for giving men rotational training experience in management. Properly planned management development is needed to make the most of this capacity, for the men concerned and for the company as a whole. In any company, it will be found that managers reach differing levels of seniority by ability, by taking advantage of opportunities and by sheer chance. Management development aims, as far as possible, to provide fair opportunities, and above all to determine, at an early date, men's abilities and their potential ceiling in the company. Each selected individual is then developed to his full capacity.

THE NEED FOR EFFECTIVE APPRAISALS OF PERFORMANCE

How to appraise effectively is still one of the most difficult problems of modern industry. Many large firms have systems involving appraisal interviews and rating forms. Since these are inevitably operated by the immediate manager and/or the manager at one level removed from the individual, they have been proved to be of doubtful value. This is because they inevitably perpetuate subjective judgments and attitudes rather than scientific assessments of performance.[8]

It is increasingly being recognised that personality differences and difficulties can very frequently develop in a hierarchical organisation between a supervisor and a non-conforming subordinate and that these differences can cause seriously distorted assessments to occur. These in their turn produce or deepen bitterness and antagonisms. This is one of the direct consequences of nineteenth century concepts of authoritarian management. It is a direct consequence of what happens, for example, when a less capable man wields power over a more capable man.

Even today, many managements are in the process of initiating appraisal schemes of this kind which are almost certainly doomed to failure or potential failure.

Enlightened firms of the 1970s will almost certainly abandon this type of appraisal procedure in favour of much more systematic measurements of performance. These will be based on carefully defined job descriptions and job specifications which pave the way for the setting of measureable objectives and measureable work targets. In these circumstances, management appraisals will automatically be forward looking and exploratory rather than retrospective and recriminatory. Estimates of potential will be more realistic when based on these foundations rather than on arbitrary rating judgments coloured by likes and prejudices.

This process is not an easy one and demands time, knowledge and skill and constant vigilance to ensure that the grooming of crown princes does not undermine job satisfaction of others involved in the process.

A doctor is not expected to become qualified unless he has completed his studies satisfactorily and has had an adequate period of practical experience in a hospital. Likewise, the modern manager needs both 'theory' and 'practice'. The corollary of this is that every manager should look upon the training and development of his subordinates as one of his most important functions. Sometimes this idea has to be 'sold' to the board before management training and management development can be a real success. In the smaller firms, this is where a well-informed training officer can help to formulate policy and procedures.

Management training and development should therefore result in the following:

Improved performance by existing managers.
Improved company structure.
Each man being developed to his maximum potential.
Replacements being available for management posts which become vacant.
Development of managerial reserves for forecasted expansion.
Immediate and long term objectives being more efficiently achieved.

The steps which a training director has to take through his training manager, and in consultation with his fellow directors are:

1. Draw out a company organisational chart.
2. Analyse existing authority and responsibilities and eliminate overlapping.
3. Prepare detailed descriptions of managerial jobs.
4. Prepare lists showing anticipated expansion of managerial appointments.
5. Compare the qualifications and experience of managers with job requirements both existing and anticipated.
6. Outline areas of training need.
7. (*a*) List managers whose background and career would indicate that they might go far.
   (*b*) List managers needing training in particular subjects and areas.
8. Produce a preliminary management training and development plan.
9. Prepare lists of external and internal courses to be used and developed.
10. Search for potential managers from within the company, and draw up a management audit.
11. Draw up an integrated management selection procedure.
12. Prepare a continuing plan of training and development based on management by objectives.
13. Finally, implement a full 'management by objectives' scheme which establishes task targets, periodically measures performance and communicates results to the individuals concerned.

These headings are easily listed but their achievement presents a formidable challenge requiring hard work and tact on the part of the training manager and his staff. For him, management training can be a two-edged weapon. Before embarking upon it, the training manager has to ensure that the objectives of top management and the senior training executives are clear and practicable. To start management training and development within a firm which has not previously indulged in the exercise invites criticism of the education and acumen of existing managers. Conversely, a well integrated plan will stimulate managers who go on courses. It can result in a complete reorganisation of the company and mean a new outlook for the firm. Such a plan needs delicate, intelligent and analytical consideration of the present situation and of expected results. It may even be advisable to call on the advice of expert training consultants before embarking.

METHOD OF IMPLEMENTING MANAGEMENT TRAINING AND DEVELOPMENT

If there has previously been no management training within the organisation, the real problem for the training manager is how to start. If there is a family tree showing the organisation of the company and the duties of individual managers, this should be checked against the actual situation from which it may differ considerably. But it would obviously be tactless to point this out to the managing director without suggesting a remedy. A better solution might be to take a country hotel and run weekend seminars for senior management at which each of the departmental managers would be invited to report on activities of his department, defining its objectives, its methods of operation, its links with other departments, its future manpower development, as he sees it. Any problems or divergences from the theoretical structure of the company should quickly emerge.

Management training and development in the skills required in the areas indicated on Fig. 18.8 can show early financial gain to a company.

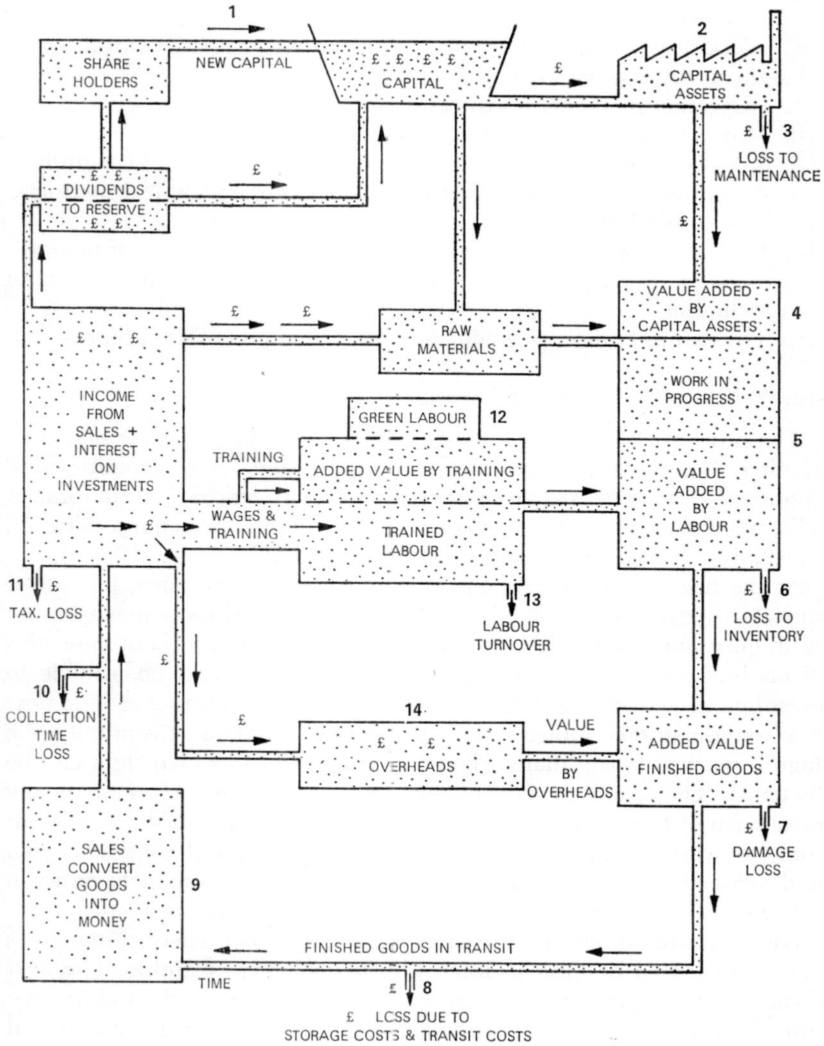

Fig. 18.8 Time—Resources—Costs

Key to Training Areas in Fig. 18.8

1. Training in financial planning.
   Use of computers in financial planning.
   Profit setting.
2. Utilisation of capital resources.
   Random observation of floor and machine utilisation.
3. Methods of accounting for plant purchase and write off.
   Planned plant maintenance.
4. Critical path analysis.
   Planning and production control.
5. Work study, human relations, value analysis, quality control.
6. Inventory control methods.
7. Inspection methods, statistical methods.
8. Materials handling, transport methods, insurance.
9. Sales management, market research, sales records and accounting.
10. Accountancy, invoice management, debt control, clerical methods.
11. Tax regulations, clerical supervision.
12. Induction training, personnel administration, education of management in training methods, job evaluation, merit rating, appraisal technique.
13. The exit interview, payment systems, economics of welfare, counselling.
14. Staff utilisation, effective research and development.

The processes of learning are always complex, but in the management training situation these processes are complicated by personalities and emotional overtones, much more than in most learning situations. A helix spiral dominated by the basic concepts of effective communication is probably the most effective system (see Fig. 18.9).

At the first circuit of the cone of management knowledge, functional subjects are dealt with at the level of broad appreciations of management techniques and knowledge. Every manager will think about how this affects him and his department. At this stage, it should be possible to assess how effectively individuals co-operate and communicate. The next time round can take subjects to greater depth and detail. Eventually, the stage is reached when 'management' as a subject in its own right can be discussed. The whole process of management training, more than any other form of training, should proceed from the known to the unknown: from the whole to the parts where skill, knowledge and confidence can be developed in their component parts and then be brought back again to form a whole and integrated picture of modern management.

Weekend discussions and the dispatch of one or two selected managers on courses should develop an interest in management training. The next job of the training manager is to prepare a management audit and budget. From information about the company's future plans for expansion and from an analysis of existing staff, an estimate can be made of the long-term demands for managers. On this forecast, a rough long-term training plan can be prepared to show the numbers required every six months. The

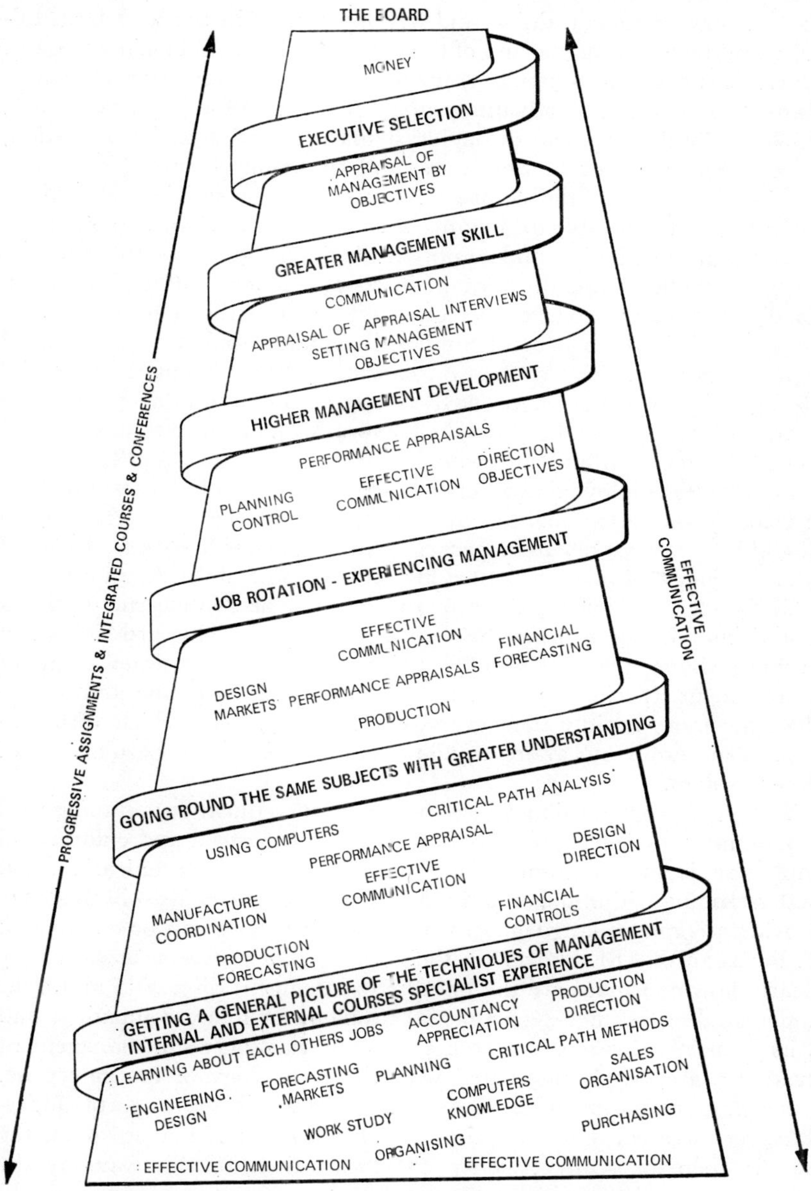

Fig. 18.9 The Repetitive Spiral of Management Development

decision on how far to use external and how far to use internal courses may depend on the policy of the particular industrial training board, under whose auspices the company comes. However, the requirements of the company will always be paramount, and every effort should be made to persuade industrial training board staff that a company undertaking its own management training and development is in the long term fulfilling, in the best way, the aim of the particular industrial training board. Where it is easier to obtain a grant from the board for external courses than for internal, full advantage may be taken of this. Nevertheless, management development through systematic appraisal and job rotation, through management by objectives, and by management selection can best be done within the company. Costs of these activities should be recorded so as to be able to convince a board of the full extent of management training activities, and so obtain maximum grant for the purpose.

One of the most useful external courses of a general nature is that run usually by regional colleges of technology to meet the requirements for the Diploma in Management Studies. These and sometimes more local area colleges run short courses on executive development for potential managers, planners, etc., as well as courses in accountancy for non-financial management, Law and the Factory Act for engineers, and general courses for foremen and supervisors.

Companies employing over 3,000 people or with capital assets in excess of £5,000,000 or an annual turnover or throughput in excess of £1,000,000 should have at least one manager who, within the last five years, attended a longer course of management study such as those of the Administrative Staff College at Henley, those of the London or Manchester business schools and in some firms the courses run by the Harvard School of Business Management. Ideas of management are in a continuous state of development. Technology, finance, politics and economic trends are changing more rapidly than ever before. No company which wishes to remain successful can afford to allow its management to be unaware of these changes.

The great contribution which management education, training and development makes to the effectiveness of an industrial and commercial enterprise is in the immediate improvement of communication and understanding within the organisation. External courses assist in that they provide a common language in which the discussions can be conducted. T. Burns and G. M. Stalker[7] in their *Management of Innovation* show very clearly how enormous are the barriers to intelligent discussion and clear understanding of inter-departmental problems in many companies, and Miss Joan Woodward's[6] researches indicate that successful management structures are closely associated with the technology of the enterprise. Once most managers of a company have reached the stage of understanding these concepts, techniques and languages of management, the way is open for really effective communication. In this way, by the systematic training and 'development' of its managers, a company will prosper.

MANAGEMENT BY OBJECTIVES

Management by objectives consists of getting each manager to suggest, discuss and agree with his superiors what his exact objectives and work targets will be for the next month, quarter, half year, year, two years, etc.

It is one of the most effective methods of ensuring measurable performance standards by which to operate management development.

An increasing number of today's successful organisations now use this system of training and development. Probably one of the most sophisticated applications is that of Smiths Industries Ltd., described by Marcus Knight in his short descriptive pamphlet *Management by Objectives*.[9] In this company, internal management analysts are trained and employed in carrying out the interviewing and in helping to set the targets. Because the analysts are less emotionally involved in the immediate situation than are the managers, the analyst staff can be objective in their judgments. Success is achieved very largely by the careful defining of key result areas and the recording of agreed problem areas and by establishing through discussion with individuals immediately concerned realistic targets and exact definition of the action required to achieve these targets. Small companies are unlikely to be able to follow exactly the procedures outlined, but the approach can be adjusted to suit any organisation. For management by objectives to be most effective, there must be a clear understanding that it is the job and NOT the man which is being analysed and improved. In the process, the individual's performance is improved, but this is as a result of the clarification of his tasks leading to effective measurement of achieved targets.[10]

MANAGEMENT TRAINING AND THE INDUSTRIAL TRAINING BOARDS

Ultimately, all boards will require companies, as a condition of grant, to have effective management training and development schemes. To satisfy the industrial training boards that management training is being carried out to their satisfaction and to get training grants, it is essential that firms keep full records of their training effort and achievements. In particular, boards will expect to see a continuous system for revising management job specifications, and for up-dating organisation charts. Companies will have to demonstrate that young managers who have been on courses are given an opportunity to apply their skill and knowledge. To do this, details of courses attended and job rotation training must be recorded. The reasons for sending certain individuals on certain courses should be clear. If the company does not use external courses but relies on internal training, this way must be seen to be soundly based. Cost records of internal courses will have to be maintained, for it may be difficult for a board to justify paying a grant unless an accurate picture of this cost to the company is available. Some industrial training boards may reduce the grant to companies whose management training is clearly unbalanced. For instance, if a company sends top management on short appreciation courses, but does not back this up with management training and development all the way down the line, it could be said that the costs of such training did not meet the board's aims. On the other hand, a company which sends young managers on diploma courses, but does not give these men an opportunity to implement their knowledge, because of the rigidity of top managers who have not undergone recent training, could easily find that the grant from the I.T.B. was less than anticipated.

Such a sophisticated approach may take some time to develop. All companies wishing to gain grants for management training should ensure that they have:

1. Full job specifications.
2. Detailed records of managers' qualifications and experience, showing courses attended, rotational experience and future planned courses and experience.
3. Regular management appraisal schemes based preferably on measurable performance.
4. Recorded details of the regular assessment of the needs of the company and of individual managers.

The activities covered should include:

(*a*) *Induction* courses for management trainees.
(*b*) *Broadening* experience—for members of staff whose experience has been largely restricted to a professional or specialist field.
(*c*) *Individual and group projects* where managers in training have been required to carry out specific well-defined tasks under planned supervision and to report and discuss their findings.
(*d*) *Visits* of managers under training to other plants or customers at home and abroad. These visits are of little training value unless the individual has been given a specific project to do or area to study, and been required to write a detailed report for this.
(*e*) *Attendance* at meetings, conferences and lectures as part of the activities of professional trade or research organisations. These are a valuable method of keeping young managers up-to-date, but part of the value of these activities is lost if the delegate does not write a report for later discussion within the company.
(*f*) *External* courses both specialist and of a general nature.
(*g*) *Internal* management discussion seminars (weekends, days or half-days).
(*h*) *Appraisal* sessions.
(*j*) *A detailed system* of management by measurable objectives.

CONCLUSION

What has been said indicates the complexity of management and of management training. It indicates how to start management training within a company and it suggests the steps necessary to ensure that a company gets a grant from a board for its management training. All management is concerned with people. Effective management means the co-operation of people joined together with a common purpose, namely the success of the enterprise. It involves the smooth working of a number of different specialists. The young accountant is responsible to an experienced accountant, the junior engineer to a senior engineer, the new salesman to a sales manager. Each before becoming a manager had been able to converse in his own professional jargon to someone who understood the exact meaning of every technical or specialist expression, but on becoming a manager he has to understand the whole field of management. Now the engineer has to explain his problems to an accountant, the sales manager to a production man or a designer—and in this situation, where each still tends to use the specialised language of his own profession, misunderstandings and frustrations arise.

Management training and development improves communication across specialist barriers and up and down and across the various management levels. Appraisals, management by objectives, role playing, group discussions all have a part to play in developing effective communications. The aim of this training is more effective management and this means making the best use of all the resources of an organisation. To this end management training provides the climate in which

TIME——→ MEN——→MACHINES——→ MONEY

are so used in each situation, whether MONEY dominated, TIME dominated, or RESOURCE dominated that maximum results are achieved.

Machines are difficult to design, money is hard to come by, time flies, but men can rebel. Human beings are the most intractable, variable and valuable resources of a company: without the support of its people, a business cannot prosper for long. The aim of management training and development is to ensure the best use of human and material resources in relation to time and cost.

REFERENCES

1. TAYLOR, F. W., *The Principles of Scientific Management*, Harper (1913)
2. FAYOL, H., *General and Industrial Management*, Pitman (1949)
3. URWICK, L., *The Elements of Administration*, Pitman (1947)
4. DRUCKER, P., *The Practice of Management*, Heinemann (1955); *Management for Results*, Heinemann (1964)
5. MACGREGOR, D., *The Human Side of the Enterprise*, McGraw-Hill (1960)
6. WOODWARD, J., *Industrial Organisation Theory and Practice*, O.U.P. (1965)
7. BURNS, T., and STALKER, G. M., *The Management of Innovation*, Tavistock (1966)
8. ROWE, K. H., *J. Mgmt Stud.*, 'Appraisal of Appraisals' (Mar. 1964)
9. KNIGHT, M., *Management by Objectives*, Smiths Instrument Co. Ltd. (1966)
10. HUMBLE, J. W., *Improving Management Performance*, B.I.M. (1965)

BIBLIOGRAPHY

KING, D., *Training within the Organisation*, Tavistock (1964)
GILBRETH, F. B., *Applied Motion Study*, Macmillan (1919)
CURRIE, R. M., *Work Study*, Pitman (1963)
VALENTINE, R. F., *The Appraisal Interview*, American Management Association (1964)
HEYEL, C., *Appraising Executive Performance*, American Management Association (1958)
TREDGOLD, R., *Human Relations in Modern Industry*, Duckworth (1963)
FOLLETT, M. P., *Dynamic Administration*, Pitman (1941)

19

Supervisory Training

V. T. STEVENSON, F.I.T.O., F.R.S.A., F.R.G.S., F.R.ECON.S., M.I.S.M., A.M.B.I.M., A.R.AE.S.

A Vice-President of the Institution of Training Officers

INTRODUCTION

This chapter is addressed to all who may be associated with the training of groups of supervisors of either sex, and it aims to deal, step by step, with the provision of such training.

A check-list of salient points is provided at the end of certain sections and the chapter concludes with a brief bibliography.

GENERAL

In this country insufficient recognition has been given to first and second line supervision, i.e., to the changing but vital role played in industry by the foremen and their counterparts. Before the Second World War, their function was relatively clear. This was to preside over a department or section and in so doing often to exercise absolute authority not only over staff, but also over operating systems, raw materials, services and even over the design of a product.

From about 1940, there has been an increasing need and tendency to employ specialists to help and to advise the line supervisor. Today there are personnel specialists, production engineering specialists, design specialists and more recently for many firms the training specialist. These functional specialists produce problems for line supervisors—particularly older foremen who are often unhappy and frustrated because they feel that their former position and authority have been undermined.

It is both good sense and good business to ensure (*a*) that these supervisors are assured that their work is vital to the efficiency of a company and an industry and (*b*) to help them understand and embrace new management concepts. These are the two ends which supervisory training should achieve.

MOTIVATION

In most firms, three factors, separately or in combination, lead to supervisory training. Firstly, the enlightened employer sees the need for it and asks either his training officer or, if he does not exist, asks some suitable member of staff to do what is necessary: secondly, supervisors as a group may seek training; thirdly, the individual foreman, aware of his shortcomings, may ask for help.

Each request requires different handling. To be fully effective training must *have the active and fullest possible support from management at all levels.* During the training itself, senior members of management, preferably headed by the managing director, should make it obvious to all by their presence and views that they strongly support the work being done. Whenever a phase of training is finished, every opportunity should be given for each supervisor to put into operation the new 'know-how' he has learnt. To fail to do this not only invalidates much of the training, it also induces frustrations even greater than those felt before the training started.

Supervisory training presents other special problems. Prospective course members may be middle-aged, with set habits of thinking and action. They are often very sensitive to the criticism of their inadequacy which they think the training implies and possibly resent having to 'go back to school'. No individual supervisor should ever be given the impression that he or she has been singled out for special correction. By the same logic, it is vitally important that the training officer should create a favourable climate of opinion towards training. He can best achieve this through the quality and high standards of the training provided. There should also be a clearly understood 'open door' provision, preferably within line management itself, whereby any supervisor may without embarrassment seek counsel and advice on training for himself. In this way, he can come to regard training as a means of self development.

TRAINING OBJECTIVES

If supervisory training is to be effective, it must have a clear purpose. This can be established through analysis of information derived from a number of sources—from critical observation of the actual work done by the particular supervisors, by discussion with them and their immediate superiors, and with members of senior management and, where pertinent, with selected workers under their control. Quite often there are considerable gulfs between what a particular supervisor thinks he ought to do and what management considers should be done, and what actually is done. (See *Supervisory Training*, p. 13, by the Industrial Training Service.)

Four objectives may be expected to emerge:

1. To clarify the function and duties of particular supervisors in a firm.
2. To assist supervisors to be more efficient and to develop their skills and knowledge to meet the demands likely to be made of them at present and in the foreseeable future.

3. To facilitate the recruitment and selection of potential supervisors for particular posts.
4. To prepare existing supervisors to undertake work of greater complexity and responsibility.

CHECK-LIST

A. Does the supervisor need to apply new knowledge and skills?
B. How good are his or her relations with colleagues and with the specialists? What recurring problems arise?
C. Should the training be only for certain existing supervisors or for prospective supervisors, or a mixture?
D. Are the supervisors aware of the need for adequate communication upwards, downwards and horizontally? Is this need met?
E. Do they understand fully the company's organisation and management structure?
F. Are they aware of the significance of satisfactory worker-leader relationships? Are there shop steward facilities?

THE SUPERVISOR'S JOB

When the supervisor's function is examined a variety of titles for it are encountered, from manager to foreman. It is not the title, however, which matters, but what he (or she) does.

He is often stated to be outside of policy making and therefore in that sense is not a manager. This is not wholly true since in practice in implementing policy, his interpretations may in small but significant ways modify it.

Functionally he is the individual who bridges the gap between senior management and the worker on the shop floor. Hence he is usually seen from below as the most junior member of the management team of a company to have *direct contact* with the workers and executive authority over them.

A specimen job specification for a first line supervisor might have the following skeletal framework. The reader could give it substance with details appropriate to his own organisation:

1. Within the framework of the function assigned him in his company, to be responsible for putting its operating policy into effect.
2. To ensure that the effort of those who work for him is of requisite standard, and that he justly treats all members of his team and maintains good labour relations.
3. To appoint and train a team of subordinates (e.g., chargehands) competent to carry out the work he requires of them.
4. To establish an order of seniority amongst these subordinates so that his work is done satisfactorily when he has to be absent.
5. To set standards of performance for his subordinates consonant with company policy and to make these standards clear to them.

6. To evaluate the standard of performance of all who work for him and to deal appropriately with those who, after fair notice, fail to reach standard.
7. To provide appropriate induction to all who work for him, whether new to his department or whether they are established workers required to perform new duties.
8. To give active support to all training and retraining schemes affecting his team, welcoming and applying new concepts and techniques.
9. To anticipate future equipment, building, material and labour requirements by making them known to higher management.
10. To communicate satisfactorily in his organisation upwards, downwards and horizontally.
11. To collaborate effectively with all specialist departments.

CHECK-LIST

A. Is there a 'family tree' in the company?—to whom is the supervisor answerable and how many individuals are directly answerable to him?
B. Precisely what relationships does he have with functional specialists, i.e., what service relationships does he have?
C. What does management require of him? What precisely are his duties and responsibilities? Have these been made specific and have they been clearly communicated to him?
D. How much help does he need from management?
E. What precisely is his authority? Is he empowered to 'hire and fire'?
F. Is he only an effort co-ordinator and/or a technical adviser to his staff?
G. What is the precise nature of the work within the general pattern of mass production, batch production, unit production, batch process or continuous process work?
H. How highly organised and complex is his own work and what particular problems does it hold?
I. How conflicting are the demands made of him by design, manufacture, production, maintenance and sales?
J. How do these demands affect him? What help does he need?

TYPES OF COURSES

At least three types of course are generally operated:

1. The course operated within a company.
2. The technical college course.
3. The 'follow-on' course.

Each has its particular uses and sometimes an amalgam of course types is the best solution to a training need. 'Within-the-organisation' courses can provide detailed attention to issues pertinent to a firm. Hence greater penetration of specific needs can be achieved through such a course. On the other hand, parochialism of outlook can occur and, unless special arrangements are made for outsiders to participate either as lecturers or trainees, the valuable broadening effect resulting from contact with views and experience outside a firm can be lost.

Over-generalising and dealing with matters seemingly irrelevent are criticisms levelled against the second type of course, the so-called 'official' external one, such as that leading to the award of the Certificate in Supervisory Studies. These courses nevertheless can provide a very useful background of essential information as well as knowledge of basic principles.

Quite obviously, the third type of course, the 'follow-on', cannot be attempted satisfactorily unless individuals have already received sound basic training in supervision. In colleges the 'follow-on' is a valuable 'refresher' facility and a means of dealing with new specialised knowledge and techniques. In the factory, follow-on courses can be used to good effect to introduce new processes and philosophies.

CHECK-LIST

A. Has management approved the course content?
B. Has adequate consultation occurred at all appropriate levels concerning the content of the course?
C. What can be done to create positive motivation?
D. How well will the course members know each other beforehand?

CONTENT OF COURSES

Obviously, this must depend on the nature of the course. If it has to satisfy an official body, the syllabus followed must conform to that required by that body. If it is an *ad hoc* course, it could deal specifically with one or all of the following, and/or with special topics appropriate to the specific need of a firm and the particular supervisors involved:

Principles of Supervision,
Practical Communication,
Job Instruction Techniques,
Interviewing and Counselling,
Factory Organisation,
Production Planning and Control,
Work Study and Method Study,
Product Design,
Elementary Statistics,
Industrial Law,
Human Relations,
Personnel Management,
Economics of Industry,
Industrial Relations,
Techniques of Joint Consultation,

Trade Unions and Shop Stewards,
Industrial Finance,
Cost Accounting.

CHECK-LIST

A. How long shall each session take?
B. What form of participation will be woven into various sessions?
C. What priority should be given to which subjects?
D. In what ways can special company insights be incorporated?

COURSE LOCATION AND TIMING

The dates, duration, timings and location of a course can each have an important bearing on its success. So many different issues have to be taken into consideration here that any arrangement made is inevitably a compromise. The only advice that can be given is that every effort should be made to ensure that the arrangements made are kept and that they satisfy the largest possible number of requirements.

CHECK-LIST

A. Shall the course be 'off-the-job' within a firm?
B. Should it be held at a local technical college or at a local hotel or recognised educational centre?
C. Should the training take place in or out of working hours?
D. When is the 'best' time in the year to hold it?
E. Has sufficient time been made available to prepare for it?

TEACHING STAFF

As with all teaching, one of the most important contributors to the success of a supervisory training course is the quality of the teaching provided. Wherever possible, choose the enthusiastic experts. What they lack in expertise of teaching can well be compensated for by their capacity to inspire.

CHECK-LIST

A. If the course is an internal one shall the company's training department staff be used for the lecturing and discussion leading?
B. Will special lecturers from within the company also be needed?
C. If so, will they require training in teaching method?
D. Who will be responsible for briefing lecturers so as to avoid unnecessary overlapping and duplication and to ensure that specific problem areas are adequately covered?
E. Will the training officer be present throughout to act as a catalyst and to ensure continuity and integration?

TEACHING AIDS AND TECHNIQUES

Every appropriate modern teaching aid and technique should be employed. Whenever possible, emphasis should be given to visual presentation (see Chapter 13). Excellent case-study material exists on films and film strips and provides a valuable means of developing analysis of supervisory attitudes and problems. The range, scope and availability of instructional films is very extensive. This source, too, can be put to excellent use provided the films are carefully integrated into the programme and are, for example, subjected to analytical discussion. There are also many and varied excellent films on industrial safety.

Training, especially supervisory training, is a process and among the earlier courses each of the Training Within Industry (T.W.I.) programmes (see Chapter 20) should find a place in all in-company training schemes. Project studies and the practice of communication skills should also be employed.

Perhaps the most useful and continuing technique for in-company training in the future will develop from regular but informal supervisory group discussions on common problems and on task objectives, supplemented by information on latest trends and developments in supervision.

CHECK-LIST

A. What audio or visual aids can be used to good effect?
B. Are any special arrangements or rehearsals necessary?
C. Should discussions on specific problems take place and if so should they be recorded and played back?
D. Should groups be sub-divided and given like topics on which to report back?
E. What special techniques are appropriate and necessary, e.g., case discussions, project reports, communication exercises?
F. What amount of documentation is necessary? Will it be prepared beforehand or will its preparation be part of a course project?
G. Should course members each in turn operate as chairman-secretary for a day?

VALIDATION OF COURSES

A very important requirement of any training scheme is that its effectiveness be assessed. With supervisory training this should be done from two standpoints: (*a*) the immediate effects and (*b*) long-term effects. Criteria of effectiveness are:

1. An improved understanding of a supervisor's duties and their relationship to the company's hierarchy.
2. Improved performance by those who work for him as measured by:
   Better-quality work.
   Increased rate of output.

Reduction of scrap.
Reduction in machine breakdowns or performance hold-ups.
Reduction in the labour turnover.
Improvement in workers' timekeeping.
Reduction in absenteeism and accident frequency.

CHECK-LIST

A. Has each supervisor's effort been rated by his immediate superior and by at least one member of higher management? Where appropriate, can assessments be deduced from his subordinates?
B. How has each supervisor reacted to his training? Did he find it helpful?
If so in what ways?
C. Could the training have been improved? If so how?
D. Has the supervisor learnt how to delegate more effectively?

CONCLUSION

Supervisory training is without doubt one of the most urgent priorities for attention by the training officer. It is also one of the most complex and difficult. It may be that the pace of technological change will ultimately change the role and job specification of the foreman. His role as a first line autocrat is clearly obsolete and in its place is the present need for supervisors who can stimulate potential and co-ordinate team effort whilst they themselves are subjected to pressures from multiple sources. In this situation, effective training and the training officer's advice and support play a vital part.

BIBLIOGRAPHY

DE C. MEADE, J. P., and GREIG, F. W., *Supervisory Training*, H.M.S.O. (1966)
The Training of Supervisors, E. I. T. B. Booklet No. 1 (1967)
THURLEY, K. E., and HAMBLIN, A. C., *The Supervisor and his Job*, H.M.S.O. (1963)
DRUCKER, P. F., *The Practice of Management*, Heinemann (1955)
JAQUES, E., *Measurement of Responsibility*, Tavistock (1956)
KING, D., *Training Within the Organisation*, Tavistock (1964)
THURLEY, K. E., and HAMBLIN, A. C., *The Basis of Supervisory Policy*, Pergamon (1967)

20

Training Within Industry for Supervisors

CLIFFORD FEE, M.I.S.M.

Deputy Senior Training Officer (T.W.I.), Ministry of Labour

INTRODUCTION

The Training Within Industry Scheme for Supervisors (T.W.I.) has as its aim the training of supervisors in the broad principles of good management.

T.W.I. defines a supervisor as 'anyone who is in charge of people or who directs the work of others', though the main concern and purpose of the scheme is to inculcate the principles of effective supervision in those who are in direct control of employees at the point of production or service.

THE NEED FOR SUPERVISOR TRAINING

Before the Second World War, few industrial managements saw the need for the training of first and second line supervisors in the arts of supervision.

Indeed, there were not many who recognised that any particular skill was required to be a chargehand or foreman. Certainly there had been little attempt to define the responsibilities and the skills required to discharge them effectively at those levels of supervision.

Usually it was thought that promotion of a worker to the task of supervising the work of others could be based on his knowledge of the work done and his own skill and character as a workman. Managements in pre-war years tended to base selection and appointment of supervisors on a philosophy of appreciation.

They liked the way a man did his work, they appreciated his observation of rules and customs so they promoted him when a vacancy arose. In effect, managements were inferring that a good workman, capable and skilled in his own particular job, should automatically qualify to control a number of other employees and bring them to his standard of workmanship and behaviour. Apart from general advice when a problem arose, little guidance and no formal training was given to the newly appointed supervisor. He was expected to 'learn by experience'. Seniors would expect him to make a few mistakes at first but gradually settle down. The fact that making his early mistakes on the employees under his control might, in many cases, have far reaching effects, seemed not to have been appreciated.

THE INFLUENCE OF SUPERVISORS

Today there is a greater realisation of the influence of a first or second line supervisor and a better understanding of how his approach can affect the performance and working habits of the employees he controls. It is more readily seen that the supervisor is the vital link between senior management and employees at the point of production.

It is he, the supervisor, who interprets the policy of seniors and gives information to workpeople. It is to him the employer turns when in doubt or difficulty. In training people to do jobs, the supervisor can inculcate working habits, good or bad, that remain with the employee the whole of his working life.

Never were the influences of a shop floor supervisor made more manifest than in the early 1940s when, because of world war, thousands of people, completely without industrial experience or background, were recruited or drafted into the munitions, armaments or engineering industries.

THE ORIGIN OF T.W.I.

In the U.S.A., industrial work forces grew apace. A large number of those employees with craft skill developed over many years had to become supervisors or instructors of the new entrants to industry.

It was then that it became apparent that a good craftsman is not necessarily a good instructor and that many with acquired skills are incapable of transferring their own techniques and manual dexterities to others.

The problem was a national one and thus the Government provided the solution.

A number of prominent executives in industry were invited to devise and introduce a short course in instructing which could be used in any kind of industry or organisation. By late 1940, the T.W.I. programme of Job Instruction was launched as a service to American industry. Other courses giving training techniques followed and by 1945 more than two million supervisors had attended T.W.I. Courses in the U.S.A.

British industrialists, who had seen the development of the T.W.I. scheme from its inception in the U.S.A., had meanwhile persuaded the Ministry of Labour, through the National Joint Industrial Council, to introduce T.W.I. to British industry. Since 1945, the Ministry has provided the T.W.I. service.

THE ESSENTIAL NEEDS OF SUPERVISORS

Every training scheme should be related to the needs and work of those who are to receive its benefits. T.W.I. service to supervisors is based on a recognition that their essential needs are as follows:

Knowledge of the work.
Knowledge of responsibilities.
Skill in instructing.
Skill in leading.
Skill in improving methods.
Skill in preventing accidents.

Knowledge of the work means knowing the output required, the machines and equipment used, the essential skills and which workpeople possess them, the materials needed and available, and the time factors involved.

Knowledge of responsibilities means knowing the rules, regulations, customs and practices of the organisation broadly and of the individual's own section in particular. It means knowing the individual's own authority, knowing agreements with trade unions, safety rules and inter-departmental relationships.

Skill in instructing means an ability to transfer knowledge and manual dexterity to employees and thus develop individual skills and the capacity of a section.

Skill in leading means a personal approach which promotes co-operation and team spirit within a working group, considerably reducing the probability of human problems developing or being poorly handled.

Skill in improving methods means a technique of recognising and bringing about improvements in the use of time, space, men, materials and machines available.

Skill in preventing accidents means a systematic method of discovering and removing unsafe conditions and acts in the work area, the work methods and workers.

THE T.W.I. COURSES

The knowledge of the work and the knowledge of responsibilities required for effective supervision differ widely from supervisor to supervisor. Such knowledge is acquired from apprenticeships, technical training and experience on the job. No national programme could cover the knowledge required by all supervisors.

The position is quite different with the skills. The principles of good instruction, the foundations for good relationships, the analytical approach to a human problem, the techniques involved in improving methods and the systematic removal of hazards are the same for all supervisors irrespective of the type of firm, organisation, product or service with which he is concerned.

It is possible, therefore, to devise training courses to present and teach basic principles of good supervision on a national scale.

The T.W.I. scheme provides four courses related to the supervisory skills. They are Job Instruction, Job Relations, Job Methods and Job Safety.

Each of these can be given in 10 hours, usually in five two-hour sessions on successive days to groups of up to 10 supervisors within the firm.

The group leader, or trainer as he is usually described, may be a member of the firm who has been trained in the presentation of the T.W.I. courses by the Ministry of Labour, or he may be one of the Ministry's T.W.I. training officers. Large firms usually have their own trainer, the smaller firms rely on the Ministry of Labour T.W.I. staff.

Each course has been carefully tested and prepared and a training manual is made available to all trainers. This ensures a uniformity of approach and outline of the basic principles of the supervisory skills to all supervisors. Uniformity is justified since principles do not differ whatever the work in which a supervisor is engaged.

HOW T.W.I. TRAINING IS PRESENTED

Unless a supervisor attending the courses can recognise the relationship between the work he does daily and the techniques being taught, much of the training is wasted.

Thus, the courses include considerable discussion of problems and practice in the application of principles to the work for which the supervisor is responsible. Learning by doing is the basis of every T.W.I. course for supervisors.

In the Job Instruction programme, for example, good and bad methods of instruction are demonstrated and examined in the first session. Agreement is reached within the group on the principles of good instruction.

During the remaining sessions, the supervisors being trained themselves provide the practical demonstrations of jobs from their own sections. They learn how to prepare instruction for a job with which they are familiar and how to present it to a trainee. Under the careful coaching of the T.W.I. trainer, the skill in instructing begins to develop in the group.

As a reminder of the lessons learned and approaches to be made, printed cards are issued during each course. Each card outlines the main points of the method to be used whenever the supervisory skill is needed.

RESULTS EXPECTED AND GAINED

T.W.I. has always been seen as a first course for supervisors which must be supplemented by other forms of training specifically related to the work and responsibilities of the individual supervisors.

Yet, for many firms, the use of T.W.I. has produced results substantial enough to warrant the claim that T.W.I. is the most rewarding national training plan for supervisors that industry and commerce has so far encountered.

Job Instruction training has meant for many firms reduction in training time, more skilled use of materials and equipment, and fewer accidents.

Settled labour forces with fewer human problems have followed when supervisory approaches have been based on training given in Job Relations.

An understanding of work study techniques and an ability to apply method study as a result of attendance at Job Methods courses have brought about thousands of improved methods which have increased productivity and made considerable savings in manpower, materials and machines.

The Ministry of Labour is modest about these results, maintaining no statistics about them but satisfied that industry and commerce must be continuing to gain good returns from T.W.I. since there has been little fluctuation in demand for the service in more than 20 years of availability.

FOLLOW-UP

Since there is little more frustrating than being taught something which one is never called upon to put into practice, it is management's responsibility to ensure that supervisors have opportunity to use the skills they have acquired through T.W.I.

Commensurate with their responsibilities and authority, supervisors must be given opportunity to develop their skills by practice in planning and giving instructions, handling human problems and improving work methods in actual work situations in their own department.

Without this follow-up, supervisors would rapidly forget their training and the value of the introduction of T.W.I. would be lost.

A planned development of the supervisor on T.W.I. lines coupled with support and guidance from his seniors can and has resulted within a few months in an efficient supervisor fully adequate to his task of developing an effective section.

NATIONAL DEVELOPMENT OF T.W.I.

T.W.I. was originally seen as a valuable aid to the engineering industry in Great Britain but it was not long before other industries recognised its potentialities in the training of their supervisors.

Industries such as textiles, chemicals, steel, coalmining and allied industries adopted T.W.I. as their basis of supervisor training. The building trades, transport organisations of all kinds, wholesale and retail distributors saw the possibilities of T.W.I. and integrated the courses into their training programmes.

Naval dockyards, Army technical branches, the Royal Air Force, police forces, the Civil Service, banks, insurance companies and hospitals are other adherents of T.W.I.

In some cases, it was necessary to modify courses to meet specific needs. As a result, there are special T.W.I. training manuals for supervisors in hospitals, in offices, in the distributive trades and in building, while for the benefit of trade unions, a training course for shop stewards has been adapted from the Job Relations manual.

At this time, 1,000,000 supervisors in Britain have attended T.W.I.

courses and more than 8,000 trainers have been trained. Over 12,000 firms and organisations use the scheme for training supervisors.

During the first 17 years of its availability to industry and commerce, T.W.I. was a free service from the Ministry of Labour. In 1962, it was decided to invite organisations to help defray the costs and moderate charges for the service were introduced. Management's appreciation of the value of the service may to some extent be gauged from the increased demands for the service since charges were made.

RESEARCH AND DEVELOPMENT

It is axiomatic that any training programme, however successful, should be examined for possible amendment not less than every two years. The Ministry's T.W.I. staff are always concerned to keep the courses up to date utilising current instructing techniques as they become known, tested and successful. In recent years, methods of role playing, syndicate work, charts, film and filmstrips, and shop floor practice have all been added to the scheme.

The latest course introduced by the Ministry as part of the T.W.I. scheme is the Training Development Service.

The course includes the training of seniors in how training can be planned in a department and the training of selected experienced employees as operator/instructors.

These operator/instructors will give instruction to new entrants and transferees mainly in semi-skilled operations.

CONCLUSION

There can be little doubt that T.W.I. pioneered formal supervisory training in this country. It has been, and continues to be, of great benefit to industry and commerce.

Although today, a case is being made for supervisory training schemes to be tailor-made to the needs of the individual, it will be a long time before T.W.I. is superseded as the major supervisory training scheme available in Great Britain particularly for the smaller organisations.

Further information on all T.W.I. programmes may be obtained from Ministry of Labour, Ebury Bridge House, Ebury Bridge Road, London S.W.1, or from any local Ministry of Labour office.

21

Training Technical Graduates

J. R. BADDILEY, B.SC. (MECH.ENG.), M. INST.PET.

Technical Training Co-ordinator, Fawley Refinery, Esso Petroleum Co. Ltd.

INTRODUCTION

The purpose of this chapter is to help employers use planned training as a means of obtaining the maximum return on their investment in technical graduates.

There are many sides to the training of technical graduates. A detailed discourse on them all would require considerable space. This chapter deals with training for specific jobs in manufacturing industries but it is hoped, with the aid of examples, to impart a systematic approach to the subject which training officers could apply elsewhere. No attempt is made to lay down hard and fast rules. It is assumed that the employers have not previously carried out planned training and that the training officers, therefore, are starting from scratch.

The average young graduate entering industry is blessed with a reasonable general education, an academic training designed to help him progress in his chosen career, an active mind and, in his attitude to work and life, a flexibility often missing in his elders.

On the other hand, he has little idea of the means by which to apply his academic knowledge in an industrial world, has only a general idea of the business in which he has taken employment and has yet to learn how to communicate with people from any walks of life. Here then is a young man (or woman) with a great deal to learn but suitably prepared to absorb rapidly the new teachings and skills of the industrial climate.

VARIETY OF WORK DONE BY TECHNICAL GRADUATES

The first job for the training officer is to consider the work expected of technical graduates in his organisation.

It is unlikely that any two industries make use of precisely the same kinds of graduates. It is certain that within an industry no two companies use graduates in precisely the same way. Again, a small firm might employ one or two graduates of one kind whereas a large chemical factory would probably employ mechanical, electrical, chemical, civil and instrument engineers together with chemists, physicists, mathematicians and others. Within a manufacturing organisation, a technical graduate may be employed in research, development of new products, construction, maintenance, or perhaps with the performance of manufacturing equipment (by manufacturing equipment is meant whatever is used in a factory to manufacture products for sale. This could be a relatively simple metal forming machine; on the other hand it could be made up of highly complex equipment involving heat, pressure, chemical reactions, etc.).

For illustration, consider the technical graduates who are concerned with the design or performance of manufacturing equipment. Depending upon the size and complexity of the factory, they could be employed on any or all of the following:

- Directing the operation of manufacturing equipment.
- 'Trouble-shooting'—investigating and overcoming manufacturing problems.
- Efficiency improvement—reducing operating costs.
- Development of existing equipment to produce either greater quantities of existing products or a greater variety of products.
- Carrying out studies to decide upon new plants and the capital investment necessary to meet changing market requirements.
- Designing manufacturing equipment (in this case design means process design, i.e., size, shape, limits of movement, pressure, temperature, etc. It does not refer to the mechanical, electrical, etc., design needed to convert process designs into manufacturing equipment).
- Planning the sequence of operations on the manufacturing equipment.

THE NECESSARY EXPERIENCE AND SKILLS

Having listed the jobs done by graduates, the next task for the training officer is to detail the experience and skills necessary to perform the jobs efficiently. In a small factory, the training officer may be fully familiar with the requirements of the various jobs; in a large factory he will almost certainly need to consult people with experience in the various fields.

Continuing with the example of the graduates who are concerned with the design or performance of manufacturing equipment, requirements could be amongst the following:

Familiarity with the operation of the manufacturing equipment. This is required, not only to enable the graduate to direct operation of manufacturing equipment, but also to make an efficient job of 'trouble-shooting'.

The ability to carry out process designs. This, of course, is essential to the man employed on design work. In addition, it can be of considerable help in appreciating the causes of equipment operating problems. Again, work on efficiency improvements can involve the use of design calculation, as can work on the development of existing equipment.

A knowledge of economics. Anyone working on the development or improvement of existing plant, or on the study of new plants, may benefit from a knowledge of the economics of the industry.

People concerned with the economics of capital investment need to prepare accurate estimates of costs for presentations to directors. This necessitates a special knowledge of the economics of the industry and the ability to make true calculations of return on investment.

A knowledge of materials of construction, of construction methods and of maintenance problems. Such knowledge is, of course, primarily the concern of those employed in construction or maintenance. It can, however, be of benefit to others. For example, it may have an effect on the design of a manufacturing process. Again, those who direct manufacturing operations may use the knowledge to obtain maximum service from the equipment.

TRAINING METHODS

The foregoing lists a few examples of the experience and skills necessary, in varying degrees, to graduates working in one field. Every firm should, therefore, aim to ensure that precise job analysis leading to clear definition of job requirements precedes the appointment of its technical graduates. Now consider means by which an individual may acquire the additional knowledge necessary to do a particular job.

There are several ways of training new staff. Some of the recognised methods are:

'On-the-job' training. By this is meant learning in the course of doing a job. It is, of course, the normal method by which many people at all levels learn new jobs and other training methods are employed only in a supplementary role. It is customary to attach the trainee to an experienced man. One shortcoming of this method is that work must be dealt with as it comes along so that the tuition can be in an illogical order and therefore be protracted.

When used as a means of learning about other people's work and problems, the trainee will normally be employed on work of no great complexity. Amongst those with experience in training, it is generally agreed that the trainee must do a useful job of work, if only a minor one. Experience shows that as an observer he would not be fully accepted and much of his time would be wasted.

Lectures. At graduate level, this is a quick and effective means of imparting specific information. It has the advantage that the information can be carefully marshalled beforehand and can often be used repeatedly with only slight modification. However, it should be remembered that no audience can, for long, just sit and absorb information. If several talks are involved, it may be advantageous to give them at, say weekly, intervals; an alternative would be to keep individual talks short and allow ample time for audience participation.

Courses designed to simulate actual practice. Potentially, this is a most effective method of instruction in that the exercises can be designed to accomplish the maximum training in the minimum time. In addition, without loss to the company, it allows the trainee to benefit from his mistakes. Commercially run courses are available to meet several needs. When run by the company itself, such a course has the big advantage that company philosophy and generally agreed methods can be woven into the exercises. However, many man-hours are required to prepare, present and tutor such a course; the time and cost can be justified only where there are several people to train.

Regular technical discussions. These are not for the beginner. The welfare of a company and of the individual depend upon the graduate staff keeping up-to-date with developments both inside and outside the company. One way of keeping up-to-date is to have regular technical discussions on subjects of topical interest. Discussion groups can vary in size from small groups within departments through to large meetings of experts from various factories and other parts of an organisation. Allied to this are (*a*) seminars at which experts from outside the organisation speak on chosen topics and (*b*) nation-wide or world-wide conferences.

CHOICE OF METHOD AND DECISION ON EXTENT OF TRAINING

Having dealt with the training needs and the methods available to meet the needs, the training officer is faced with the problems of selecting training methods and of getting decisions on the extent of training. This could well present him with his most difficult tasks.

Choice of training methods and the extent to which training is carried out depend upon a variety of factors. Included amongst these factors are the size of the organisation, the number of graduates employed, the annual intake of graduates and pressure of work.

There are also management factors. Management must attempt to balance the direct and indirect costs of training against the long-term benefits. They have the conflict between the day to day need to get work done and man-hours necessary for training. This is not a matter of precise economics; it is to a large extent a question of which option to take. Another factor is the extent to which management consider it their duty to develop the individual to the limit of his capabilities.

In the end, management must make the decisions; the training officer can only make recommendations. In some cases, he may be obliged to offer a variety of suggestions so as to give management an opportunity to select the arrangements which they believe will suit them best.

When it comes to choice of training method, it is fairly obvious that the very large organisation, compared with the small organisation, is far better placed to select the best method to meet a particular requirement. It normally has sufficient staff to be able to release people to organise technical training; to give lectures and demonstrations, and to guide the development of the individual. There will probably be a variety of work in the same factory; this can help widen the scope of training. There could be sufficient staff requiring the same training to make it worthwhile comparing the merits of simulation exercises in a classroom with those of 'on-the-job' training. Again, when it comes to implanting information as distinct from work methods, it may, due to the numbers needing the information, be advisable to compare classroom lectures or in some circumstances, the use of available programmed learning courses with 'on-the-job' learning. At the other end of the scale, the small organisation must, generally, be much more limited in scope. The intake of graduates may be so slow that, even if an instructor could be made available, the setting up of classes for simulation exercises would occur so rarely as to be of little value. Here, too, programmed learning courses, if suitable and available, could be an expedient method.

The training officer in a small organisation will need to look outside for help. He may find that technical colleges and other institutions run special courses with industry in view. He may be able to persuade business contacts to help; for example, the contractor who has designed and installed a company's manufacturing equipment might be persuaded to accept a trainee on a construction site elsewhere.

Even if training is limited almost entirely to 'on-the-job' training, there is no reason why the training officer should not carefully plan the sequence and extent of the training. The above may appear to be vague. In fact, it is difficult to apply rules to training. No one formula will suit. No two organisations have precisely the same requirements and it would seem to be reasonable to say that within an organisation, no two people are in complete agreement on training needs and methods.

TIMING OF TRAINING

The timing of training is almost as important as the training itself. There is divergence in opinion on timing—some believe that a man should attempt a job and make a few mistakes before being taught the job properly. The majority hold the view that training should be given just before the knowledge is needed in a man's work. It is also generally agreed that early follow-up after actual performance on the job has started is a vital stage in the reinforcement of knowledge.

In practice, of course, timing is frequently far from ideal. For example, a course involving classroom instruction may have as students people recruited at various times; this could mean that training is late for some and early for others. Again, 'on-the-job' training in a specific job may have to await the convenience of the department concerned. In most cases, timing must be a compromise between the ideal and the practical.

CONCLUSIONS

In this chapter an attempt has been made to indicate a logical approach to the problems of helping the young graduate to become a productive member of a manufacturing organisation. Training for specific technical needs has been emphasised.

Training and tuition in other fields can however have a profound bearing on career success. Training in management skills and techniques is of first importance (see Chapter 18). In particular, attention must be given to the social skills so essential if a graduate is to work in harmony with people of widely different backgrounds from his own. He must be able to communicate effectively at all levels. If he has bright ideas but cannot communicate them, they and he are wasted.

Not least important in terms of both training and timing is the implementation of the training recommendations of the professional bodies given by progressive firms and organisations to their newly appointed graduates. Institutions such as the members of the Council of Engineering Institutions (C.E.I.)* issue booklets dealing with their respective schemes of training. These are commended to employers in each of the industries concerned.

Meanwhile the 'Bosworth' report, which recommends postgraduate courses linked to specific industrial needs, may well result in another and very significant bridge to quicker achievement in the future of full competence by the technical graduate.

BIBLIOGRAPHY

DYER, F. C., *Executives Guide to Effective Speaking and Writing*, Prentice-Hall (1962)

STIRES, D. M., and MURPHY, M. M., *Modern Management Methods PERT and CPM*, Industrial Education Institute Ltd. (1964)

GUNNING, R., *New Guide to More Effective Writing in Business and Industry*, Industrial Education Institute Ltd. (1964)

'Bosworth' report on *Education & Training Requirements for Electrical and Mechanical Manufacturing Industries*, H.M.S.O. (1966)

*Addresses of the constituent members of the C.E.I:

The Royal Aeronautical Society, 4 Hamilton Place, London, W.1.
The Institution of Chemical Engineers, 16 Belgrave Square, London, S.W.1.
The Institution of Civil Engineers, Great George Street, London, S.W.1.
The Institution of Electrical Engineers, Savoy Place, Victoria Embankment, London, W.C.2.
The Institution of Electronic and Radio Engineers, 9 Bedford Square, London, W.C.1.
The Institution of Gas Engineers, 17 Grosvenor Crescent, London, S.W.1.
The Institution of Marine Engineers, 76 Mark Lane, London, E.C.3.
The Institution of Mechanical Engineers, 1 Birdcage Walk, London, S.W.1.
The Institution of Mining and Metallurgy, 44 Portland Place, London, W.1.
The Institution of Municipal Engineers, 22 Eccleston Square, London, S.W.1.
The Royal Institution of Naval Architects, 10 Upper Belgrave Street, London, S.W.1.
The Institution of Production Engineers, 10 Chesterfield Street, London, W.1.
The Institution of Structural Engineers, 11 Upper Belgrave Street, London, S.W.1.

22

Instructor Training at Letchworth

J. SIMS.

Senior Lecturer, Ministry of Labour Instructor Training College, Letchworth. Assistant Manager, Government Training Centre, Letchworth

INTRODUCTION

Since the 1964 Industrial Training Act, there has been a growing need for qualified instructors and a growing awareness on the part of industry of the need for more instructor training. The number of students passing through the Ministry of Labour's Instructor Training Colleges at Letchworth, Hertfordshire and Hillington, near Glasgow, has steadily increased. In November 1966, the Minister of Labour announced to the House that the number of places (1,500 per year) would be doubled in the next 12 months. Many branches of industry are now anxious to take advantage of the service offered to them by the Ministry. However, very few of those responsible for the selection of instructors realise what is required to make a good instructor. Their choice is probably governed mainly by considerations of technical ability. Personality and attitude to training may be taken into account as an afterthought, if at all.

It is possible for a skilled craftsman to leave a tool room on a Friday evening and start work in a training school on the following Monday morning totally unskilled as an instructor. An instructor must, of course, be a good craftsman since, if he is to impart skills to others, he must possess those skills himself. What also needs to be recognised is that he must have the ability to instruct. Few people are born instructors. Instructing is a skill in its own right and like other skills has to be learned. Employers would not ask an unskilled operative to tackle a skilled job and they should not expect any person to take on the demanding task of instructing without suitable training. At its two colleges, the Ministry offers an intensive two-week training course in instructional techniques and it is the aim of this chapter to outline briefly the methods used to teach students some of the main techniques of instruction.

WHY AND HOW PEOPLE LEARN

If a person is to learn a subject he must have:

1. The desire to learn.
2. The means of learning.
3. The incentive to learn.

The embryo instructor has to be made aware of the fact that nobody can be compelled to learn. The learner has to be positively motivated towards the learning process. If he begins to lose interest, it becomes increasingly difficult for him to learn. If the basic desire to learn is analysed, it will become apparent that it varies greatly from person to person according to age and responsibility and other factors. A boy may join a firm as an apprentice, not necessarily because of a strong desire to learn, but because of parental influence or the example of his friends or because of guidance from a youth employment officer. An adult may be seeking advancement or security or some other aim in connection with his home or family. The instructor's first logical step in helping a person to learn is to find out what are his student's motives for learning and whenever possible to extend and deepen them.

Individuals learn through the use of sight, hearing, touch, taste and smell, and learning is made easier if two or more of these senses, for example sight and sound, are employed at the same time. Learning involves the acquiring of knowledge and is primarily a process of registering and retaining practical and theoretical information, to which the senses and the powers of reason and understanding combined with constant practice all contribute.

METHODS OF INSTRUCTION

It is essential to differentiate between instruction appropriate to the workshop and instruction appropriate in a classroom.

THE CLASSROOM LESSON

Although the classroom lesson, even for adults, is a very old method of teaching, to many instructors it is something quite new. Its object should be to foster enthusiasm and secure a closer relationship between instructor and student by encouraging the latter to participate as actively as possible. In this way, the instructor can get to know which among his students are the bright, the average, the backward or the apathetic. The formal lesson can be one of the most thorough methods of instructing. Throughout it, the instructor should aim to rivet attention and exercise his students' reasoning powers by posing carefully selected questions during a logical build-up of relevant information. This activity, where appropriate, should be interspersed with suitable exercises. After asking a question, the instructor should always pause to allow his students time for thought and only when necessary should he name a person required to answer.

Should he then receive the bare answer 'Yes' or 'No' he can immediately follow up with 'Why?' or 'Why not?'. The student should always be able to deduce the answer to a question from the information previously supplied by the instructor.

The procedure at the Instructor Training Colleges is that an exposition of the lesson technique and an example lesson are given by a college lecturer. Each student then prepares and presents to his fellow students a lesson on a subject of his own choosing within his trade. At the end, he receives from the lecturer a commentary containing constructive criticism.

A suggested pattern for a classroom session of 45 minutes is:

Introduction—2–3 minutes	Arouse interest; link with previous subjects; state objective.
Revision—2–3 minutes	Where necessary, revise material dealt with in a previous lesson.
First stage—10 minutes	Give clear exposition with class participation.
Summary of first stage—2–3 minutes	Reiterate main points so far covered.
Second and third stages with summaries	As for first stage and summary.
Final stage—10 minutes	As for first stage.
General questions—2–3 minutes (longer if necessary)	Give students the opportunity of asking questions on the total range of subject-matter covered.
Final summary—2–3 minutes	Re-emphasise the main points of the session and link each stage to form a coherent whole.

The number of stages in a session will, of course, vary according to the time at the instructor's disposal and the extent to which he needs to develop the subject.

INSTRUCTING THE INDIVIDUAL WORKER (DEMONSTRATION)

As well as being able to impart information in the classroom, the instructor must be able to give practical instruction in a practical setting, i.e., in the workshop. Individual instruction is particularly valuable where students are at different stages of training or when the job is such that it can be seen by only one learner at a time. The trainee instructor is taught that the essence of individual instruction is the breakdown of each skill or operation into short stages or units, which, when assembled, form a coherent whole. This technique may be compared with building a ladder. After each rung has been inserted, pegged and wired, the student should be able to mount it and stand on it in complete security.

A complete job which has to be performed as a continuous process can still be dealt with unit by unit, provided that the instructor remembers that his objective is to train the learner progressively until full competence is achieved. It is also stressed that throughout the learner's performance, his work must be checked and supervised by the instructor. It is explained that errors can thus be corrected as soon as they occur and as soon as the learner's mastery of the instruction can be verified. After this initial instruction, the next question considered is how best to determine the way to supervise the further development of a learner's skill, and it is explained that correctly supervised practice will encourage enthusiasm and bring the learner closer to full competence.

At the Instructor Training Colleges, after an exposition of the techniques of individual instruction and an example breakdown of a subject by a college lecturer into five-minute demonstrations and explanations, each student first prepares (under supervision), and then practises, a similar breakdown of his own chosen subject. After his practice session, he receives constructive comments from the lecturer. It may be thought that five minutes practical instruction is not long enough for teaching any operation, but if the learner is given too much information to absorb at one time, he will become mentally saturated. When he is receiving this type of instruction, he is watching the example, listening to explanations on it, and trying to foresee and understand the next move. Above all, he is aware that he is expected to commit all he sees and hears to memory. To him, this is a big task and, moreover, it is often an entirely new experience. Very few people can absorb knowledge for longer than five minutes without a break and a session for discussion, revision and consolidation.

The use of a job-breakdown sheet is also essential in preparing and presenting individual instruction in progressive stages. Suggestions for the layout are as follows: head a sheet of paper with the title of the job, and divide the paper into two columns, the left-hand column headed 'Stages' and the right-hand column headed 'Key points'. 'Stages' will be each part of the operation which substantially advances the work—in other words, the number of operations to be done in order to complete the job. 'Key points' will be any points occurring in a stage which might make or mar the work, injure the worker or make the work easier to do. Examples are a knack or trick of the trade, special timing or a special piece of information. It is important that the instructor should understand how to present the job to the learner. The steps suggested are:

1. See that the learner is in the right position to observe the job.
2. Outline the subject by means of a short introduction and thus promote interest in it and in mastering it.
3. Proceed to do the job stage by stage, explaining as you go and emphasising the key points. Slow down your actions where appropriate and repeat if necessary.
4. Direct attention to points that might pass unnoticed.

INSTRUCTING A GROUP OF WORKERS (DEMONSTRATION)

The lesson technique may be used whether instructing a group of students by demonstration of the skills of a trade or of a single operation. The number of students per group who can be effectively taught by this method will vary with the type of job and situation. Small complicated jobs—possibly in cramped conditions—will permit only a limited number of learners. With larger jobs in more open situations the number can be increased.

At the Instructor Training Colleges, after an exposition of the techniques involved, a practical demonstration is given by the lecturer to a group in the workshop. One person is then chosen from the group and put to work. He is stopped at appropriate places by the instructor and his method is confirmed or corrected by the remainder of the group. The instructor gives more than one learner an opportunity of participating in the work by changing the learners around at suitable intervals. When the job has been completed, the instructor summarises the work done and supervises individual practice in the way advocated for individual instruction.

As with individual instruction, a job-breakdown sheet is essential. Each stage of the operation must be analysed and questions carefully formulated. It is emphasised that the instructor should never lose sight of the fact that he has to encourage the reasoning powers of his students, and guide them in logical deduction. It is suggested that the instructor should present the task to the learners as follows:

1. See that the group are in the best position to observe the job.
2. Outline the subject by means of a short introduction.
3. Do the job stage by stage, posing questions, pausing and allowing discussion or nominating persons to answer. Maximum group participation should be the aim.

CONCLUSION

As a final thought, it is emphasised that to improve the learner's skill, firm yet sensitive guidance and advice is far more effective than adverse criticism. Praise is synonymous with encouragement, and the course aims to demonstrate that both provide re-inforcement to the learner.

There are indications that for some time to come, demand for places on Instructor Training courses by industry and commerce will outstrip supply except where arrangements are made sufficiently far in advance. This is in spite of expansion of the number of places at Letchworth and Hillington Colleges for Instructor Training and the places provided through the setting up of Instructor Training Units attached to G.T.C.s in various parts of the country, e.g., Perivale, Cardiff, Leicester and Killingworth. By the end of 1968 there will be a total number of 4,000 places available—eventually this will be increased to 5,000.

23

Technician Training

D. E. WHEATLEY, M.A.(CANTAB.), B.SC.(LOND.)
Director (Further Education and Training), City and Guilds of London Institute

INTRODUCTION

This chapter describes what is meant by the term 'technician', gives an account of the special characteristics of work at technician level and builds on that to illustrate the distinctive features of schemes for the recruitment, training and further education of technicians. Based primarily on practice in the technical field, the chapter also includes a note on technician-level occupations in commercial and administrative departments. It concludes with a note on the importance of a technician training scheme to any firm which seeks to recruit a fair sample of the more able school-leavers in its area.

TECHNICIANS: DEFINITIONS—CHARACTERISTICS—EXAMPLES

Technicians are key people, vital to the efficiency of any modern industrial enterprise. For a variety of social, educational and industrial reasons, British industry has been backward in recognising the contribution which properly trained technicians can make to productivity. Great Britain's performance in the international league is doubtless due to a combination of factors; but the scarcity of technicians to provide technical stiffening to the labour force and to form the indispensable link between the professional scientist and technologist on the one hand and the shop-floor worker on the other is certainly among them.

The word 'technician' is a generic term, used to identify individuals doing responsible jobs—jobs which involve a higher level of scientific and technical knowledge than that needed by an operator or craftsman, but below that of a technologist, engineer or scientist. A technician's education and training enable him (her) to exercise technical judgment; this judgment is based on an intelligent application of the general principles underlying his (her) work as compared with the reliance upon accumulated skill and experience which is characteristic of the craftsman. Whereas the term 'craftsman' applies to a fairly easily defined and generally recognised degree of skill, the term 'technician' applies rather to a broad band of abilities, ranging from jobs which are only slightly more technical than those of craftsmen, right up to others where high-grade technicians are rubbing shoulders with technologists and scientists.

THE RANGE OF TECHNICIAN OCCUPATIONS

There are of course thousands of technician occupations. The following are a few examples by way of illustration: draughtsmen in engineering or construction; design assistants; jig and tool designers; inspectors; electronic maintenance technicians; analytical technicians (physical, chemical, metallurgical and biological); production planners and controllers; plant or site supervisors; estimators; quantity surveyors; contract managers; testers; technical salesmen; technical service staff; fabric designers; carding overlookers; spinning overlookers; quality control technicians; work study officers; print designers; service managers (automobile); industrial photographers.

In general, technicians are normally found among workers in the following fields:

1. The detailed design and development, or the manufacture, erection or commissioning of equipment and structures; drawing, estimating; inspecting and testing equipment; use of measuring instruments; operating, maintaining and repairing machinery, plant and services and locating defects therein; process control; activities connected with research and development; testing of materials and components; quality control; technical advice to customers; servicing of equipment; data processing; work study.
2. Assisting qualified scientists in such activities as physical measurements; collection and evaluation of experimental observations; the devising and setting up of experimental apparatus; the preparation of chemicals or biological cultures, or similar preparations in other fields; photographic work; the taking and routine testing of product samples; chemical or metallurgical analysis, biological assay, etc.

WHAT IS REQUIRED OF THE TECHNICIAN

The special characteristics of technicians' work determine the nature of their training and further education. First and foremost, technicians need to have mastery of those techniques which are directly relevant to their immediate employment. Proficiency in these techniques requires an understanding of the science and mathematics on which they are based. This understanding need not be so profound as that of the professional scientist, but it should be sufficiently firm to permit its further extension as new applications of the relevant science and technology are introduced, and also as the technician himself encounters situations which present new requirements for specialisation.

Most technicians also find themselves concerned technically with the work of others at craft or operative level and they must, therefore, be aware of its nature and problems. This awareness is preferably acquired through the medium of a broad basic training and planned experience which is specifically designed to give direct acquaintance with the work in question.

Technicians in all fields need to be able to communicate with colleagues at senior and junior levels, for example, in making reports, issuing instructions and so on. Therefore they need fluency in the use of the written and spoken word and this can best be built on the basis of a sound general education.

The relative extent of the technician's requirement for manual as opposed to mental skills will depend very largely on the field in which he is employed—in design, for example, it will be minimal, in servicing it may be high. It is, however, not likely that mental skill will ever be less than 50% of his total skill and the overall balance of his education and training must take this into account.

IDENTIFYING THE NEED FOR TECHNICIANS

Only a few training officers will be fortunate enough to be appointed to firms in which a well designed scheme for the recruitment and training of technicians is in operation. This is due partly to the relatively recent and, as yet only partial, recognition given to the technician function. In part also, it is due to the fact that many of the technician training schemes which do exist are not really suitable, being merely minor adaptations of traditional patterns of craft apprentice training. Most training officers will therefore find that they have a substantial task on their hands and that they need to embark upon the following programme:

1. Identify functions in their firms which could best be carried out by technicians.
2. Produce job specifications for technicians as appropriate.
3. Estimate the scale of recruitment required and identify the sources from which recruits may be obtained.
4. Devise patterns of recruitment and schemes of training for new entrants (i.e., school-leavers), upgrading of existing trainees (e.g., craft apprentices), and for the retraining of adults, as appropriate.
5. Assess the basic educational requirements for recruits to technician training and identify the technical college courses that would provide suitable complementary further education.
6. Alert schools, technical colleges and youth employment officers in the firm's catchment area that an education and training scheme for technicians is to be introduced.
7. Instruct assistant training officers and instructors and also advise

line supervisors on the policy to be implemented and inform them of the distinctive requirements of technician training.

8. Convince the management and get authorisation for necessary expenditure.

SOURCES OF RECRUITMENT

It will readily be appreciated that the large theoretical element in the work and training of technicians has a decisive influence on selection requirements. This is especially so because of the importance of success in the appropriate course of further education. The situation is, however, complicated by the fact that technicians do not form a single, narrow band in the spectrum of ability which could be defined in terms of a single entry requirement. On the contrary, the range of ability found in technician occupations is matched by a corresponding range in the sources of recruitment. These may be somewhat arbitrarily summarised as:

1. *School-leavers*

   (*a*) Age 16+ with five years (in Scotland four years) progressive secondary education including science and mathematics up to the level of the Certificate of Secondary Education (C.S.E.)

   (*b*) Age 16+ with success in the General Certificate of Education (G.C.E.) or Scottish Certificate of Education (S.C.E.) at 'Ordinary' level in science, mathematics and at least two other subjects.

   (*c*) Age 18+ with success in G.C.E. at 'Advanced' level in a relevant subject, having also studied at least one other relevant subject to that level (with supporting successes at 'Ordinary' level).

2. *Up-graded trainees*

   Typically these will be craft apprentices of age 18 or thereabouts who have demonstrated in training and in further education that they possess the ability to profit from technician training.

3. *Retrainees*

   Adults, e.g., ex-servicemen, but also including personnel from other industries, whose training and work has been at technician level. Such recruits require a relatively short conversion course to enable them to apply their previously acquired knowledge and experience in a new field.

TRAINING NEEDS

Technical college courses for technicians are in the main geared to the three types of school-leaver indicated under 1 above and the principal types of course are as follows:

For category 1(a) above. Technician courses devised by the City and Guilds of London Institute for technicians in specific industrial fields, e.g., telecommunications, mechanical engineering, electrical engineering, industrial measurement and control, metallurgy, construction, printing, clothing, milk processing and control, foundry work, textiles, mining, computers, chemical, shipbuilding, papermaking and many others.

For category 1(b) above. The well-known schemes for Ordinary National Certificates (O.N.C.) (part-time) and Ordinary National Diplomas (O.N.D.) (sandwich or full-time) which are two-year courses administered by 'Joint Committees' of the Government Education Department(s) and the relevant professional body or bodies. The standard reached is regarded as equivalent to that of the G.C.E. at 'Advanced' level. Schemes exist in engineering, construction, business studies, metallurgy, mining, naval architecture and textiles.

For category 1(a) and 1(b) above. General courses for intending technicians. These are one- or two-year part-time courses which enable school-leavers of age 15 or 16 to study with a view to entering City and Guilds Technician Courses or O.N.C./O.N.D. courses according to their attainment in the general course examination. There are seven of these courses—in engineering, construction, textiles, mining, shipbuilding, science and printing. General courses are not available in Scotland.

For category 1(c) above. Part-time Higher National Certificate and full-time or sandwich course Higher National Diploma courses, administered as for the corresponding schemes at Ordinary level. The standard reached is approximately that of a university ordinary degree though on a narrower front. The Higher National Diploma qualification is at almost professional level; in practice, most students in H.N.D. courses have full professional qualification in view and many of them go on to achieve it, after further study and with appropriate industrial experience. There are schemes in building, chemistry, business studies, metallurgy, applied physics, civil engineering, engineering, electrical engineering, chemical engineering, naval architecture, textiles, mining, foundry technology and printing.

PATTERNS OF TRAINING

Technician training schemes, taking four or five years in all, are generally built on the following pattern:

(*a*) *Induction* (*a few weeks*). Familiarisation with the firm, employment conditions, working environment, the training scheme, opportunities for further education, welfare and safety arrangements.

(*b*) *Basic training* (*1–2 years*). (This usually takes one of two forms, according to whether the field is craft-based or non-craft based.)

In craft-based industries basic training generally is broad practical training, designed to give experience of the principal materials and production techniques used in the industry or sector. Typically, this is found in the field of engineering and in the training of engineering technicians in other industries, e.g., iron and steel, or mining.

The training has some similarities to the initial basic training of craft apprentices, except that it usually has wider scope. It differs also in that the emphasis is less on the acquisition of skill through extended practice than on gaining an appreciation of the characteristics and limitations of a wide range of processes and materials used in the industry, in relation to the production or operational requirements. Training of this kind is often 'off-the-job' except where this is particularly difficult to arrange, for example, in the construction industry.

In non-craft industries, the basic training of technician apprentices generally is 'representative practical training'. This is practical training in a situation selected for its suitability for the introduction of new recruits to a broad and representative sample of the operations of the firm concerned. This pattern is found particularly in the process industries, for example, in a large chemical works or a steelworks, a general analytical laboratory might serve this purpose, for chemical/metallurgical technician trainees, who might ultimately find employment in a variety of different fields. Alternatively, a period in a technical service department would bring the trainee wide contacts within the firm and also with its customers; it would also provide excellent opportunity for acquiring product knowledge and an awareness of the firm's general policy. There are of course very many other situations of this kind, depending not only on the type of industry but also on the individual firm's arrangements.

(*c*) *Secondary training* (*1–2 years*). In this stage, the trainee spends periods of weeks or months training in various departments, with or without periods of off-the-job training, as appropriate. The aim of secondary training is to broaden and deepen the industrial experience of the trainee so that he may be able to see as a whole the operation and inter-relationships of the various departments with which he is likely to be associated in the future. In addition, the fairly lengthy periods in each department give opportunity to the trainee and his employer to decide where his aptitudes and interest can be used to best advantage. During this period, it is particularly important that all concerned should be fully aware of the purpose of the trainee's stay in a particular department, so that he may be given full opportunity to concern himself with technical considerations in varied situations rather than spend long periods of time doing routine tasks or in acquiring skills that are not relevant to his intended technician function.

(*d*) *Specialised training* (*1–2 years*). This final period of practical training aims to give the trainee a thorough knowledge of those functions with which he will be predominantly concerned after completion of training. The programme should be individually devised for each trainee. Although normally based primarily on the department for which the trainee is intended, it need not, and often should not, exclusively be so; for example, a future engineering inspector might usefully spend a period in laboratories with a wide range of testing facilities, coupled with another period in a design office, and yet another in a production department, before finishing off with a lengthy spell in the inspection department itself.

COMMERCIAL TECHNICIANS

The foregoing has been written exclusively in terms of technician training for technical occupations. The concept of technician-level occupations is, however, equally applicable to the commercial and administrative fields. Regrettably, well designed schemes of recruitment, training and further education for commercial and administrative technicians are even scarcer than their counterpart in the technical field. There is no good reason why this should remain so. As long ago as 1958, the Association of British Chambers of Commerce introduced its Commercial Apprenticeship Scheme based mainly on recruitment at 16+ with G.C.E. or S.C.E. at ordinary level in English and two other subjects from a specified range, with training over a four-year period analogous to what has been described above for technical occupations. During these four years, study is by day-release for the O.N.C. in Business Studies (two years) followed by *either* H.N.C. *or* a course leading to the examinations of an approved professional body (such as the Chartered Institute of Secretaries). There is also a variety of comparable schemes devised by individual firms, some of which call for recruitment at the Advanced level of the G.C.E. It seems likely that the immediate future will see considerable development in this field as industrial training boards produce training recommendations, based on the 1966 report of the Commercial and Clerical Training Committee of the Central Training Council.

CONCLUSION

The contribution that good technician training can make to a firm's competitive efficiency has already been referred to. To the training officer and the personnel department, it has even wider significance in terms of recruitment. No firm nowadays can expect to get a fair cut of the more able school-leavers in its area unless it is able to offer them clear prospects of training and education to responsible levels. In all, this difficult sector of the training officer's task bristles with problems as well as with opportunities—it presents a challenge which cannot be ignored without peril.

BIBLIOGRAPHY

Better Opportunities in Technical Education–Cmnd. 1254, H.M.S.O. (1961)

The Education and Training of Electrical Technician Engineers, I.E.E. (1963)

Recommendations for the Training of Technicians and Technologists, Iron and Steel I.T.B. (1966)

Recommendations for Technician Training, Joint Committee on Training for the Foundry Industry (1966)

Building Technician Training Scheme, Eastern Federation of Building Trades Employers (Jan. 1964)

Construction Technician Training Scheme, Scottish National Building Trades Federation (Employers) (Jan. 1965)

Scheme for the Training of Civil Engineering Technicians, Institution of Civil Engineers (1965)

Training for Commerce and the Office, Report of the Commercial and Clerical Training Committee for the Central Training Council, H.M.S.O. (1966)

Guide to Commercial Training, British Chamber of Commerce (1961)

Commercial Apprenticeship Scheme (2 pamphlets), British Chamber of Commerce (1961)
DEPARTMENT OF EDUCATION AND SCIENCE, *Further Education for School-leavers*, H.M.S.O. (1966)
Further Education for Technicians, City & Guilds of London Institute (1964)
Report on the Commonwealth Conference of the Education and Training of Technicians, H.M.S.O. (1967)
DEPARTMENT OF EDUCATION AND SCIENCE, *Reports on Education–No.* 26, *National Certificates and Diplomas*, H.M.S.O. (Nov. 1965)
CITY & GUILDS OF LONDON INSTITUTE, *Institute Publications*, C. & G. (Current)
YOUNG, J. T., *Technicians Today and Tomorrow*, Pitman (1965)
Report on Recruitment and Training of Technicians, British Iron and Steel Federation (1961)
Technicians in Laboratories and Quality Control Departments, British Iron and Steel Federation (1962)
Training Recommendations for Building Technicians, Construction I.T.B. (1968)
The Training of Technician Engineers, Engineering I.T.B. (1968)

24

Clerical and Administrative Training

A. G. BEVERSTOCK, B.SC., M.A., PH.D., C.ENG., M.I.E.E.

Education and Training Officer, Southern Electricity Board

INTRODUCTION

The aim of clerical training is to produce clerks who are competent in required office procedures, skills and services, who have a thorough understanding of the work of the sections in which they are employed, who possess a knowledge of and are skilled in the application of the principles of communication, and who have the necessary background and attitude of mind to be effective as members of the firm and as citizens.

The aim of administrative training is to produce persons who can interpret and execute the relevant policies and procedures laid down by management.

THE NEED FOR PLANNED TRAINING

Effective and efficient communications within a firm and between the firm and the customer have become recognised as two of the most important aspects of the firm's operations especially in larger organisations. In terms of communication, clerks are the craftsmen and craftswomen, and administrators are the technologists, of the communication system. It is odd, therefore, that the majority of employers have given so little attention to the systematic training of office staff. The first report of the Central Training Council on this subject indicates that no more than 8% of office staff under the age of 21 are being trained at present and only about 7% are being released to attend a course of further education. This lack of training may well account for many inefficiencies in industry and commerce.

INITIAL CLERICAL TRAINING

Clerical training needs to be organised in a systematic and planned way similar to modern apprenticeship training. There should, therefore, be a period of full-time induction comprising introduction to the firm, instruction in company procedures and information, followed by a period of training in basic skills.

Clerks could join other new employees for the initial stages of induction. They should then separate for a basic skills course which ought to be full-time. A firm which is large enough can set up a clerical training school for this purpose. With small firms, a course could be arranged conjointly with local technical colleges.

The course should consist of training in those skills and procedures which are to be found in the normal office. These are mainly registry operations, postal and filing methods and procedures; preparation and maintenance of records; abstracting; collating and presenting information; the principles of good business correspondence; the use of the telephone; office machines; and office calculations including statistics.

A number of firms have experimented with courses of this type and their experience indicates that 10 or 12 weeks are necessary to cover the content.

An interesting recent development has taken place in Denmark and is referred to in the Report mentioned above. For some time, the Danes like the Germans have instituted clerical apprenticeship systems. The experimental introductory course for apprentices in Denmark consists of 240 hours instruction spread over 10 weeks. The majority of the subjects mentioned above are covered in the syllabus and the course is arranged at the state commercial schools, the employers meeting part of the cost and paying apprentice wages during the period.

PHASED CLERICAL TRAINING

Immediately following the period of induction, the trainee should be allocated to his first section for on-the-job experience and training according to a planned programme which will include the introduction to the work of the section and its relationship with other sections and the role of the department in the firm. Allowance should be made in the programme at appropriate points for job-rotation, not only for reasons of affording the trainee additional experience and a sight of the contribution of other jobs to the work of the section, but so that a flexible clerical force will be available when contingencies arise such as illness or absence from duty for training or other reasons.

During this phase, opportunities should be provided for the clerk to see something of the technical side of his department. For example, a clerk in an engineering department should be shown production work and the equipment which is made, or a sales clerk should observe the methods of selling and problems arising with customers. Such experiences give width and depth to the training programme.

Proper supervision of training during this phase is essential. The Commercial and Clerical Training Committee of the Central Training Council recommend that a senior manager or director be nominated by firms to have responsibility for training and he should have the advisory services of a specialist training officer. Office supervisors should, as part of their duties, carry responsibility for trainees in their charge. Training itself should be an important subject in courses for such supervisors.

Subsequent phases of training should be concerned with preparing the clerk for work in other departments. In modern industry and commerce, a mobile and flexible force is needed. This will necessitate periods of off-the-job instruction similar to the 'module' concept in craft apprentice training.

Finally, there is the question of the preparation of selected clerks for advanced work. This requires a substantial period of off-the-job training which can extend to as much as two years. The programme will include training in various departments and at different management levels.

In all phases of training the use of training diaries and progress records is advocated.

CLERICAL EDUCATION

As with other properly planned training schemes, a clerical training scheme should have a component of further education. Young office workers who possess less than three 'O' Level G.C.E. subjects can undertake study for the Certificate in Office Studies—a part-time course lasting for two years. Those who possess the necessary 'O' level passes can follow the two-year course leading to the Ordinary National Certificate in Business Studies. On successful completion of this course, studies can be undertaken on a part-time basis for the Higher National Certificate in Business Studies, a valuable qualification in its own right. Clerks who enter a firm at 18 years of age or over and hold two 'A' level G.C.E. passes in subjects relating to business can enter this course direct. Similar arrangements exist in Scotland for business studies courses.

A CLERICAL TRAINING FRAMEWORK

Detailed programmes of training can only be arranged according to the special needs of a firm. The following framework, however, illustrates the principles which might be followed:

1. Entry — G.C.E. 'O' level passes in at least three subjects including English language. Age 16–17 years.
2. Induction — Full-time course—company information and basic skills—3 months.
3. Phase A — Allocation to working section, on-the-job instruction, limited job rotation in section—9 months.
4. Phase B — Modules of job training and experience in at least two other sections of the department—12 months.
5. Phase C — Modules of job training and experience in another department—6 months.
6. Phase D — Allocation to the department in which the trainee will undertake initial permanent work. Final on-the-job training—6 months.

Total—3 years.

Educational course	O.N.C. in Business Studies.
7. Advanced Training	Job training and experience in several departments—selected clerks—2 years.
Educational course	H.N.C. in Business Studies or course for professional qualification, e.g., A.C.I.S.

ADMINISTRATIVE TRAINING

An administrative training scheme is designed to produce professional administrators and executives capable of devising and applying the procedures necessary to implement the policies of the company.

Entry would be at two main age levels—at about 18 years of age from school or as transfers from a clerical training scheme—or at about 21 years of age from a university or college of technology. In the former case, entrants should possess five G.C.E. passes at 'O' level including English language, mathematics and a foreign language plus two appropriate subject passes at 'A' level. In the latter case, entrants should hold a degree or diploma in business studies or equivalent and/or have satisfied the academic requirements for a professional qualification.

It is recommended that all entrants should follow a scheme of training in accordance with the basic principles of planned induction and modules of training and experience. The Report referred to above recommends, however, that in the induction course, emphasis should be placed on the understanding of office organisation and systems rather than on learning basic clerical skills.

Modules of training should be arranged in at least three departments and include the study of principles of organisation and their application in the company; work control including planning, budgetary control and statistical systems; electronic data processing, sales and marketing; production control and purchasing. Trainees at this level should also undertake projects and assignments.

Trainees entering at about age 18 should follow educational courses leading to a professional qualification or the H.N.C. or a degree in business studies. In planning training programmes, firms might consider the advantages of block-release courses or degree sandwich courses of the type recognised by the Council for National Academic Awards. The total length of training needs to be about four years for young entrants and two years for graduates.

SUMMARY

Formal recruitment and training arrangements for clerical and administrative staff are just as important to industry and business as engineering or other technical training schemes. Attention to off-the-job training, avoidance of casual recruitment methods and planned training from the start of employment will improve the standards and productivity of the office and the general efficiency of the company. Use of some of the interviewing techniques advocated in Chapter 6 together with clerical and other aptitude tests will improve methods of selection and placement.

The appointment of specialist training advisers within larger organisations or to service the needs of groups of small firms, and the development of formal practical training courses at local technical colleges should be considered.

BIBLIOGRAPHY

CENTRAL TRAINING COUNCIL, *Training for Commerce and the Office*, H.M.S.O. (1966)
CENTRAL TRAINING COUNCIL, *Training for Office Supervision*, H.M.S.O. (1968)

25

Training of Salesmen

N. B. SUTCLIFFE, F.I.T.O., M.INST.M.S.M.

Personnel and Training Manager, Vickers Division, Sperry Rand Ltd.

INTRODUCTION

'We want you to start a sales training programme!' The training officer, given such a basic brief, may well ask himself 'Can salesmanship be taught or learned?' The facts are an adequate answer. They are that many large companies now have their own schools for sales training, and that consultants hold hundreds of well subscribed courses on salesmanship every year.

Certainly the principles of salesmanship can be taught and learned as effectively as those of engineering or accountancy. However, just as the engineer or accountant needs to be trained and given practice and experience to become competent in his profession, so too the training officer has to consider certain basic training elements which must be employed in order to create properly trained salesmen.

PRODUCT KNOWLEDGE

People buy from a salesman who knows his product, and can answer their questions on what the product can do for them. So an essential first step in the sales training programme must be to ensure that every salesman is given the chance to know his products thoroughly. It has been said that instruction in product knowledge can be overdone, but if the instruction is presented in such a manner as to stress the benefits of the product to the customer, then any time spent on product knowledge training will be valuable and will be seen to be valuable by the sales trainee.

SELLING SKILLS

The salesman's knowledge of his product must be communicated to his client. How this knowledge is used when talking with him is clearly of prime importance. A considerable part of the training programme will, therefore, consist of training in the skills of selling—how to make an effective approach to the client, how to gain his interest and confidence, how to answer questions and objections, and how to close the sale.

ON-THE-JOB TRAINING

Armed with good product knowledge, and having learned the principles of selling, the trainee salesman needs job experience. It must not be thought at that point he can be left to his own devices—he will need active training and support in the field. He will need to know not only how to plan his day, but how to plan each call, how to keep his records, how to analyse his results, and how to find new prospective areas for his selling efforts.

ATTITUDE TRAINING

The salesman's attitudes to his company, his fellow employees, his job and his customers are important particularly if his efforts do not produce immediately successful results in the early stages. Since sustained enthusiasm is essential for success in selling, individual morale is a factor which the training officer must aim to promote throughout the programme. Attitude training does not belong in a specific category on its own, it is an integral part of all the activities which form the total training programme. Improvement of the trainee's attitude is achieved through the manner in which he is taught his product knowledge and selling skills, and through the attitudes of those who give him training on the job. For these reasons, the need for very careful selection of the trainers at all stages of the programme is very important.

The elements of the training programme have been examined briefly, but how much of each element is required and how does each element apply to selling the particular product?

DETERMINING SPECIFIC TRAINING NEEDS

The specific training needs of the salesmen must be clearly established at the outset on a systematic basis. As with other forms of training, the first step is to analyse the job concerned. From this analysis, a job description should be written. With this document and with knowledge of the recruitment level of the trainees, the likely subject matter for training then becomes clearer and the training officer is able to formulate preliminary plans, and place emphasis where it is most needed.

If there is no job description available, it is the responsibility of line management to provide essential information. It will be necessary for the relevant line manager to give this in sufficient detail to the training officer, who in collaboration with him will collate the necessary information. This involvement of line managers is important, since it is they who will be actively involved in most if not all of the elements of the training programme. All too often the relationship between line management and staff training functions is confused. It would be interesting to consider, for example, how many sales training officers are still given to understand that they are the training men—that they are responsible: that they should go ahead and get on with it—as if the line manager could wash his hands of any responsibility for training of his salesmen.

SALES QUESTIONNAIRES

When determining required sales training for existing salesmen, it is prudent to consider a particularly useful method of identifying and establishing their needs. The training officer and sales manager can devise a questionnaire for circulation to all salesmen. It should be designed to obtain from them their own view of any difficulties which they commonly experience in the course of their job. Careful wording of the questions can provide information on the depth of understanding of selling skills amongst the salesmen. This approach makes individuals think of their selling problems and helps to condition them to being receptive to sales training since they should welcome training which will help them overcome these difficulties. Furthermore, this gives identification with the programme, since they realise that much of it is based on their problems.

LOCATION OF TRAINING

(*a*) *For Product Knowledge.* In many firms where there is a manufactured product, it is clearly evident that product knowledge will be best gained through a planned programme of work for the trainee in the factory or where relevant by on-the-spot experience of the service available. Product knowledge can also be imparted in the classroom, but this training would normally be additional to exposure to work on the manufacturing process itself.

(*b*) *For Selling Skills.* Training in selling skills can be achieved initially in the classroom. This is a very specialised part of sales training, and the inexperienced training officer would be well advised to seek the assistance of one of the well established sales training consultants in this field. One vital point is that if the training officer has not had personal experience of selling, he will almost certainly find himself in an extremely difficult position if he intends to participate personally in the teaching of selling skills. He may, therefore, choose to involve his line managers in this activity, whilst acting as course adviser in his capacity as the training expert.

(*c*) *On-the-Job Training.* On-the-job training will obviously take place in the field. Whilst this is the last of the clearly defined elements in the training programme, it must not be thought that it ceases once the trainee has reached a reasonable degree of competence. The training officer should always seek to promote the attitude amongst line management that on-the-job training for their staff is a continuous process, however experienced the individual may be.

DETAILED PLANNING OF THE PROGRAMME

When the main elements of the sales training programme have been clarified and when specific needs have been established, and when the locations for each part of the training have been determined, the detailed structure of the programme itself can be planned.

Training for newly recruited personnel will obviously start with a planned induction course, and the length of this will depend on the product—for example, length of training can vary in different companies from a few days to several months. The course will consist of exposure of the trainee to the work of the departments with which he will have contact in the course of his job—such as the order section, publicity department, main sales office. In addition a considerable amount of the trainee's time will be concerned with the acquisition of product knowledge: and with manufactured products he will spend time in the production shops. Too many induction courses consist of the trainee 'watching Nellie'. The programme should be planned so as to incur as far as possible actual involvement in the work of appropriate departments.

Additional product knowledge training in the classroom will probably follow the initial induction programme, although some trainers may prefer to plan formal classroom sessions into the induction course itself. Classroom training in selling skills, however, will probably take place as soon as the trainee has acquired a reasonable product knowledge.

THE SELLING-SKILLS COURSE

What should be the content of a selling-skills course? A natural starting point would be to arrange a talk by the sales manager on the company sales policies coupled with an explanation of the need for skilled salesmen and hence the need for the training course.

The corner-stone of good selling skill is a thorough understanding of human relations and this subject will naturally occupy a considerable part of the total course time. Most people acquire an understanding of human relations through their daily contacts in life, but the salesman under training needs to take a sideways step to look at the main elements which constitute good human relationships.

The likeable salesman will probably be a successful one—but the question then arises why are some people more likeable than others? Sometimes personal characteristics can irritate—buyers are not likely to be drawn quickly towards a salesman who does not look open and frank, who cannot look the other man in the eye, who just talks incessantly (how many salesmen do this?) or who is overbearing or a story-telling bore. In the rush and pressure of business life, courtesy is often overlooked: yet the salesman must learn to be courteous always. He cannot afford to be without knowledge of everything that the word implies.

There are many other aspects of human relations which the trainee salesman must study—the need for enthusiasm, the need for loyalty to his company, and tact towards the customer, the capacity to avoid arguments and 'knocking' of competitors.

All these points form a part of training in good human relations and the wise training officer knows that considerable teaching expertise is required to communicate effectively on this subject. There is a number of sales training specialists in this country who have such expertise in varying degrees, and their advice can be invaluable. However, before using the advice of specialists, a careful review should be made of the proposed methods of several consultants. This enables a careful and informed choice to be made.

INTERVIEWING THE CUSTOMER

Most programmes for training in selling skills include sessions on how to conduct each interview in a logical, planned manner. Each call will require some prior planning and preparation, and advice on this is needed by the trainees. The call itself should present a logical sequence. In solving a mathematical problem, certain logical steps must be taken to reach the correct solution, and if any one vital step is missed, the solution will be wrong. In the same way, the salesman who omits an important step in his presentation may well reach the wrong result, i.e., he loses the sale. The number of steps varies with different sales training specialists, but the sequence itself is always fundamentally the same. The salesman must make his approach and gain the interest of the customer. He has to establish customer confidence in his capability, his company and his product. He creates in the customer the desire to buy and ultimately has to know how and when to close the sale.

Nowadays, very few companies believe in the use by the salesman of a well learned and rehearsed monologue. On the contrary, the sales sequence should be so devised as to allow for the maximum use of the individual salesman's personality, and for him to proceed and sell in a conversational manner, asking and answering questions naturally as they arise. Although a good selling interview follows a sequence, the steps are not so absolutely defined as to be recognisable in any given selling conversation by anyone but a highly trained salesman.

Other subjects which should be included in the first selling-skills course would be handling objections, use of the salesman's time and possibly a specific topic such as selling quality.

SALES TRAINING TECHNIQUES

Sales training methods in the classroom follow the lines of well tried methods in other areas of training. Films should be available as an introduction or support to talks or lectures. They can also be used to start analytical discussions. Role playing, too, can be widely used, but the training officer should so organise the sessions as to avoid embarrassment sometimes suffered by the sales trainee victim in a two-man role play of a selling situation.

There are three stages to on-the-job training. Firstly, the trainee makes calls with an experienced salesman and only observes. Then the trainee begins to conduct the interview himself—but in the company of a senior salesman. What the trainee is taught at these stages is of vital importance, since the impressions he gains will greatly affect his efficiency for a long time. Thirdly, however experienced the salesman, he can still benefit from an occasional training day with the sales training manager. Weaknesses and bad habits can be detected, diagnosed and cured.

During the training-on-the-job, it is important that only one man should do the talking at any interview. The trainer should not be introduced as the senior man or the customer will switch the conversation to him. The trainer should always avoid interrupting unless the situation deteriorates seriously.

Each call should be analysed afterwards by the trainer and trainee together, perhaps whilst sitting in the car before moving on to the next call. Informality should be the keynote of such conferences—the trainer should avoid making a ceremony of each investigation, and should try to establish an atmosphere in which the trainee is prepared to criticise his own efforts.

CONCLUSION

All this is work which is a test and a challenge for the trainee salesman. In all its aspects the work of building a sales training programme is equally a challenge and a test of the capacity and dedication of the sales training officer.

BIBLIOGRAPHY

RUSSELL, F. A., BEACH, F. N., and BUSKIRK, R. H., *Textbook of Salesmanship*, McGraw-Hill, Revised (1955)

TACK, A., *A Thousand Ways to Increase Your Sales*, The Worlds Work Ltd. (1958)

NATIONAL SOCIETY OF SALES TRAINING EXECUTIVES, *Handbook of Sales Training*, Prentice Hall (1961)

TACK, A., *How to Train Yourself to Succeed in Selling*, The Worlds Work Ltd. (1964)

26

Safety Training

A. E. G. GROOMBRIDGE, C.M.INST.I.S.O.

Safety, Fire, Security, Training and Research Supervisor, Fawley Refinery, Esso Petroleum Co. Ltd.

INTRODUCTION

Safety in industry should be an integral part of every human act and operation. In every context, safety performance is directly related to efficiency. Safety in industry is good business. Safety should, therefore, be planned into industrial operations as an effective economic investment as well as for obvious social and humanitarian reasons.

In every industry, it is essential that some system of training be arranged to ensure that all personnel, employed directly or indirectly by a company, are conversant with that Company's policy on safety. All employees should be aware of the hazards which they may meet and be conversant with safe operation of equipment which they may use. It is, therefore, necessary to ensure that management as well as employees are precisely aware of statutory as well as personal obligations concerning safety of the individual and the safe operation of equipment.

BASIC PHILOSOPHY

1. Safety must be planned into and integrated with job training.
2. Training should aim to secure practical personal involvement.
3. There must be a follow-up to training to ensure that right habits are being cultivated.
4. Some periodic training should be arranged.

HOW TRAINING NEEDS ARE DETERMINED

NEW EMPLOYEES

On joining a firm, new employees must be given sufficient verbal information on safety to ensure their immediate personal safety until detailed instruction can be given concerning their job. This induction information should include the company policy on safety, safety rules and emergency drills. The aim should be to ensure that this information has been absorbed during the first day at work. As a check, it will be necessary to have a follow-up session several weeks later; to include practical use of fire extinguishers and with many firms a site visit in order to point out special hazards.

APPRENTICES

Particular emphasis should be paid to all apprentice training, to ensure that safety is planned into every act and every job. Only in this way can safe working habits be formed at this crucial stage of development. Discipline, to ensure the enforcement of safety rules, is necessary and requires to be emphasised during instruction.

Specialised lectures and practical work will be required outside normal workshop routines and should be planned at regular intervals over the entire period of apprenticeship.

Safety advisers can use a special lecture aid to explain accident sequence to apprentices and other new employees. This aid consists of the use of five standing 'dominoes' which are caused to fall in 1–5 sequence. They are labelled:

1. BACKGROUND OF PERSON
2. PERSONAL DEFECTS
3. UNSAFE ACT OR UNSAFE CONDITION
4. INCIDENT
5. INJURY

As each dominoe falls it strikes the next and causes it to fall. By removing No. 3—the Unsafe Act or Unsafe Condition—the injury sequence does not proceed. The demonstration emphasises the crucial point that without an unsafe act or an unsafe condition, injury cannot occur.

JOB TRAINING

Training to ensure that men possess the task skill to carry out their work efficiently (and therefore safely) is an obvious necessity. When programming any such instruction in how to handle tools and equipment and operate machinery and plant, particular emphasis must be placed on the correct methods to avoid injury or damage to operator, plant or equipment. It will, therefore, be necessary for the instruction to be given by someone who has experience, to underline correct and incorrect procedures, good and bad habits. The breakdown of the instruction to elemental functions is of primary importance. It should be possible to answer the questions why, when, how and what in reference to the use of basic tools, for example:

Why is a particular tool used?
When is it the best tool to use?
How is it used?
What is the best and therefore safest way of using it?

These questions can be asked of tools and equipment for any job and will ensure a more efficient and safe operation.

After training follow-up, to make certain that correct selection of equipment and procedures are being carried out, must be undertaken by supervision on the job. Compliance with regulations must be enforced and bad habits must not be allowed to develop.

TRAINING TO OVERCOME SPECIAL HAZARDS

Special training may be required in addition to that already advocated in order to enable the individuals to carry out their work safely. This training should be given by a person experienced in the particular subject, since it may require more developed skills than those possessed by general job training instructors. With, for example, instruction in the use of *breathing apparatus*—the programme should also be based on the questions why, when, how and what. In addition, practice in the wearing of this equipment should be given and, if at all possible, under actual controlled gaseous conditions. A suitable medium is smoke or tear gas. It is surprising how many men tear off their masks or opt out once they are put into such a practical situation; it is better to learn who these persons are before a real life situation occurs. A similar situation, of course, arises amongst persons required to work at heights. Follow-up of practical training every 18 months is strongly recommended.

The same approach applies with *fire training* where emphasis on the practical use of equipment, and some information on fire prevention and how to act in an emergency, should be given at least every 18 months.

Training in *lifting and handling* is a subject requiring specialised training by a person who understands not only the anatomical effect of incorrect body loading but who can also diagnose lifting problems and relate these to correct body movements and/or the use of mechanical aids. In a large works, a suggested and effective method of training is to obtain the services of a specialist for one week. During this week, up to 20 persons can be trained as specialists in lifting and handling. They should be selected on grounds of their recognised skills and should include hourly paid personnel as well as supervisors. They should cover as wide a geographical area and selection of jobs as possible. Subsequently, all persons at risk should attend a one-day course, to be given by one or several of those persons now qualified, whose skill as instructors in this work has been recognised. The aim should be to give practical exercises within the work situation. All the specialists should ensure that the principles and practices learnt on the one-day course are employed and practised and that any problems arising in particular work sequences are also tackled.

There are, of course, many other special hazards in particular industries and these may be recognised as needing the advice of a specialist. Excellent courses are operated by the Ministry of Labour, by the Royal Society for the Prevention of Accidents and others; particularly for industries where specially hazardous material is being handled, such as at the Atomic Energy Authority at Harwell, where instruction on the safe use and handling of radiation sources is given. Special hazards should also be quickly diagnosed whenever new equipment and new tools are employed. These too may require specialised job instruction for safe use and handling.

Other areas of desirable instruction may be identified where no suitable media for passing information exists. For example, a firm might recognise that its design and engineering division is not ensuring that equipment is being designed and located to the best advantage of the person who has to operate it, thereby hazarding the safety of the person—and probably reducing efficiency. A specialised talk on ergonomics may be useful but would not necessarily relate to the real life situation. A member of the design and engineering division could be seconded to the safety division for a period; during this time he would be able to obtain a deep understanding of safety and ergonomics and relate this to observed practices and discussions with the individuals doing the work. Finally, he might conduct a session with his colleagues and their management. In this context, it is useful to illustrate by slides and real life examples what can happen as a result of unsafe design practice. Examples are operating equipment and valves placed in difficult or impossible positions; light fittings incorrectly placed, etc. Training by such example and involvement has real impact on the attitudes of the design engineer.

NEWLY APPOINTED SUPERVISORS

When a man is appointed in a supervisory capacity, it is important that he receives instruction in his duties and his task objectives. One of these objectives should be to ensure the safety of his men and the avoidance of damage to plant and equipment. His training programme will depend on his particular responsibilities. Safety should be planned into it as in job training. Emphasis should be on the statutory obligations related to the industry or work in which he is involved, and the work regulations and how he is required to ensure that these are carried out. He should be made accountable for all injuries sustained by his men. He will need to receive instruction on how to carry out investigation into the cause of an incident, and how to ensure that any recommendations he makes as a result are followed up effectively.

SAFETY CONTACT

No job can be broken down too far in order to ensure that a person fully understands what is to be done and how to do it safely. The system often described as 'safety contact' is an excellent means of ensuring that persons are adopting safe working habits. By this method each week or month the first line supervisor discusses with each person the complete breakdown of a particular job. This could even include such simple items as how to use a screwdriver and on the other hand how to stack a load on a fork lift truck, etc. This system may well require initial training of supervisors in how to make the correct approach and the most effective use of this particular instruction. If carried out regularly and reported to management, this can be a most effective method of improving work habits and reducing injuries.

TRAINING EXERCISES

Training exercises are sometimes the most effective way of anticipating an unplanned, unsafe event, for example, a serious fire or emergency. This training is arranged in two separate training sessions. First, the exercise would be discussed by those persons responsible for control; in this case the manager of the area and senior supervisor, the chief fire officer of the works and a senior representative of the local fire service. The object of the exercise should be made known, the information gathered, problems discussed and plans made.

The second training session could be a practical field exercise using persons who would be required to use fire equipment, isolate machinery, shut valves, etc., and put into action those plans which had been evolved at the first session. It may be preceded by a general discussion on what is to be done. As a result of an analysis of this training, new problems will arise which can then be planned into future exercises and the safety and fire procedure for that area. A general discussion on the results with everyone present is worth the effort. By this practical involvement, many lessons will be learnt and it may well show a need for more specialised training, such as in the use of fire extinguishers, etc.

There are many other applications for this type of training to meet unplanned events by theoretical and practical involvement. It should certainly be regularly planned into the training of persons operating equipment when hazardous materials are handled.

An analysis of the causes of all injuries, fires, near misses and incidents will enable management to determine other safety training requirements. For those premises included under the Factory Acts, the provision of the law requires details of those persons who are injured and are absent from their place of work for three days or more. Statistics on this kind of information are published annually by the Ministry of Labour and give some indication of cause trends for serious injuries. However, much more detailed information is required to determine where real efforts should be directed.

The Royal Society for the Prevention of Accidents has devised a reporting procedure on time losing injuries, for which the definition is—'a man unable to attend work the next day or next shift'. A frequency rate can be produced making comparisons possible between groups within departments, or between departments, or between industries, so taking into account the difference in numbers of persons employed. The formula is:

$$\frac{\text{Number of time losing injuries of those persons at risk} \times 100{,}000}{\text{Number of man-hours worked}}$$

This method of calculating frequency rate can be extended to calculation of frequency rate for injuries. The calculation of this last figure, to promote comparison between working groups, is of the utmost importance. Similarly, the investigation of all minor injuries by supervision will enable an analysis to be made of where the problem lies and consequently help determine where training needs are required.

Likewise, a formalised system of investigation into all fires and serious incidents will add to the information available.

HOW TRAINING POLICY IS MADE

Company and management policy should include a number of objectives and one of these should be to operate safely. In order to implement this objective, departments will need to determine the type and amount of training which is necessary to carry out a safe operation. Some of this training will be inter-departmental, some will be specialised.

It is recommended that a five-year broad training plan be established to enable management to see at what stage in the plan they are at any time, and to assess the effect of this training on performance. The assessment of results could mean a change in the training plan.

As training time must be allocated and may well be at the expense of production time, it is necessary that the safety programme, which will include safety training, should be directed by management. A management safety committee, comprising the general manager, departmental heads and the safety adviser, meeting regularly, is essential to control the effort which will be required and to ensure that the objectives and targets which are set are being met.

THE PERSONAL FACTOR—ATTITUDE

Analysis of all industrial injuries will show that the greatest single cause of incidents is personal attitude. This is true in Europe, the Commonwealth and the U.S.A.—and undoubtedly throughout the world. In general terms, it has been found that between 70% and 80% of all incidents are caused by 'unsafe acts' and the remainder by 'unsafe conditions'.

However, it should be recognised that:

Unsafe Acts are *committed* by PEOPLE.
Unsafe Conditions are *created* by PEOPLE.

In other words, the common factor is PEOPLE and *not* things; PEOPLE are the main cause of incidents.

Given this information, what then should be the training policy, in what direction should the main efforts be directed? How is it possible to set about training people to improve their attitude, how to stop carelessness, thoughtlessness and disregard of instructions, etc.? How is it possible to set about training supervisors to recognise, understand and be dedicated to a progressive safety policy? The answers lie in a policy of involvement.

A POLICY OF INVOLVEMENT

This section includes a suggested training programme to influence *attitude* which is the largest single factor in causing incidents, injuries and fatalities, and is, therefore, the most important feature of the safety training programme. Whereas considerable data on safety training already exist for other problems, too little time is at present spent in endeavouring to overcome problems associated with attitude. The following is a suggested programme which has been found to be extremely effective not only in improving safety performance but in highlighting problems associated with work.

Concern, involvement, commitment and follow-up are the principle sub-headings. Each is an important phase of the programme and must be properly understood if the training is to be successful.

Concern for the safety of all employees at the board of directors and top management level is the first step. Training should not be necessary, but may be required, for top management to be convinced that good safety performance is not only a moral, and usually a statutory obligation, but also that it is another measurement factor of efficient operation.

Top management involvement and commitment to improved safety performance could be obtained by the following training. The director or general manager would give an opening address to his senior management group (*a*) on his *concern that injuries are occurring*, (*b*) on his *conviction* that there is a *problem*, (*c*) on his conviction that *improvement is possible*, and on his *commitment* to safety improvement and (*e*) on his *determination* that this will be achieved.

Useful information to demonstrate that the present safety performance is poor and that improvement is possible and justifies proposed training costs should be compiled. This can be obtained for similar industries in this country, Europe, the Commonwealth and the U.S.A. and can be presented in comparative industrial injury frequency rates. It is never too difficult to find firms safer than your own and yet recognised as being highly efficient. In this way, it will be demonstrated that provided management and supervision are competent, they should be able to control safety in the same way that they control production, maintenance and quality.

The management group should be given an opportunity to debate safety, and it will be useful to break it down into five basic questions:

1. Is there a problem?
2. Where is the problem?
3. What is the problem?
4. What must be done to overcome the problem?
5. What can I do?

A statement of their case to the general manager following this group discussion will reveal many deep-rooted problems mainly of attitude at all levels. Frank answers will be required. Argument will occur around such subjects as 'this department has more hazardous or dangerous work than that; this plant is more recently constructed and better designed than that; maintenance personnel are more at risk than operators; boilermakers are more prone to injury than electricians'. Is this true? Need it be so?

SUPERVISOR'S INVOLVEMENT AND COMMITMENT

Once the senior management group is committed to a programme of safety improvement, a safety training programme for supervisors should be arranged to follow a similar pattern. The initial address should be given by the general manager with at least one departmental manager present at the general discussions. This involvement is essential for success. Management must demonstrate commitment to an improved safety performance by demonstrating concern and willingness to give up time to discuss the problem. Supervisors' groups should consist of five to nine persons in each group, preferably selected from several departments and occupations in order to give a balance of experience, and they should include several levels of supervision from foremen or first line supervisors to senior supervision. Preferably a supervisor directly responsible to another should not be included in the same group. Two or three groups attending the initial presentation and final discussion at the same time, but having their group discussion in separate rooms, will enable a better and more balanced final discussion to take place. The same five headings will be the items for the discussion in groups followed by a presentation of the result of these discussions in general session at which senior management will be present.

FOLLOW-UP

Follow-up to these safety studies for supervisors will be necessary, to ensure that they are committed to and are in fact taking action on their own recommendations following group discussions. This could take the form of about three hours in the same groups as before, with questions to be answered such as:

1. Has action been taken on previous recommendations?
2. If results have not been satisfactory, why not?
3. Is there anything that management can do to assist in making better progress?

A short verbal report to senior management from each group on these problems, accompanied by recommendations, should be followed-up by a general discussion.

ALL-PERSONNEL INVOLVEMENT

Finally, all personnel from the shop floor should be given an opportunity to share this safety study experience away from their working environment, in the same way as the supervisors. A senior member of management should give the initial presentation with other levels of supervision present. The groups should be representative of a cross-section of trades and preferably be inter-departmental. The subjects for discussion would be the same as for supervisors and the presentation would follow a similar pattern. Follow-up to these studies is preferably done on the shop floor by supervision and they must demonstrate their concern by action.

It will soon be seen from this method of instruction that safety is a very personal thing and that the problem is not somebody else's but primarily is with individuals, *you* and *me* and all of *us*. It will also be found that everyone becomes more committed to the safety programme since the future programme should be planned from some of the recommendations made during the safety studies.

HOW TRAINING POLICY IS IMPLEMENTED

It is necessary to ensure that all persons receive the training which they require. Consequently, the central records office should enter the training which has been carried out on the individual's personal record card. Also on this card should be noted when the time falls due for further training. For example, a man receiving a special course will require follow-up training every 18 months. To ensure that this is done, his supervisor will need a reminder, particularly if the person has been moved in the meantime. Only in this way can continuity and effective training be assured. In a very large complex organisation, this information can be set up on a computer.

Similarly, a definite programme should be followed for all new employees, all persons moved from one job to another and all persons promoted to more responsible positions. This will ensure that they can be effectively trained in their personal responsibilities for safety.

As previously stated, not all the facilities for such training in safety will be available at every firm. The Ministry of Labour and the Royal Society for the Prevention of Accidents run a wide variety of courses for specialised needs and if they are unable to provide the training, they will be able to recommend where training can be obtained. For example, training facilities are provided by the principal manufacturers of certain equipment and specialised products, at their own works.

FILMS

Films can be used most effectively provided that their use is planned into a training programme. The indiscriminate showing of films to fill in time should be avoided. For example, a film showing the hazards of not wearing eye protection needs to be planned into a discussion on 'What are the eye hazards in that particular environment, and what is the correct type protection to wear': then to try the equipment on, and see that it is comfortable and is right for the wearer.

The most effective eye film known is called *It's up to you* and is obtained from Harvest Films, Riverside Drive, New York, 100, U.S.A. This is of general interest to everyone who should wear eye protection at work and is a 'shock treatment' type of film.

A general film on the guarding of machinery is unlikely to show exactly the same machines as the worker may be using. Such a film therefore must be followed by practical demonstration.

There are of course many safety films available through the various film libraries and, in addition, large concerns usually have made one or two films related to their own particular subject. For example, the oil and chemical industries have a vast range of specialised films. One film which is of good general instruction to laboratories personnel is called *Black Monday*, which was made by the Imperial Chemical Industries Ltd. This film is a typical example of the type which needs to be followed by a group discussion of all the lessons learned.

Finally, there are some very dull films where the message is extremely complex or not suitable and severe vetting should be exercised before they are programmed. It should be noted that a film does not have to cost thousands of pounds to be of training value. One of the most effective safety films used costs less than ten pounds. It was made by an industrial engineering section of an oil refinery and shows with sub-titles the effective use of dry chemical powder to extinguish fire on a person if petroleum contamination occurs.

SUMMARY

Safety training should be included in on-the-job instruction wherever possible and be integrated with any special job training which may be planned.

As the largest single cause of injury is the human factor and, in particular, the attitude of the person to his/her work, this is the area in which training will be found to be most effective in reduction of injuries. To be effective, safety training is best carried out by group activity and it is essential that senior management and supervisors are seen to be personally involved.

Finally, safety, morale and efficiency are so dependent on each other, that the effect of one on the other cannot be exactly assessed; but one thing is certain, only with good safety appreciation and effective control, will those in authority demonstrate that they have met their moral responsibilities. An operation is not efficient if it is paid for by human misery. Moreover, safe firms *are* efficient firms.

USEFUL ADDRESSES

Addresses from which safety information may be obtained are listed below.

R.S.P.A., Terminal House, 52 Grosvenor Gardens, London, S.W.1.

H.M. Inspector of Factories (and Safety Health and Welfare Exhibition), Horseferry Road, London, S.W.1.

Ministry of Labour—T.W.I. Safety Course, Training Dept., Ebury Bridge House, Ebury Bridge Road, London, S.W.1.

27

Retraining the Older Worker

JOHN W. BARBER, M.A. (OXON.)

Senior Lecturer in Management Studies, Portsmouth College of Technology

INTRODUCTION

The purpose of this chapter is to discuss the retraining of older workers (the over 35s) in the context of rapid technological change, and to show how with insight and vigour staggering returns on investment might be achieved. Certain assumptions and attitudes will be examined. Employers and older workers are often diffident about retraining. The role of the training officer is critical in this situation. He may have the difficult task of convincing both sides that training is desirable and possible before even attempting to prepare or implement it. He himself must also be completely convinced of the need and value of essential preparation and must be able to justify it. He must also be prepared to modify his training methods.

OLDER WORKERS AND TECHNOLOGICAL CHANGE

The accelerating pace of industrial change makes the retraining of older workers an important issue. Evidence of this change is visible everywhere. New materials usually require new techniques of work, and new machine techniques make former manual skills obsolete.

In the cities, the new giant buildings are being built with prefabricated units. Completed walls are swung into position and fixed. Crane operators and component assemblers are replacing the old craftsmen. Construction is being revolutionised.

In steel mills, hot strip steel is now being produced by remote push-button control and much of the drama and labour is going out of steel smelting with the introduction of magnetic overhead cranes and electrically-operated carbon arc furnaces. In the engineering workshop, huge multi-purpose lathes can operate automatically. Metal forming machines may no longer have their cutting flames guided by a skilled hand. In the radio industry, conventional circuit wiring is rapidly being replaced by printed circuit panels. It takes only a few seconds for hundreds of components to be securely soldered on to a circuit. Formerly skilled radio mechanics might have taken a day to complete such a job by hand. In another factory, soldering can have been completely abandoned and joints will be made by special pressure guns which twist one wire upon another. These are a few indications of the variety and pace of current technological change which is going on throughout industry.

Side-by-side with the reduction and sometimes the total extinction of former traditional skills, there is an increasing need for new skills or for new flexibility within existing crafts. In this situation, craft demarcation tends to become more acute. The redundant worker, his job status damaged and sometimes destroyed is often left stranded, puzzled, resentful and afraid. The steam engine drivers for instance found it hard to accept redundancy and to decide what to do for the next 20 years.

It is the older worker who finds himself in this kind of predicament. Often he cannot accept the possibility of learning a new skill 'at his time of life'. He may be in despair. He faces humiliation and defeat. He looks round desperately for something he believes he *can* do well and he usually decides on general labouring. For men in their late fifties who have been unaccustomed to heavy work, this is seldom a wise decision.

In theory he might expect to find new opportunities for training and retraining as a result of the 1964 Act. As yet, however, the new training boards have had little significant influence on the training of the older worker. This is partly because the training of the young worker and of other employees at all levels has been regarded as more important. It may also be because industry has not asked for grant awards for the design and implementing of retraining courses for older workers. Many employers and training officers have so far failed to see the value of establishing such schemes, and many older workers are reluctant to respond to offers of retraining. Despite the excellent training provided in the expanded Government Training Centres which specifically offer training for new craft skills, the number of older workers taking up places is very small. This contribution to the total problem is significant but small. Ultimately each industry and each board will have to face the problem of how to make the best use of its older workers and in particular how to ensure their effective training or retraining.

PRESENT ATTITUDES

There are both bleak and bright aspects to the problem of the retraining and employment of older workers. One of the less favourable aspects disclosed by research is that in most regions of the world there is discrimination against the employment of older workers. The same trend prevails in Great Britain at all levels of employment. This discrimination probably stems from a number of widely held but often erroneous assumptions about age. The age of employees is frequently given as the reason for any ineffectiveness. Complacency and intolerance and resulting inefficiency are frequently attributed to middle age. Today, seniority in years, even at supervisory and management levels, carries less weight than hitherto. In the last few years in many industries, skilled and unskilled men at all levels have been invited to retire at 55. Meanwhile, even solidly organised trade unionists like miners and railwaymen find that the application of the seniority principle offers little protection when technological change makes jobs obsolete and causes job redundancy.

Attitudes to age and employment are heavily influenced by the pace of change. At all ages, for example, employees resent change unless it happens to involve them in promotion. It is usually assumed, however, that as employees become older they hate change more than ever, that they become more set in their habits, less flexible both mentally and physically, more difficult to teach, slower to learn, less confident, less tolerant, less patient, more difficult to handle, harder to change, and that their labour is, therefore, much less marketable.

Few people over 35 believe these things about themselves. Most employers believe them, or some of them, to be true of some of their existing staff—usually only of those older workers whom they dislike. When, however, the question arises of recruiting and training new employees, it appears that the unfavourable assumptions made about some of their own older workers are transferred to all unknown older candidates.

Since half of the workers in British industry are now over 40, it is obvious that these attitudes and assumptions if wrong can have serious consequences, for the national economy in loss of skill or of potential skill, and for the individual in society, in prolonged human anxiety.

THE NEED TO RE-APPRAISE

As yet many employers are not prepared to change or even to re-examine their attitudes towards the recruitment and training of older workers. They may, however, be prepared to consider the retraining of their existing older workers and success in this area combined with support and advice from the industrial training boards could lead to re-appraisal.

In terms of what can be done, it is worth mentioning the retraining which B.O.A.C. undertakes with its captain pilots. The professional association of the pilots ensures that promotion to the senior flying position is open only to those with long experience. Before a new fleet of aircraft is commissioned, their more experienced pilots are invited to apply for lengthy retraining courses to equip them to fly and command the new machines. This retraining is prepared and implemented with great thoroughness. It always involves the learning of many new skills, procedures and techniques and the unlearning of older ones.

Over the years, B.O.A.C. has consistently operated this policy. Today they are renowned as one of the most efficient and profitable of the world air lines. Their safety record is a tribute to the entire staff of the organisation. It demonstrates too the effective response made by Captain Pilots to their training. *These Captain Pilots are all in the 42-55 years old age group.* Witness, for example in April 1968, that 47 years old Captain who, with amazing skill, successfully landed a crippled and blazing Boeing aircraft full of passengers.

B.O.A.C. believe that it will be well within the competence of Captains in this age group to command the supersonic transport aircraft of the 1970s, in spite of these aircraft being considerably more complex, and operating at a higher tempo with a smaller crew.

Once employers and training officers begin to examine facts like these they will begin to ask whether industry itself can also provide similar evidence of high performance after training. Research workers have been sifting the evidence from industry. This shows that many misconceptions exist about the older worker's capacity to train. By definition, an older worker is over 35. No definition in terms of age can measure vigour, consistent effort, high motivation, dedication, determination, mental agility and those other attributes which usually contribute so much to quality and performance in any particular skill. Francis Chichester at the age of 65 provided an impressive example of consistent performance of *skills learnt late in life.* Industry has already provided much less dramatic but convincing examples of high quality performance by retrained older workers. Part of this story will be told in an impending book of case studies edited by Dr Eunice Belbin, Director of the Cambridge Research Unit into Problems of Industrial Training.

THE WAY AHEAD

Industry at all levels is seeking consistent effort and a good average performance. At workshop level, this is usually referred to as experienced worker standard (E.W.S.). If this only were achieved, many productivity targets would be shattered. There is no evidence to prove that reasonably fit older workers are not capable of achieving this standard in all but the most severely paced and strenuous occupations. As yet older workers are seldom trained to new occupations.

Three possible remedies are suggested for removing prejudice. The first concerns the climate of work relationships. Authoritarian supervision still tends to persist and this creates anti-organisation primary group pressures. The older worker is not immune to these 'we they' pressures. He, too, is not prepared to 'spoil the job' by excessive effort.

If, however, supervisors adopted a more liberal approach, this could stimulate healthier attitudes and much more discussion of tasks and responsibilities. This in turn could produce recognition of the value of training as well as acceptance and help for the newly trained. At present, indifference, suspicion and sometimes open hostility are often shown especially by older workers to management proposals on training.

The second remedy stems partly from the first and concerns the general absence of a reassuring tradition or public image of retraining. Dorothy

Wedderburn's book on *Redundancy and the Railwayman* shows clearly how little the railwaymen knew or wanted to know about the retraining opportunities offered by the Government Training Centres. The implication is that very few redundant workers have any confidence in retraining. *They identify training with defeat instead of with opportunity.* This is one field in which mass media could help. Television, in particular, could bring home in dramatised form the conflict, fears, doubts, excitement and achievements which training for a new occupation can evoke in the older worker. There is urgent need for employers guided and encouraged by the industrial training boards to stimulate an effective public relations policy on retraining of older workers.

The third remedy lies in the provision of better training. At present, there is often indifference towards, and unawareness of, the methods and techniques of training required. On the shop floor 'working alongside Nellie' has usually been accepted as the only method. Since the 1964 Act, new trends have begun to develop. Industry is beginning to recognise the value at all levels of task analysis, of isolating specific problems in a job, of systematic and detailed job analysis, of job descriptions and job and personnel specifications. All these are at last being seen as the essential preliminaries for effective analysis of training needs. A systematic approach to these activities would make for a general improvement in training standards and results. Here, too, the I.T.B.s can show the way.

The successful training of older persons demands in addition the understanding and application of certain special principles and techniques of learning. These are neither complicated nor abstruse. They constitute a straightforward challenge both to the training officer in an individual firm and to the training adviser of an industrial training board. The remainder of this chapter will outline some of these principles and techniques.

PRINCIPLES AND TECHNIQUES

GOOD SELECTION

Careful selection of older trainees is of prime importance, whether for purposes of retraining existing employees or for recruiting and training of new employees. Initial interviews should be conducted by training specialists. Normal checks of sound selection procedures should be adopted, including assessments of aptitude, ability, previous experience and attainments, intelligence, personality and physical fitness. These can, moreover, give extremely useful indications about the real ageing of a person.

There are some who age quickly and others who are dynamic at 60. An attempt must be made to gauge this important factor not necessarily to disqualify but rather to clarify and estimate whether personnel specifications will be met for defined manpower requirements, and to ensure that the intended training can be fitted to particular needs. For similar reasons, it should be found out when each candidate last did any consistent study or training.

TRAINING THE TRAINERS

The role of the trainer is crucial with older students. His social skill in dealing with their fears, doubts and inhibitions, especially in the early stages of training, is probably more important than his instruction. He must be able to win their confidence quickly. He himself may be an older worker, though this is not essential. He should be respected for his knowledge and ability in the subjects he teaches. He should have received a course of systematic training in job instruction and in techniques of training including the use of visual aids. In addition, he should at first observe and at later sessions should take part as an interviewer when candidates are being selected for training. He should similarly observe and later participate in all forms of job analysis and training needs analysis. These activities should be accepted as an essential part of his training and subsequently should become part of his normal responsibilities. Finally, he should be given intensive training on how to conduct individual instruction and counselling interviews with older workers.

COUNSELLING

Individual counselling should be an integral part of the training programme. It ought to take place before, during and after the training. If training has been intensively planned, provision will have repeatedly been made for trainees to undertake specific training tasks. Not all of these will or should require supervision. These activities can also be planned to provide opportunities for individuals to have private sessions with their trainer. A room free from interruption in which to hold such sessions is essential. Before, during and after training, tensions, apprehension and feelings of insecurity are inevitable. In the beginning, the older worker who is to be retrained often has misgivings and may feel undermined. He asks himself why he was chosen . . . was there something wrong with him? He may already have received repeated reassurances and yet be unconvinced. He can be harassed by fears of failure and by thoughts of what his friends or his family will think of him. Similar fears exist in the older worker who undertakes new occupation training. He may appear to have exercised more freedom of choice and might be expected, therefore, to have stronger motivation but he too is often worried about whether he will be successful and if so whether he will be accepted into the work environment by existing employees.

Both types of individual need further reassurance and reinforcement. Both need to be able to talk things over in confidence and in privacy with someone whom they respect and whom they know can and will willingly make time to listen. The trainer must exercise great skill and tact. This mainly involves getting the other person to talk and helping him to disclose and explore his problems. In doing this, the trainer will place himself in the strongest possible position to stimulate effective learning. Instead of hiding their faults and difficulties and most of all their fears, his trainees will discuss them with him and seek his confidence and help.

In this situation, the dividing line between individual instruction and counselling should disappear. A good trainer will be a competent practitioner in both of these skills. Individual attention of this kind can produce astonishingly successful results. It provides exactly the reinforcement which so often is absent.

THE LEARNING PROCESS*

It is obviously incorrect to expect a rapid pace of learning from an unintelligent older worker who has not practised learning since he left school. At the other extreme, one can expect the opposite result from persons of high intelligence who have continued to learn. Dr. R. M. Belbin states that learning is a skill which benefits from practice; he also suggests that for people with very high I.Q.s intelligence which is used increases with age. Below this level, all older persons undergoing training may experience confusion in the early stages if the pace of instruction is too fast. At this stage, they learn more slowly than younger people. (Speed of performance can be developed at a later stage in the course.) Learning is the product of three activities—registration, retention and recall. Too fast a pace causes a blockage between registration and retention. The individual forgets what he has just registered because something else intervenes before he has time to assimilate the new learning.

Older people have to integrate new knowledge and new skills with existing skills. For this reason they not only like plenty of time in the early stages—*they must have it.* Furthermore, the unlearning of former skills or procedures may also have to take place. At first, all this can be most disturbing and unnerving. If a real change of approach has to occur, time to ponder and to discuss it is essential.

For these reasons, the registration process will almost certainly occur slowly at first. It is, therefore, crucial at the outset that trainers should proceed logically, patiently and progressively. Side issues and distractions (interference factors) should be avoided. Talking down should never occur. If training has not been planned with adequate foresight and imagination, it can cause mental blockages in the trainees through their inhibitions and inability to establish continuity and logic.

The amounts of instruction given should be carefully balanced with the most effective means of consolidating it. Trying to go too far too quickly creates 'interference'. The firm consolidation of simple basic concepts is the easiest way to move successfully towards more diffuse skills and procedures. The consolidation process is often most effective when the individuals are put on their initiative to participate either by doing or by discussing or by both and above all to *discover* for themselves. This occurs most successfully when learners are faced with a progressively

*For a fuller and much more lucid explanation the reader is referred to the 'Review of Reports and Discussions' by Hilary M. Clay which is the first section of the O.E.C.D. Final Report of its Management Seminar held in London in October 1964 on the subject of *Job Re-design and Occupational Training for Older Workers*; and to the O.E.C.D. publication *Training Methods for Older Workers* by Dr. R. M. Belbin (consultant to the Manpower and Social Affairs Directorate of O.E.C.D.).

designed and challenging task which can be undertaken and practised in pairs or threes at their own pace in a relaxed, informal way, preferably not in a classroom atmosphere—and certainly not in the workshop 'alongside Nellie' until the final stages of the course.

In the early stages, involvement by every individual is essential. It is especially important to give maximum time to the putting into practice of what has been taught. Failures of short-term memory by older persons do not occur with skills which have been effectively learnt at this stage in this way through repeated practice.

Especially in these early stages, it is often a mistake to employ formal lecturing as a method of instruction. In effect, it can amount to an acute form of pacing which does not allow the individual time to integrate what he is hearing with what he knows or believes he knows. Here and in the early stages of most formal instruction, the young person with fewer preconceived ideas often has a temporary advantage.

On new occupation training for redundant craftsmen's mates, I described a method aimed to secure maximum involvement by older trainees. The account appears as Appendix D of *The Fawley Productivity Agreements* by Allan Flanders. Though this is basically a discussion method it is very largely an extension of the T.W.I. Job Instruction Course.

On teaching methods (p. 335) I wrote:

> 'In the prevailing circumstances we decided that lecturing as the main method of instruction would be useless. Somehow we had to design a form of instruction which would allow the trainee *to participate actively for most of the time.* This, we felt, would be a way of capturing and retaining his interest and of ensuring his learning. We decided, therefore, to break up the instruction and yet arrange it so that even an inexperienced instructor would have all his material logically marshalled before he ever began his course.
>
> 'Behind this approach lay another concept. It was that the training should be broken down into very small self-contained items. These items would be set out in the form of instructor's notes in such a way that each item would probably be contained within one or two pages. The object of this laborious break-up of the work was twofold. It was to ensure participation by the trainee and at the same time to prevent headaches or major upsets for the instructor who might otherwise be thrown off his stride by discussion.
>
> '*Participation in the form of item by item discussion by the trainees was the key-stone of the courses.* In practice all the instructors quickly grasped this concept and made it work. They saw that each self-contained item was a logical step towards the acquisition of all the skills required in the particular occupation.
>
> 'Quite deliberately however we often took one pace back in order to take two paces forward. We planned deliberate overlapping and repetition as part of a streamlined process of effective learning.'

Another part of the training problem is the question of how to secure effective communication, by other means than the use of discussion methods. New information can be presented in many ways. All of them contain difficulties. Oral instructions are often misunderstood or incorrectly remembered. Written instructions are preferable provided they are brief, simple and clear. Training documents and manuals need to satisfy the same standards of clarity (many fall short). Appropriate tests can provide re-inforcement for the trainee and valuable information for the trainer. All these methods of communication depend for their effectiveness on the skill of their initiators. All of them when badly used can provide interference factors for the older worker who almost always blames himself for the semantic difficulties which he encounters.

Similar comments can be made about detailed procedures for acquiring skills. In the past, probably too much emphasis has been put upon 'How' a skill is to be acquired and not enough on the overall concepts of 'What' and 'Why'? With the training of older workers, much more emphasis could be given to placing the trainee in a position where he and perhaps a colleague can together *discover for themselves* the answers to carefully planned sequences of task problems. With any adult, this is far more likely to produce effective understanding than any methods of learning by rote, or of being told 'How' by someone else.

CONCLUSION

The retraining of older workers offers a challenge to every training officer. The basic concepts involved are simple motivational ones which are easy to grasp. It must be increasingly recognised that interest and perception are key factors in a learning situation, and that with every individual both are intensified by success. This is a hidden reason why quality and standards of training are so important. They help the learner in all the various ways they are employed, they can create and strengthen motivation within the individual. With re-inforcement of this kind, the future scope and development of training in new skills for older workers could bring immense returns to British industry in terms of high performance and low labour turnover, and to the older half of our working population—job satisfaction.

BIBLIOGRAPHY

CLAY, H. M., *The Older Worker and His Job*, H.M.S.O., (D.S.I.R.) (1960)
BELBIN, E., *Training the Adult Worker*, H.M.S.O., (D.S.I.R.) (1964)
WELFORD, A. T., *Ageing and Human Skill*, O.U.P. (1958)
GRIEW, S., *Employment of Older Workers–No. 1 Job Re-design*, O.E.C.D. (1964)
BELBIN, R. M., *Employment of Older Workers–No. 2 Training Methods*, O.E.C.D. (1965)
WEDDERBURN, D., *Redundancy and the Railwayman*, C.U.P. (1965)
SCHNEIDER, B. V. H., *The Older Worker*, Inst. of Industrial Relations, University of California (1962)
Job Re-design and Occupational Training for Older Workers, Final Report and Supplement to Final Report on O.E.C.D. International Management Seminar, O.E.C.D. (1965)
OWEN, R., (ed.), *Middle Age*, B.B.C. (1967)

28

Vocational Training and Retraining in Government Training Centres

C. P. RAWLINSON, M.B.E., C.ENG., A.M.I.MECH.E.

Deputy Chief Inspector (Technical), Ministry of Labour—Training Department

INTRODUCTION

The object of this chapter is to describe the concept and principles of accelerated vocational training as carried out in the Government Training Centres, and to explain the Ministry of Labour training services that are available to assist firms (and thereby training officers) to play their part in creating a well-trained labour force that will both suffice for their own needs and harmonise with the national objective of securing a steadier and more rapid rate of economic growth—the case for which was outlined in Cmd. Paper 1892 *Industrial Training: Government Proposals.*

Firms with a well established training organisation will already be aware of external sources from which educational or training assistance can be obtained. Firms which are creating a new training organisation or appointing training officers for the first time may wish to know of the training assistance that can be obtained from the Government Training Centres. In development areas, for example, provision exists for direct assistance to employers with the training or retraining of employees, and in locations throughout the country, provision exists for individuals on their own initiative to apply for courses of full-time accelerated training.

WHY THE GOVERNMENT PROVIDES TRAINING OPPORTUNITIES

It is accepted by the Ministry that training at all levels below professional is primarily the responsibility of the industries and trades concerned. With the coming into force of the Industrial Training Act and the setting up of industrial training boards, the Government has a bigger interest in industrial training than ever before. (On this see Chapters 1 and 2—see also Chapter 20 on T.W.I. Courses and Chapter 22 on Instructor Training.) Industry will nevertheless *still retain* the overall responsibility for its own training.

Training is provided by the Government Training Centres for two reasons:

(*a*) *Economic* To help meet the shortages of skilled labour particularly in the engineering and construction industries where there is an important contribution to make to economic growth. (This also assists the redeployment of labour.)

(*b*) *Social* To give opportunities for resettlement in permanent jobs to disabled persons, ex-regular members of H.M. Forces and unemployed persons with special resettlement problems.

This training is provided by the Ministry of Labour under powers derived from the Employment and Training Act, 1948, and the Disabled Persons (Employment) Acts, 1944 and 1958.

THOSE ELIGIBLE FOR TRAINING

Government vocational training courses are available to selected persons aged 18 or over, with no upper age limit, who are not already in skilled work with prospects of regular employment. Those engaged on a skilled job in a declining industry or anyone likely to become redundant through their work being superseded by a new process are eligible for training in a new skill or for retraining in their own trade to adapt their existing skill to the needs of modern industry.

There are also special facilities for training people with special resettlement problems, ex-regular servicemen and disabled persons.

TRAINING AT GOVERNMENT TRAINING CENTRES (G.T.C.s)

The training for which the Ministry of Labour is responsible is provided at Government Training Centres, which generally are established in modern factory-type buildings with typical workshop floor layout. Hours, time-keeping and other routines follow industrial practice as far as possible.

There are at April 1968 38 Centres, but this number is being increased to 42, capable of producing 17,000 trained men a year by the end of 1968. The Ministry offers courses in over 40 different trades and most courses are available to all categories of persons eligible for training. Applicants are normally selected for training by panels on which representatives of employers' associations and trade unions are invited to serve. Standards of suitability have been agreed with each industry concerned. Individual applications for training are made through the local Employment

Exchange. (Details are given in Table 28.1 on p. 291 of facilities extended to firms who sponsor their own employees.)

For trades with a high technical content (e.g., draughtsman), a pre-entry test must be taken, but the first three weeks of every trainee's course is a period of assessment of aptitude, during which suitability to continue is determined. At the end of the course, the trainee has a good grounding in the practical skills of his trade with the necessary related 'theoretical' knowledge. Courses last 6–12 months and for some trades there is a further period of training with the employer, during which the trainee is paid at the skilled craftsman's rate, or an agreed percentage of it, and the Ministry pays the employer a small training fee.

In addition to the standard courses under the vocational training scheme, refresher courses for those who already have a skilled trade which they have not practised for some time and conversion courses for those whose service trade is not fully equivalent to its civilian counterpart, can be arranged.

THE CONCEPT OF ACCELERATED VOCATIONAL TRAINING

Several titles have been given to G.T.C. training. Accelerated Vocational Training (A.V.T.), Rapid Training, Shortened Training, Adult Vocational Training, Vocational Re-adaptation or Rehabilitation, Retraining or Conversion Training—these are the principal titles that have been given to a system of training which, under Government auspices, helps adult workers to acquire in a few months knowledge and skill which should enable them to occupy posts demanding a defined level of qualifications. Accelerated Vocational Training (A.V.T.) is the most commonly used title.

A.V.T. courses last a certain number of months, rather than years; most are for six months' full-time training. This is the main distinction between A.V.T. and other shorter or longer forms of training.

A.V.T. developed mainly, but not exclusively, in the basic manual crafts in building and metalwork. Even in its earliest days, its principles were applied to help teach the elements of a 'real trade' which could be followed immediately, with a certain level of qualifications, comparable to that which could be acquired through apprenticeship or in other formal training schemes. Its field of activity now covers not only the traditional trades or skills but is becoming increasingly concerned with training for the 'new' skills and with the retraining and redeployment of labour. Its particular value in a transitional period is to minimise the effects of unemployment on the skilled labour force.

The necessity for skilled craftsmen to transfer from one job to another or from one area to another, used to mean that in many cases these craftsmen ceased to work at their crafts and were lost to industry as skilled workers, but experience now shows that this loss can be reduced by offering re-training which facilitates the quick return of the craftsman to skilled employment. The flexibility of the system allows advantage to be taken of the skill, knowledge and experience of the individual so that a course of suitable content and length can be arranged for him.

The system can be applied to most of the increasing number of jobs requiring quite a high degree of skill within a narrow range, because the processes of production in large sectors of industry often do not call for the continuous and full use of the varied skills of a fully qualified craftsman.

This form of training, limited in its duration, intended for a particular purpose and pursuing restricted aims, calls for special methods adapted to meet these particular training and retraining needs.

PRINCIPLES OF ACCELERATED VOCATIONAL TRAINING

The Ministry of Labour has provided training courses in Government Training Centres for almost 50 years and has built up its own techniques of accelerated vocational training. This training is intensive and is geared to a progressive syllabus covering the highest common factor of employers' (production) requirements in the various trades. It is a systematised form of training and is entirely different from the traditional methods used by industry and is unique in this country. The main essentials are as follows:

> The courses are full-time, highly intensive and specialised. A comparatively short full-time course is considered more suitable for an adult than a long part-time one. The length of the course is set at the essential minimum considered necessary to enable an 'average' trainee to absorb the job knowledge and skills associated with the training trade. During the course, the basic skills of the trade are taught and the trainee is introduced to production speeds.
>
> The 'theoretical' part of the course is reduced to the absolute minimum necessary for an understanding of the principles involved in the job process. For trades with a high technical content, for example, instrument maintenance mechanics, there may be one or two full-scale lectures a week, but for the majority of training trades, the necessary related knowledge is given in short workshop talks and demonstrations. The classes are small, ranging from 8 trainees in wood-cutting machining to 12 in radio and television servicing and 16 for draughtsmen. These small numbers permit a good deal of individual tuition and the intake to the majority of the classes is staggered. Thus, for certain periods in the trainee's course he is in the ideal learning situation of 'one trainee to one instructor'.

The A.V.T. scheme is intended to produce a good workman in a particular trade with sufficient background knowledge of the industry he is entering. It is not a substitute for the long-term training given to the apprentice, whose training includes a high proportion of technical education intended to produce a craftsman with the appropriate wide degree of knowledge and skill which will sometimes help him to qualify for promotion to executive and higher posts.

THE SYLLABUS OF TRAINING

The syllabus of training in each course includes informal talks on the shop floor on day to day problems arising during practical training and educational instruction in basic arithmetic and the reading of engineers' drawings related to each trade. Progressive practical exercises serve as tests and at the appropriate time the trainee is moved away from basic training exercises on to selected industrial production work, enabling him to acquire experience (e.g., in machine operating) of different materials, cutting tools, jigs and fixtures. Work obtained from employers develops a knowledge of job breakdown and production processes and helps the trainee to learn to meet the production times required by industry. The principle of training on production work applies where possible to all training trades and is the means by which employers can have training 'biased' to suit their requirements for a particular trainee.

Training in all Government Training Centres conforms to a national syllabus compiled by the specialist technical staff of the Ministry of Labour Training Department Headquarters in consultation with representatives of both sides of industry. This procedure is adopted in order to provide the same standards of training throughout the country. Syllabuses are planned in considerable detail with a progressive series of exercises prescribed for each week of the course. They are kept under periodic review to ensure that redundant techniques are eliminated and new ones included.

A trainee does not progress from any exercise until he has thoroughly understood and mastered it, and a weekly record of progress is maintained by the instructor. The chief instructor checks on progress by visits to trainees in the trade sections at intervals throughout the course, and a certificate is awarded to all trainees who complete the course satisfactorily.

THE METHOD OF DRAWING UP SYLLABUSES AND CURRICULA

The curriculum of training used in the Government Training Centres is mainly concerned with the descriptive matter covering the job technology and drawings of exercises covering the range of practical work to be accomplished during the course.

When there is a request for a new course of training, the specialist division in Training Department Headquarters responsible for compiling syllabuses will assign the most suitably qualified technical officer to make a survey of firms at which the new occupation or skill has been evolved (or is already in use in the case of a well-established trade for which the Ministry has not previously trained).

The report on the survey will describe in detail the actual work being done and the range of skill and knowledge required of a qualified worker, the conditions in which he will follow his trade and the equipment, machines, tools, materials used. It serves as the basis for drawing up the syllabus content and for the guidance which will be provided for the instructor in the use of training aids. In compiling the syllabus, the technical officer will have in mind the importance of the practical work to be carried out as training exercises, putting in the optimum amount of work in the best learning sequence. The need, as appropriate, to introduce industrial production work will allow for biased training where advisable or necessary. The theoretical content or related knowledge will be limited to what is just sufficient to enable the trainee to understand 'what' he is doing and 'why' he is doing it. The system of training relies heavily on a talk and demonstration by the instructor to the trainees as each new element of skill or technique is introduced, with continuing guidance and encouragement designed to inculcate initiative and confidence. The instructor is the 'key' man—there is no question of learning from 'Nellie'. The instructor frequently gives individual guidance.

When the syllabus has been completed, decisions are taken concerning the major items of machinery and equipment required and what electricity, gas and water services are necessary for the training to be carried out. A 'class layout drawing' is prepared and this layout is fitted into the space available at the Government Training Centres concerned.

TECHNICAL CONTROL OF THE EFFICIENCY OF TRAINING

Standard national syllabuses and class layout drawings facilitate the overall technical control of the efficiency of training which is the responsibility of a separate division of Training Department Headquarters. Technical surveys of all Government Training Centres are regularly carried out.

In the course of a survey, the 'inspector' is concerned with all technical aspects of training activities at the centre, which cover:

(*a*) *Educational work.*

Effectiveness of lectures, shop talks and demonstrations. Reading of workshop drawings—safety precautions.

(*b*) *General training matters.*

Effectiveness of the three weeks assessment of suitability procedure —application of T.W.I.—use of training aids—standard of instructors' weekly training reports of progress of trainees—biasing of training where appropriate—upgrading—safety—general discipline. Quality of training in all trade classes.

(*c*) *Production work.*

Suitability—quality—quantity in relation to syllabus requirements. Control and progress of customers' orders.

(*d*) *Conclusion of the inspection.*

Summing-up with centre manager. Assessment of the overall efficiency of training.

OTHER TRAINING SERVICES

UPGRADING TRAINING

Training is given at Government Training Centres to men sponsored by their employers who continue to pay wages while they attend the course. This service is on a fee-paying basis, the fee varying according to the trade concerned. Usually such training is for specialised parts of the training courses which the employer cannot provide himself, e.g., upgrading from capstan operator to setter. This form of training is limited and can be given only where it is possible to fit in with the Ministry's other training commitments.

GRANTS TO FIRMS FOR THE TRAINING OF LABOUR IN DEVELOPMENT AREAS

The Ministry of Labour provides grants towards training costs incurred by firms on first moving into, or expanding in, development areas, or who are retraining workers as part of general measures to prevent a substantial reduction in the size of their labour force. Development areas are designated as those where a high unemployment rate exists, or is expected to develop and be likely to persist.

The financial benefits towards the cost of the training of additional workers by employers in development areas have been improved as from October 16th, 1967. Direct assistance with such training has been provided since July 1st, 1966, by the Ministry in Government Training Centres and by lending Ministry instructional staff to get training started in firms' own establishments.

RECRUITMENT, SELECTION AND TRAINING OF INSTRUCTORS IN GOVERNMENT TRAINING CENTRES

(*a*) *Recruitment.* Instructors are recruited from the ranks of craftsmen in industry. Vacancies are advertised in the national and trade press. Applicants are required to have served a normal apprenticeship, followed by at least five years employment as skilled craftsmen.

(*b*) *Technical qualifications.* Possession of an Ordinary National Certificate, City and Guilds Certificate or equivalent qualification is to a candidate's advantage, but lack of such qualifications does not exclude him from consideration. In the preliminary selection process breadth of practical experience is a factor of primary importance.

(*c*) *Selection and trade testing.* Candidates who appear to be suitable are invited to take a theoretical trade test and those who are successful in this test appear before a selection panel. The purpose of the panel is to assess:

1. General educational standards and intelligence.

2. Personality and bearing.

3. Potential instructing ability.

Success at the panel is followed by a practical trade test.

Both the theoretical and practical tests are of a standard sufficiently high to guarantee the technical ability of successful candidates. There is, however, a further safeguard in an obligatory three-months probationary period.

(*d*) *Training the newly-appointed instructor.* During the first few weeks of his service, the newly-appointed instructor attends a two-weeks Departmental Course on Instructional Techniques at one of the Ministry's instructor training colleges situated at Letchworth (Herts.) or Hillington (Glasgow). This course includes the study of subjects such as human problems concerning trainees; why and how people learn; lecture techniques; lesson techniques, use of visual aids, assessing and reporting on progress; workshop instruction for individuals and groups. The primary purpose of the course, however, is to provide an opportunity for each member to practise the various instructional techniques in a realistic setting but under observation and with guidance from the lecturing staff. A report on progress at this course is supplied to the manager of the centre to which the instructor has been allocated. In this report, attention is drawn to any weaknesses displayed by the instructor which may need corrective action at the centre and he is closely guided on taking up duty so that he may integrate himself effectively into the established pattern of training activities.

Additionally, the new instructor attends a one-week course which is intended to familiarise him with the procedures he will need to follow as an instructor in a Government Training Centre. At the earliest opportunity, he takes the four courses in the 'Training Within Industry for Supervisors' scheme, namely 'Job Instruction', 'Job Safety', 'Job Relations' and 'Job Methods'.

PLACING TRAINEES IN EMPLOYMENT

Placing officers at the G.T.C.s work in close collaboration with the Employment Exchange service. They aim to place each trainee in employment before he leaves the centre. If a trainee has been trained in a trade where a personal tool-kit is usual, a free kit is provided. If the trainee has not been placed in employment by the time he has completed his course, the Employment Exchange nearest his home, under the supervision of the Regional Office which accepted him for training, continues placing action. About 95% of trainees secure employment in their training trade either immediately at the end of their course or within a short time afterwards. This high placing figure shows that training is being given to a standard which is readily accepted by Industry.

TRAINING ALLOWANCES

Training is free and allowances and earnings-related supplements are paid to trainees. These allowances are not subject to income tax and there are free credits of national insurance contributions. The allowances also vary according to whether training is given locally or away from home. In addition, a good midday meal is provided free. For trainees living at home, basic allowances range from £8 to £11 per week and supplements are available on reckoned earnings for the previous tax year ranging from 1*s* 0*d* per week on £450 reckoned earnings to £7 per week on reckoned earnings of £1,500 or more.

Daily travelling expenses are paid for able-bodied persons where the daily journey is over two miles each way and for disabled persons irrespective of distance. Two centres have hostels attached for those who have to leave home to take training, but where there is no hostel accommodation lodgings are arranged and a lodging allowance is paid. Assistance is given towards fares home. There are also paid holidays according to the length of the course, for example, eight days (including public holidays) on a six months course. Hours of work are normally 42 per week.

AMENITIES AT THE CENTRE

Medical supervision and first-aid are available at all Government Training Centres. The centres have good canteens and most arrange sports and social activities.

CO-OPERATION BETWEEN CENTRE AND EMPLOYER

A very important feature of the scheme is that each centre not only attempts to specialise in training for the needs of the surrounding districts, but specifically invites employers to visit the centre to discuss their particular requirements of recruitment and training so that arrangements can be made to modify or 'bias' the course to meet particular needs. Very often an employer's own materials can be used for such courses so that in the later stages the trainee he has chosen can be intensively instructed on the exact work he will ultimately be called upon to deal with as an employee of the firm.

The closest co-operation should exist between the centre and the employer so that there is no waste of training effort or overlapping. After discussion, agreement should be reached on the type of course to be covered by the prospective employee, the drawings to which he will work, the materials he will use and the degree of accuracy he must attain.

Employers can also materially assist their local centre managers by being willing to accept on occasions small parties of trainees and their instructor for educational visits.

PUBLICITY

Visits to the centres by representatives of industry are regarded as an important part of the training activities and those who are 'working with the tools' as well as training officers and other members of managerial staffs, trade union officers etc., are particularly welcome. It is important that all shop floor personnel should know at first hand the nature and content of the courses, the methods and philosophy of training in Government Training Centres and the employment capability of the trainee who successfully completes the course.

What will at once be apparent during a visit is the true workshop atmosphere, to which the trainee rapidly becomes accustomed. From clocking on to clocking off, he is subject to normal factory rules and regulations and in his busy surroundings, soon overcomes that uncertainty of action which sometimes assails those new to the noise and activity of industry. Although he may be training to operate only one class of machine or equipment, he readily becomes familiar with the uses of other types and this gives him a feeling of confidence when he finally steps into industry.

Every endeavour is made to instil confidence in the trainee throughout his period of instruction at the centre; he is encouraged at the start to acquire the spirit of craftsmanship and to develop a manipulative sense rather than try to attain speed in operation, which follows only after he is well grounded in the other essentials.

CONCLUSION

With the greater publicity which has been given to Government Training Centres over recent years, the trainee knows from the start of his course that by his efforts he will achieve new status and new skill. He also knows that he is participating in a great national effort to increase and improve the quality of the labour force in this country. These factors help to explain the enthusiasm for training which so many trainees possess. (It is known, for example, that in some centres, a high proportion of the men entering the centre have given up their jobs, including some paid as much as £20 per week or more.) A trainee, equipped with the training which has given him a 'second chance' in his industrial career, expects that industry will fully utilise his aptitudes, interest and capacity. As efficient production is the touchstone of all industrial enterprise, the national interest of raising future total output can be achieved only if individual firms adopt a positive policy of investing in skill as a necessary adjunct to investing in the modernisation of plant and equipment.

Details of training schemes and current G.T.C.s are given in Tables 28.1 and 28.2.

Table 28.1 TRADES IN WHICH TRAINING IS AVAILABLE AT GOVERNMENT TRAINING CENTRES

Trade	*Number of training places per class*	*Length of course in weeks*
BUILDING AND CIVIL ENGINEERING		
Bricklaying	16	26 plus 60 continued training with employer
Carpentry	16	26 plus 60 continued training with employer
Contractors Plant Maintenance	16	26
Electrical Contracting	12	32
Heating and Ventilating Fitting	16	26 plus 78 continued training with employer
Painting and Decorating	12	26 plus 60 continued training with employer
Plastering	16	26 plus 60 continued training with employer
Plumbing	16	26 plus 60 continued training with employer
Slating and Tiling	16	26 plus 60 continued training with employer
Street Masonry and Paving	12	26
ENGINEERING		
Draughtsmanship	16	39
Fitting (General)	16	26
Fitting (Jig and Tool)	12	26
Instrument Bench and Machine Working	12	26
Machine Operating (Centre Lathe Turning, Milling Machine Setting/Operating, Capstan Setting/Operating, Precision Grinding)	10	26
Sheet Metal Working	16	26
Welding, electric arc	12	26
Welding, oxy-acetylene	12	26
MISCELLANEOUS		
Agricultural Machinery Repairing/Fitting/Welding	16	30 plus 10 continued training with employer
*Boot and Shoe Repairing	16	26 plus 26 continued training with employer
Canteen Cooking	2 or 3	12
†Furniture—Cabinet Making	12	30 plus 24 continued training with employer
Scientific (Bench) Glass Blowing	12	26
*Hairdressing (Men's)	12	26 plus 13 continued training with employer

*Training restricted to the disabled, ex-regulars and unemployed persons with special resettlement problems.

†Training restricted to persons in North Western Region.

Table 28.1 (continued)

Trade	*Number of training places per class*	*Length of course in weeks*
MISCELLANEOUS (Contd.)		
Instrument Maintenance	12	52
Motor Repairing	12	26
Heavy Vehicle Repairing	12	26
Radio Television and Electronic Servicing	12	39 plus 52 continued training with employer
Screen Process Printing	12	26
Storekeeping	1	13
*Tailoring (Retail Bespoke)	12	36 plus 68 continued training with employer
Typewriter Repairing	12	52
Vehicle Building—Body Building	16	26 plus 58 continued training with employer
Vehicle Building—Coach Painting	16	26 plus 58 continued training with employer
Watch and Clock Repairing	12	52
Woodcutting Machining	8	26 plus a period of continued training with employer which varies according to the industry in which placed
Engineering for Blind Persons—Capstan Operating, inspection with use of Braille instruments and repetition assembly work	5	8

*Training restricted to the disabled, ex-regulars and unemployed persons with special resettlement problems.

Table 28.2 GOVERNMENT TRAINING CENTRES AND REHABILITATION UNITS

Centre or Unit	*Address*	*Telephone No.*
BILLINGHAM-ON-TEES G.T.C.	Industrial Estate, Leeholme Road, Billingham-on-Tees, Co. Durham *Station:* Billingham	0642 52066
BIRMINGHAM G.T.C./I.R.U.	255 Holyhead Road, Handsworth, Birmingham *Station:* Soho and Smethwick	021–554 1998 for I.R.U. 021–554 3621 for G.T.C.
BLACKBURN G.T.C.	Philips Road, Blackburn, Lancs.	0254 52056
BRISTOL G.T.C./I.R.U.	Gill Avenue, Fishponds, Bristol *Station:* Fishponds	0272 653241

Table 28.2 (continued)

Centre or Unit	*Address*	*Telephone No.*
CARDIFF G.T.C./I.R.U.	Western Avenue, Cardiff *Station:* Cardiff	0222 40731
COVENTRY G.T.C./I.R.U.	Torrington Avenue, Coventry *Station:* Coventry	0203 66634
DUMBARTON G.T.C.	Castlegreen Street, Dumbarton *Station:* Dumbarton	0389 2247
DUNFERMLINE G.T.C.	Townhill, Dunfermline, Fife	0383 21148
EDINBURGH G.T.C./I.R.U.	West Granton Road, Edinburgh 5	031–552 5568
ENFIELD G.T.C.	Bilton Way, Enfield, Middlesex	01–804 2257/9
FELLING-ON-TYNE G.T.C./I.R.U.	Green Lane, Felling, Gateshead 10, Co. Durham	0632 693365
GLASGOW G.T.C./I.R.U. (HILLINGTON) Instructor Training College	Queen Elizabeth Avenue, Hillington, Glasgow S.W.2	041–883 8721
GLASGOW G.T.C. (QUEENSLIE)	Blairtummoch Road, Industrial Trading Estate, Queenslie, Glasgow E.2	041–774 2173
GLOUCESTER G.T.C.	Bristol Road, Gloucester	0452 27721
HINDLEY G.T.C.	Swan Lane, Hindley Green, Hindley, Wigan	0942 55277
HULL G.T.C./I.R.U.	Chamberlain Road, Hull, Yorkshire *Station:* Wilmington	0482 20738
IRVINE G.T.C.	Ayr Road, Irvine, Ayrshire *Station:* Irvine	Irvine 2688
KILLINGWORTH G.T.C.	Harvey Combe, Station Road, Killingworth, Northumberland	0632 663801

Table 28.2 (continued)

Centre or Unit	*Address*	*Telephone No.*
LEEDS G.T.C./I.R.U.	Dewsbury Road, Hunslet, Leeds 11 *Station:* Leeds	0532 76074
LEICESTER G.T.C./I.R.U.	Humberstone Lane, Leicester *Station:* Leicester	0533 68571
LETCHWORTH G.T.C. and Staff College	Pixmore Avenue Letchworth, Herts. *Station:* Letchworth	0462 6 3084
LIVERPOOL G.T.C./I.R.U.	Stopgate Lane, Liverpool 9 *Station:* Edge Hill	051-525 9331
LLANELLI G.T.C.	North Dock, Llanelli, Wales *Station:* Llanelli	0554 2 3217
LONG EATON G.T.C./I.R.U.	Wilsthorpe Road, Long Eaton, Nottingham *Station:* Long Eaton	0607 6 5145
MANCHESTER G.T.C./I.R.U.	Windmill Lane, Denton, Manchester *Station:* Denton	061-336 5451
MEDWAY G.T.C.	Hoath Lane, Gillingham, Kent	
MOTHERWELL G.T.C.	Java Works, Java Street, Motherwell, Lanarkshire	0698 2434
NORWICH G.T.C.	Mile Cross Road, Norwich, Norfolk	
PERIVALE G.T.C./I.R.U.	Walmgate Road, Perivale, Greenford, Middlesex	01-997 1451-5
PLYMOUTH G.T.C.	Strode Road, Plympton, Plymouth	0752 36671
POPLAR G.T.C.	Empson Street, London E.3	01-987 4898
PORT GLASGOW G.T.C.	Industrial Trading Estate, Port Glasgow	0475 42151

Table 28.2 (continued)

Centre or Unit	*Address*	*Telephone No.*
SHEFFIELD G.T.C./I.R.U.	Richmond Park Road, Sheffield 13 *Station:* Darnell	0742 42517/9
SLOUGH G.T.C.	119–122 Buckingham Avenue, Slough, Bucks. *Station:* Slough	75 23326
SOUTHAMPTON G.T.C.	West Bay Road, Southampton	0703 28281
TURSDALE G.T.C.	Tursdale Colliery, Ferryhill, Co. Durham	0385 77 457
WADDON G.T.C./I.R.U.	Stafford Road, Waddon, Croydon *Station:* E. Croydon	01–688 6191
WEST SUSSEX G.T.C.	Churchill Industrial Estate, Lancing, Sussex	

BIBLIOGRAPHY

The following leaflets are obtainable from the Information Branch, Ministry of Labour, Ebury Bridge House, Ebury Bridge Road, London S.W.1.

PL 392—*Supervisory Training Pays*
PL 393—*Allowances for Persons undergoing Courses of Industrial Rehabilitation*
PL 394—*Allowances for Persons undergoing Courses of Vocational Training*
PL 395—*Instructor Training*
PL 396—*Better Supervision*
PL 403—*Allowances for Persons undergoing Courses in Sheltered Workshops*
PL 405—*Are you short of Skilled Men?*
PL 406—*A Second Chance to Learn a Trade* (issued with training allowances inset)
PL 407—*Additional Training Services for Ex-Regulars*
PL 408—*Additional Training Services for Disabled People*
PL 409—*Courses for Supervisors and Instructors*
PL 416—*Invest in Training*
PL 421—*A Career as an Instructor at a G.T.C.*
DDL 1 (Rex.)—*Assistance with Industrial Training for Firms in Development Areas.*

29

Apprenticeship Training

A. G. BEVERSTOCK, B.SC., M.A., PH.D., C.ENG., M.I.E.E.

Education and Training Officer, Southern Electricity Board

INTRODUCTION

The word 'apprenticeship' is derived from the French word 'apprendre' which means 'to learn'. Strictly speaking, therefore, the term apprenticeship could indicate training for any occupation or profession. In this chapter, however, the word apprenticeship is restricted to training for the skills of a craft or trade which was the original use of the term in the days of medieval Gilds. The purpose of an apprenticeship, in this sense, is to train and educate an individual to become competent in one or more crafts, to develop his personality and to equip him with the necessary background of knowledge and attitude of mind which will enable him to play his part as an effective employee of his firm and as a useful citizen outside.

FIVE MAIN PRINCIPLES

The five main principles inherent in the modern concept of apprenticeship are:

1. The recognition that education and training, during the early years of a young person's experience in working life, form a continuous and progressive process.
2. The acceptance that apprenticeship involves not only the learning of skills but also the acquiring of allied knowledge by further education.
3. The programmes of training and those of appropriate 'formal education' should be so integrated that the various sections are clearly inter-related and lead to a 'whole' and not a 'part' concept.
4. Apprenticeship should be regarded as a training period rather than a 'direct production' period. This does not mean that an apprentice should not participate in productive work—but that emphasis should be on 'training' rather than 'labour'.

5. In the organisation of apprenticeship, industry has a social responsibility. The young working group is a most vital section of the country's manpower. The potential of apprentices must be developed to the maximum of their own interests as well as those of their employer and the country. This wide issue demands that due attention should be given to both moral and economic factors. Apprenticeship must, therefore, develop character and be an education for citizenship.

COMPONENTS

The components of a sound craft apprenticeship system are:

(*a*) Selection.
(*b*) An induction period.
(*c*) Formal and simulated training in workshops.
(*d*) Practical training 'on-the-job'.
(*e*) Technical education.
(*f*) General education and related studies.
(*g*) Skill and educational attainment testing.

SELECTION

Selection for craft apprenticeship presents a special difficulty in that the primary object is to select young people of not too high an intelligence level who possess or are likely to acquire manual skills and who have a 'feeling' for the craft job. A satisfactory procedure, therefore, takes into account not only the recommendations of school teachers and youth employment officers but also an objective assessment of the manual aptitude of the candidate. It is incumbent upon industry to make the best use of available skilled potential and this demands the application of refined methods of selection.

The majority of applicants for apprenticeship are school-leavers with little or no experience of industrial life and without very much idea of the working conditions of adult life. To many, the move from school to industry can be an abrupt and complex change. Selection procedures, therefore, need to be as simple as possible and should be conducted in an atmosphere of sympathy, understanding and tolerance.

Care should be taken with the production of the application form which should only ask for key factual information. In writing to referees or head teachers a questionnaire or a brief letter should request information on a limited number of points. This is desirable as heads of schools have to cope with a large number of such requests usually in the summer term which is the schools' busiest period. The questionnaire to the schools should be confined to standards reached in science, mathematics and

English, the degree of interest in and aptitude for practical work, and to the required personal qualities.

Special care should be taken in interviewing young people to make due allowance for nervousness and shyness on the one hand or over-confidence and the desire to impress on the other. To many young people straight from school, the interview is an ordeal. The interviewer should therefore try to bring himself to their level and continually bear in mind their inexperience and immaturity. Questions should be simple and direct. More often than not the type of boy who seeks a craft apprenticeship finds difficulty in expressing himself and the interviewer must be prepared to assist him to overcome this. The interviewing of applicants for apprenticeship should be conducted on a friendly personal basis, by the rotational method of interviewing rather than by the panel method (see Chapter 6).

USE OF SELECTION TESTS

The use of selection tests merits special consideration, not only in view of their mention in the Industrial Training Act, but also because probably more research has been conducted on this aspect of selection than any other. However, management, in general, is still not convinced of the value or validity of such tests. This is indicated by the fact that in a survey of 750 firms, it was found that only 8% were using selection tests.

Many attempts have been made to establish the structure of practical or craft ability and to isolate the factors involved. While, of course, there is considerable overlap in the factors claimed from the results of such experiments, it seems evident that the important factors are general intelligence (g), mechanical comprehension (M), spatial imagery (K), number ability (n) and mechanical manipulation (m). Existing tests, therefore, tend to be designed with a view to measuring these factors or combinations of them.

Tests of general intelligence fall into two main categories, verbal and non-verbal. The verbal test consists of similies, mixed sentences, analogies, words of opposite meanings and problems—the non-verbal subjects test thought relationships by using diagrams. Mechanical comprehension tests consist of exercises in which the results of mechanical operations are deduced either from working models or static diagrams. Spatial tests usually consist of diagrams in which the main problem is to identify relationships between isometric figures of various sizes and shapes. Mechanical manipulation tests, as the name implies, involve the use of fingers, hands and tools for simple practical tests.

Various prediction experiments have been conducted, in which attempts have been made to validate test results against criteria of educational and craft ability. They indicate that three types of tests have prognostic value. These are non-verbal tests of intelligence, mechanical comprehension tests and number attainment tests. The difficulty, however, is that most of the research work has been concerned with the selection of both craft and student engineering apprentices and very little work has been carried out on the use of tests in selection for specific trades. For example, the recent experiments of the Applied Psychology Unit at Edinburgh University tend to confirm other previous findings that the use of psychological tests in the selection of engineering apprentices, in general, commands respect but the determination of practical skill still presents a problem. Thus, the often accepted hypothesis, that a 'boy who is good with his hands or has many practical hobbies will also be good at practical work', has not yet been confirmed. Hobby ratings do not appear to have significant value in prediction for success in craft work.

The above findings do not mean that tests, hobby and practical ratings should not be used as part of a selection procedure for craft apprenticeship. While the determination of actual practical skill appears to defy solution, combinations of present known tests can be useful as an aid to predicting general proficiency for engineering and craft work and for trainability for such work. More research, however, is required, not only into the validation of present tests, but also into the skills which constitute practical craft ability. Perhaps the industrial training boards will sponsor such research on a large scale.

Whatever the method of selection, it is important to remember that the main aim of recruitment is to provide sufficient craft manpower to meet the future needs in a firm or an industry. Craft apprentices must, therefore, be selected at the *right* level and not above it although provision must always be made for the late developer to transfer to a scheme of higher education and training.

INDUCTION

Induction has a very important part to play in the development of human relations. A young person's sense of personal significance is easily lost when he exchanges the intimacy of the home and school environment for that of the industrial factory or organisation. A planned period of induction is necessary not only to assist the young person to adjust himself during this transitional period but also for the employer to help the young person to acquire the mental disciplines and attitudes which will enable him to make a full contribution to the working unit of which he becomes a member.

The process of induction should be gradual; and it should not be thought of as lasting a few days or a week but should take place intermittently over a period of at least three months. Particularly during this time, there should be a sympathetic understanding of the needs of the young person on the part of managers and supervisors.

A planned induction period consists of three elements—social aspects, company information, introductory workshop practice. These three elements embody induction to the company and the job. The social aspects include the immediate introductions to the responsible officials, supervisors, managers and the working group; conducted tours of the works or plant; and, the explanation of the conditions of employment.

Company information should include the history of the firm; the structure of the organisation; company products and markets; sports, social and welfare facilities; facilities offered for further education; the place of the firm in the national economy and its relationship with other firms and with the trade unions. An apprentice handbook containing these and other features is an essential.

The following syllabus for introductory workshop practice is suggested as a guide:

> The features of good craftsmanship and the general care of tools; the composition, physical properties and characteristics of the common workshop materials; the care, safety precautions and use of hand tools including soldering and brazing, elementary fabrication, simple heat treatment, taps and dies, wood chisels and gauge; the elementary facts of electricity supply; bench drills and simple lathe work; accident prevention including elementary first aid, use of protective clothing, principles of the tidy work-place, correct ways of lifting and handling, use of ladders, safety precautions when using hand tools, avoidance of practical jokes, personal hygiene; simple introduction to workshop administration.

FORMAL AND SIMULATED TRAINING

Formal training is that training which is carried out in workshops especially designed and equipped for the purpose and staffed by qualified instructors. Simulated training is where the field or site conditions of a job are reproduced as accurately as possible in the training workshop.

The need for training of both types is characteristic of modern thought and practice. The traditional method of training an apprentice by attachment to a craftsman has many disadvantages. Expert craftsmen are not necessarily the best instructors, bad methods and techniques can be perpetuated, and training may be spasmodic and incomplete, since the work followed by the apprentice may only consist of that carried out by the craftsmen. In addition, the method lends itself to a great deal of observation on the part of the apprentice instead of actually carrying out tasks. Conversely, there is also the danger that the apprentice may become a form of cheap labour.

Many firms, however, have sought to give an apprentice an introductory basic training in a company workshop. During recent years, this principle has been extended by the organisation of integrated one-year courses in technical college workshops.

A further extension of this idea is illustrated by the revised scheme of craft apprenticeship introduced in the Electricity Supply Industry in 1964. This scheme includes a period of formal and simulated training in each year of apprenticeship thereby reducing the total length of apprenticeship from five to four years.

Nearly the whole of the first year and appropriate periods in succeeding years are devoted to this type of training. The remainder of the time is spent in 'on-the-job' training coupled with the educational course which, of necessity, for a scheme of this kind, is by three-months block release. The formal training is carried out in Electricity Board workshops or at selected technical colleges or by a combination of these methods.

More recently the first year training scheme for engineering craftsmen and technicians was announced by the Engineering Industry Training Board for which a grant of about £500 per apprentice is given. This consists of a full-time course in workshops for the whole of the first year including induction (3 months), acquisition of basic skills (6 months) and commencement of special skills (3 months). The periods specified are approximate. This training (also referred to as 'off-the-job' training) may be carried out in a company workshop or bay, or a technical college, or in a Government Training Centre or in the Board's own training centre. The programme and syllabuses are suitably integrated with the appropriate further education course. The full details of this arrangement, together with suggested workpiece drawings, are contained in the Board's publication on the scheme. After the first year, the apprentice embarks upon specific 'modules' of training as outlined below.

The two schemes described illustrate the move towards a recognition that the best way of providing a sound beginning to the training of a craftsman is by 'off-the-job' training for the whole or practically the whole of the first year. This method enables the apprentice to acquire the correct foundations of the skills needed, and to appreciate important allied matters such as safety. It also helps him to develop the right outlook towards his craft, his place in industry and his function in the company.

The modern approach to training after this first basic year is that the rigidity of the traditional system, under which an apprentice learns one craft only, needs to be removed. The acquisition by an apprentice of more than one set of skills has been attempted in some industries through a system of dual apprenticeships but this has been of very restricted application. The Engineering Industry Training Board proposes a system of training and experience modules. The 'training modules' are based upon the time needed to *learn* the skills concerned and the 'experience modules' the time required to *develop* these skills. The minimum requirement envisaged for a craftsman is the basic year followed by at least two training and experience modules during a period of three years. Clearly, in the case of more complex craft jobs, a greater number of modules will be required. The flexibility inherent in this system is admirable and imposes a desirable change in the traditional arrangement whereby it has been accepted that all types of craft need the same amount of time for training, i.e., the British system of five-year apprenticeship for all.

Again it is thought that the training modules envisaged by the Board will best be carried out by 'off-the-job' methods with appropriate educational courses integrated with them. To permit this, some modification in the present length and content of craft educational courses will be required including more flexible provision than in present conventional apprenticeships.

Early indications are that the more recently formed Hotel and Catering I.T.B. will also adopt a modular approach to apprenticeship but that there the emphasis will be on smaller training units or 'bricks'.

'ON-THE-JOB' TRAINING

Despite the trend towards an increased component of formal and simulated instruction in the training of the craft apprentice, training on the actual job remains the most important section of the programme. It is here that the apprentice gains experience of actual job difficulties, learns to work as a member of a team or production unit, meets the stresses and strains of the industrial world of the adult worker and comes into contact with shop stewards and trade unions. Training 'on-the-job', therefore, becomes almost as much a character forming period as a skill experience period.

Training 'on-the-job' needs to be just as carefully planned as any other section of the training programme. Movement should be planned from one section to another in order to give the apprentice as wide an experience as possible and if the firm is too small for this to be done then the firm should consider grouping with other similar small firms for this purpose.

The proper supervision of apprentices during this section of training is essential. Young trainees are often left without information or proper instructions. Management should nominate clearly the persons who are to be responsible on the shop floor for the training. Apprentices need to be placed with craftsmen who are not only competent to instruct them but also have an interest in young people and their problems. Foremen and others should be given additional remuneration for these special responsibilities.

'On-the-job' training should follow naturally the period of formal and simulated training. The aim should be to implement the skills taught under instruction in the training workshop in the actual working conditions of the shop floor. Integration between these two sections is vitally necessary. The former is a period of skill-learning, the latter one of skill-experience.

TECHNICAL EDUCATION

It is now accepted that further education is one of the essential components of apprenticeship and attendance at educational courses should be a requirement in the apprenticeship agreement or indenture. The Engineering Industry Training Board already stipulates that, at least, first-year craft and technician apprentices must be released for a minimum of one day per week for study in a technical college and it is anticipated that other training boards will follow this example. This study is no less important than training in the workshop or on the job.

While the day-release procedure has become the usual method of allowing an apprentice to undertake technical college courses, a growth in block-release systems can be expected. Educationally, a block-release course of, say, three months continuous in one year or three sections of one month per year enable the apprentice to concentrate on study without the interruption experienced in the day-release system. The physical difficulties in certain industries (e.g., building), where much of the 'on-the-job' training of the apprentice is on site or in the field, can also be avoided if college is attended for continuous periods. Moreover, if there is to be a greater formal workshop training content in apprenticeship then blocks of education coupled with blocks of formal training may prove the best system for integrating these two components. In some major industries which cover wide areas, it might be necessary for some colleges to concentrate on block-release courses for certain crafts and some colleges for other crafts. Such a system might be necessary, not only to make maximum use of technical college resources, but also to provide the best service possible to industry in a region.

The main examining bodies concerned with the education of the craft apprentice are the City and Guilds of London Institute and the Regional Examining Unions.

Each union is composed of representatives of industry, commerce, local education authorities and regional councils of further education. The work of a union is carried out through examination committees covering a wide variety of subjects. The unions can examine on behalf of the Institute or conduct examinations at certain stages for which there are no other external examinations. For example, a union could provide examinations to test craft apprentices at the end of the first and second years of a City and Guilds course, the Institute conducting the examination at the third year intermediate stage. The present Regional Examining Unions are the Yorkshire Council for Further Education, the Union of Lancashire and Cheshire Institutes, the East Midland Educational Union, the Northern Counties Technical Examinations Council and the Union of Educational Institutions (Birmingham).

The courses provided by the City and Guilds Institute range over a very wide field from traditional subjects such as mechanical engineering, craft practice and building subjects, to more recent additions in fabrication engineering craft practice. The Institute encourages high standards of proficiency by awarding gold, silver and bronze medals, the highest award being the Insignia (C.G.I.A.) open to craftsmen and technicians who have gained the appropriate certificates and completed a thesis of high merit and original thought.

City and Guilds Craft courses normally require three years of study by part-time day-release for an intermediate examination, two further years for a final examination supplemented by a year for the full technological certificate. Some courses are arranged in parts, for example, the recent Electrical Craft Practice Course in which Part 1 consists of a general course undertaken by all craft apprentices and Part 2 comprising a series of specialist papers to meet the needs of individual crafts.

The syllabuses normally have four sections: workshop, craft theory, principles and related studies. The last named deals with necessary calculations, measurements, drawings and any special statutory or other regulations concerned with the craft. With a possible increase of workshop training within firms or arranged by training boards, care will be needed to avoid overlap and duplication of the practical content of technical college courses and that of courses arranged within industry.

GENERAL EDUCATION AND CHARACTER TRAINING

Apprentice training is incomplete if attention is not given to the all-round development of the individual. This involves not only training the young person to become a highly skilled craftsman or craftswoman but educating him as an individual and assisting him to develop his character.

Time is now allowed in most craft courses at colleges for broader education including the study of English and current affairs and allied to this are the activities which take place in industry. One of the themes of the Duke of Edinburgh's study conferences was the important role of industry in contributing to the standards of behaviour and character in the community. Both management and trade unions have a responsibility in this respect.

Industry can help in the sphere of the character training of the apprentice through the sponsorship of clubs and apprentice associations and by supporting schemes such as the Duke of Edinburgh's award, Outward Bound schools, adventure schools and organised visits abroad. All such activities not only give the apprentice a wider background of experience but present challenges taxing his physical and mental toughness. They help to develop confidence in the young person and prepare him for service to the community.

PERFORMANCE TESTING AND ASSESSMENT

Progressive testing and assessment of the performance of the apprentice is necessary not only to maintain good standards but also for diagnostic purposes. Tests should be given at the end of each year of training or at the end of each skill training module.

In designing such tests, care is needed to differentiate between a practical test and a skill test. A practical test involving simple workshop exercises should properly be regarded as a test of the actual work an apprentice will be called upon to perform. A skill test, as the name implies, is a test of actual skills where it is possible for these to be determined.

The following subjects are envisaged in a craft practice test: knowledge and ability to use tools and materials; practical tests of actual tasks, for example, a fitting, assembly, turning or milling job; the use of drawings and measurement; diagnosis and correction of faults as, for example, in electrical work. In devising the marking plan, it is important to assess factors such as work planning, layout of equipment and tools, safety, accuracy and cleanliness.

Since knowledge and actual practical work are to be tested, the progress test should consist of a written test in addition to the practical. For craft apprentices, it is recommended that this test should consist of short statements requiring completion to obtain the answer.

The results of tests should be combined with properly kept records and training diaries for the purposes of overall assessment.

THE APPRENTICESHIP AGREEMENT

An apprenticeship does not legally exist unless there is a deed of apprenticeship signed by the apprentice, the parent or guardian and the employer. For the purpose of certain Acts (e.g. National Insurance Act, 1946) the apprentice, even if a signatory to an indenture, is regarded as being under a Contract of Service, and the position is now more fully covered by the Contracts of Employment Act, 1963.

It is important that the indenture should be regarded as a training document and include provisions for probation, attendance and performance at education courses, the obligations of the employer in respect of practical training, the standards of behaviour expected of the apprentice, pay and conditions, length of apprenticeship and arrangements for dealing with a dispute between the employer and the apprentice.

On completion of training, the apprentice should be provided with a copy of the completed indenture and a Certificate of Apprenticeship.

CONCLUSION

Stimulated by the Industrial Training Act, and by developments in European and other foreign countries, apprenticeship in Great Britain will in the near future acquire a new look.

The traditional system of 'on-the-job' experience, with all its defects, including the use of the apprentice as cheap labour, will be replaced by a system in which training under workshop conditions and qualified instructors, integration of theory and practice, proper progress testing and assessment, training in more than one set of skills, will be the important elements. Nor will apprenticeships be confined to boys. Experience has shown that girls who have the necessary interest and aptitudes are quite capable of undertaking craft apprenticeship.

A controlled integrated apprenticeship system of the type now unfolding will produce craftsmen and craftswomen flexible in skill and outlook and trained for the needs of the technological age.

BIBLIOGRAPHY

WILLIAMS, G., *Recruitment to Skilled Trades*, Routledge and Kegan Paul (1958)
LIEPMANN, K., *Apprenticeship*, Routledge and Kegan Paul (1960)
INDUSTRIAL TRAINING COUNCIL, *Craft Apprenticeship*, H.M.S.O. (1962)

Day Release, H.M.S.O. (1964)
SCIENCE RESEARCH COUNCIL, *Training Made Easier*, H.M.S.O. (1960)
WELLENS, J., *The Training Revolution*, Evans Bros. (1963)
First Year Training of Engineering Craftsmen and Technicians, Engineering I.T.B. (1966)
Group Training Schemes Booklet No. 2, Engineering I.T.B. (1966)
Recommendations for the Training of Craft Apprentices, Iron and Steel I.T.B. (1966)
Recommendations for the Training of Apprentice Boatbuilders, Shipbuilding I.T.B. (1966)
The Training of Craft Apprentices, Electricity Supply I.T.B., 2nd Ed. (1967)
Recommendation No. 2, The Training of Instructors, Electricity Supply I.T.B. (1967)
Instructors Handbook for Part A of First Year Training for Engineering Craftsmen and Technicians, Engineering I.T.B. (1968)
Selection Tests, National Institute of Industrial Psychology, 14 Welbeck St., W.1
Selection Tests, National Foundation for Educational Research Test Agency, The Mere, Upton Park, Slough, Bucks
Selection Tests, Applied Psychology Unit, University of Edinburgh

30

Operator Training

The Late T. J. FULLER, A.M.B.I.M.

Formerly Research and Development Officer, Joint Committee for Training in the Foundry Industry

INTRODUCTION

In this chapter, an operator is defined as 'one who has not served an apprenticeship for the job he is doing and is generally engaged in repetitive production tasks which require manual skills but are not classified as skilled occupations'. (Douglas Seymour, *Operator Training*, I.P.M. (1964).)

Operators form the largest and often, in terms of training, the most neglected section of modern industry. It is on them that the profitability of a company in terms of quality and quantity often depends.

Since the war, full employment has produced a high labour turnover and consequent labour shortage. This shortage is due to companies who recruit people into jobs for which a few years ago they would not have been considered.

Individuals often have difficulty in reaching even a minimum standard. This is at a time when technological change and competition, especially in the export market, are making standards more and more stringent. There is, or should be, therefore, an increasing need to train and retrain operators especially in new skills.

This chapter aims to show that in order to provide training of a sufficiently high standard, the first essential is to establish by intensive analysis the exact nature and content of the required skill.

RESPONSIBILITY FOR OPERATOR TRAINING

Three fundamental principles should apply to Operator Training:

1. The operator should not be left to discover for himself the quality and quantity of the work demanded of him.
2. He should be trained in the shortest possible time to reach the quality and quantity standard of an experienced worker. This is as essential to management in terms of cost as it is to the operator in terms of wages.

3. He must receive systematic training based on the knowledge and skills of an experienced worker.

Many companies state that the first line supervisor is responsible for seeing that these principles are carried out. Delegating responsibility in this way is no guarantee that systematic effective training is actually carried out. This is not always the fault of the supervisor. He has many pressing responsibilities. In the larger companies, the functions of service departments are usually well defined, but the average supervisor often combines such functions as personnel management, maintenance, method study, quality control and training, in addition to his responsibility for production.

The survey carried out by Thurley & Hamblin and a smaller survey confined to the foundry industry has shown that, although in general each supervisor considered training as one of his functions, it was proved on examination that he gave very little time to it. He was aware of its importance and was often one of its firmest advocates but, because of the pressures on him, training came low on his day to day list of priorities.

If operator training is to be effective, it is not enough for higher management to delegate it to shop floor supervision and then forget all about it. Effective training in any field must have support from senior management and the impetus should come from that source. Operator training policy to be effective should start in the boardroom and end by being implemented on the shop floor only after receiving full support by management at all levels. The responsibility for training is the managing director's and he should delegate the authority to line management to see that it is carried out.

Before training begins it is essential to discover the knowledge and skills required.

COMPANY KNOWLEDGE

Every operator actively wishes to know about his working conditions. This should be the first stage of his induction. Too often induction programmes start with the history of a company and the company products. This can come later in the training period when an operator is more acclimatised.

A new operator undoubtedly wishes first to find out about method of payment, sickness benefit, hours of work, canteen, locker and washroom facilities. To attempt to instil pride of company in someone who has only been employed for a couple of hours and has yet to find out if the company has any pride in him is hardly likely to be successful. The first essential, therefore, is to have conditions of service explained by someone known to be authoritative on the subject, and not by the man on the next machine, who may well be biased.

KNOWLEDGE OF THE JOB

Once an operator is aware of what is required of him by the company conditions of service, he will next require a knowledge of the job itself. To this end, he will need to be trained for the use of the machines, tools, processes and materials. The training will obviously vary according to the type of job. It could range from lectures and job practice in a training school to demonstration on the job.

The training must be of a high quality and not just a 'sit by Nellie' approach. 'Sit by Nellie' is not always an inefficient method. Everything depends on the prevailing conditions. If the skilled worker is effective as an instructor, if interruptions are at a minimum, if he has no psychological objections to training the new recruits and there is no interference with his earnings, then the method may be effective.

KNOWLEDGE OF QUALITY

The operator of today frequently presses a button or operates a pedal or lever and then takes a component from the machine. His main function is to recognise whether the component is faulty and if possible adjust his machine to rectify the fault. If he cannot do this, he must know who to notify in order to get the fault rectified.

There are several methods of training the operator in fault recognition; these include instructional tapes, slides, films, programmed learning and the showing of a 'museum' of faults to the trainee. A very dangerous practice is to show the trainee the wrong method before showing the right method.

Maximum effectiveness will be achieved by preparing a comprehensive faults analysis chart giving the name of the fault, appearance, cause, effect, person responsible for correcting faults and method of correction or prevention.

Faults analysis should be documented in consultation with skilled operators, foremen, inspectors and other process supervisors. Often the name, interpretation, effect and even prevention of a fault will differ within a company and agreement must be reached between everyone concerned before the faults analysis document is finalised.

ANALYSIS OF SKILLS

This crucial section can be broken down into two stages. What is done must be clearly defined. This includes body, arm, hand, leg and finger movements. The second stage is to define how the skill is performed. This is arrived at by careful questioning and observation of the skilled operator.

Firstly, there is the visual aspect. What is the operator watching as he performs the task. Secondly, there is the tactual aspect. Can the feel or sense of touch involved in the skill be defined and demonstrated. Thirdly, and more difficult to define, is the kinaesthetic sense—a combination of touch and movement.

It is essential by these measures to cut through the *mystique* that often surrounds a job. This can only be achieved by a very penetrating study of the job followed by a carefully worded statement which will accurately define exactly what has been observed. All too often it is assumed that there is a *mystique* associated with the new skill whereas so-called hidden skills, when made explicit, can often be quickly acquired.

Obviously the approach to operators is of prime importance. Their co-operation is a first essential. Most skilled operators are in fact co-operative if approached intelligently and are often flattered to discover that they are being used in the investigation of the skill.

The job to be studied must be broken down in sequence into individual movements and, at this stage, only what is actually done is recorded. There is a strong tendency when doing an analysis to become involved in method improvement, but this must be avoided. If the method can be improved later, this can probably be done at the training stage.

The production of an analysis involves considerable work. Many companies have found, however, that once the heavy initial work is completed, it is a relatively easy task to add to it. For example, one machine shop producing a wide variety of jobs discovered, as a result of analysis, that there were basic skills common to all the jobs. Once these had been analysed only a short analysis was needed if there was any particular skill required for a certain job.

The final analysis sheets should have recorded left and right hand movements, vision, other senses and any particular comments relating to these.

INSTRUCTION SCHEDULES

Once the knowledge and skills have been analysed, instruction schedules can be prepared. Using the information gained from the analysis, the instruction schedules should be broken down into a number of teaching stages. Target times should be set for each stage in the schedules. It is not essential that these target times should correspond with an experienced worker standard; in some cases, it is necessary to add up to 20% to target times in order that the trainee should not be discouraged in the early stages of his training. The amount of increase in time will depend on the standard the trainee is expected to reach before being accepted in production.

AIMS OF TRAINING

During the training period there are three main objectives:

1. To build up a trainee's speed of operation to target times.
2. To make the trainee aware of the faults that occur and explain the methods of prevention.
3. Where necessary, to give trainees as wide a variation of work content as possible.

TRAINING AREAS

Whilst 'off-the-job' training is probably the most advantageous in terms of individual instruction, training to a standard method, establishing individual competence and the carrying out of special exercises, the most important factor in deciding whether to install 'off-the-job' training areas is their economic justification. Many companies do not have the labour turnover to justify the capital and running costs of a separate training area.

If training is done on the shop floor, however, in order to be well done, machines and equipment must be allocated for training purposes and a qualified instructor appointed, although neither the machines nor the instructor would be necessarily occupied full-time with training.

TRAINING PERSONNEL

It is one of the functions of the instructor or training officer to determine from line management: (*a*) who has to be trained, and (*b*) what standards are required. It is *not* the job of the training officer or instructor to set the standards, but to see that the standards set by line management are achieved by effective training.

Great care should be exercised when selecting instructors. It must be remembered that the best operator does not necessarily make the best instructor. There is also a tendency on the part of some companies to select people for instructors who are production 'dead beats' and unsuited for their present jobs. Although such people have, on occasions, become good instructors this is not to be recommended.

It is essential to recognise that once an instructor has been selected, at whatever age, he must be trained in instructional techniques either within the company if it has a qualified training staff, or on external courses.

ADVANTAGES OF SYSTEMATIC OPERATOR TRAINING

There are five main advantages:

1. The individual is trained in the method of an experienced operator.
2. He is made aware, all through his training, of faults that can occur.
3. He can be introduced in short sequences to most of the jobs he is likely to meet on the shop floor.
4. He is trained to an experienced worker standard, and his training is not complete until he has reached this standard.
5. A systematic approach to training reduces considerably the time taken to reach this standard.

The advantages of systematic operator training are threefold. They can undoubtedly help in furthering quality products, in scheduled quantities, at competitive costs; and these are a major consideration in today's need for increased productivity.

BIBLIOGRAPHY

CLAY, H. M., *Research in Relation to Operator Training*, D.S.I.R. (1964)

SEYMOUR, W. D., *Industrial Skills*, Pitman (1966)

31

Simulators and Continuous Process Operator Training

JOHN W. BARBER, M.A. (OXON.)

Senior Lecturer in Management Studies, Portsmouth College of Technology

INTRODUCTION

The first purpose of this chapter is to present a detailed case study describing how simulators are used to consolidate the understanding of certain basic concepts in the work of the continuous process operator.

The second purpose is to provide a postscript to the study which examines some of the implications for the process operator of the increasing pace of automation in this field.

The case study was originally published by courtesy of Esso Petroleum Company in the *BACIE JOURNAL* of June 1965.

IN THE PIPELINE

Today everything is either about to enter the pipeline, in the pipeline, or about to come out of the pipeline. At an oil refinery they have 'feed' in the pipeline. This feed—the incoming raw material of the petroleum industry—is crude oil, and crude oil is the starting point for this account of the use in a refinery of simulators in the training of continuous process operators.

To continue the refiner's jargon . . . the feed goes to the process operator's plant—and if anything goes wrong with the feed, the plant may 'suffer an upset'. It may even 'puke' . . . This jargon has sound sense and subtle implications. The words plant, suffer, and upset are so simple. For the operator, they have undertones which convey the idea of a very expensive complex of machinery and equipment which, like the human body, only functions well when its many variables are working harmoniously. The plant consists of pipes, vessels, valves, pumps, furnaces and the automation devices which control them and ceaselessly process the 'feed' for as long as 18 months on end.

The good process operator develops an uncanny skill—almost an intuitive skill—which enables him quickly to take a right decision if something goes wrong or, better still, to take corrective action to avoid the trouble. He is usually one of a shift of half a dozen men or less who, between them, may control a 'plant' worth anything from £1,000,000 to £8,000,000.

The case study will begin with a brief comment on the oil refining industry itself, and then define as clearly as possible the difference between the continuous process operator and the manually skilled worker. After that, three simulators will be described which have been found useful in developing the skills of the continuous process operator.

THE OIL REFINING PROCESS

First then a word on refining. Crude oil, as everyone now knows, is brought to Great Britain in great ocean-going tankers, which come mainly from the Middle East and from Libya. On arrival at the marine terminal of a refinery, they pump their crude oil into a pipeline and on into tank storage. It is there for a very short time, before being pumped out again through another pipeline into furnaces. While in the furnace, it is still in the pipeline. In the furnace, the oil vaporises. It then moves to a cooling tower—a form of condensation tower which is called a fractionating tower. These towers are between 100 ft and 200 ft high, and are so arranged internally that condensation at different temperatures can take place on trays at different levels of the tower. The base of the tower is the hot end. The vaporised crude is shot in near the base and as the vapours rise they cool and condense. Some of the trays on which they condense are fitted with draw-off pipes.

The significant thing about crude oil is that, when condensed in this way, the condensed liquid which collects at different levels—and therefore at different condensation temperatures—produces different products. Nature has all these products mixed up in crude oil. Distillation separates them. However, they are not immediately marketable products. They have to be re-refined and purified to get them to standard quality for competition with other firms. To do this, basically, the same process of distillation or fractionation is repeated over and over again. From it emerges a huge variety of products ranging from high octane gases like butane and propane, then high octane petrols, then through what are called the middle distillates such as paraffin and white spirit, down to heavy fuel oils and (with certain crudes) bitumen.

TYPES OF SKILL

So much then for the process; now to the people who operate a refinery. About two-thirds of the labour force maintain the refinery. Skills there vary from highly skilled mechanical craftsmen to semi-skilled maintenance

Plate 31.1 Explaining a Manual Skill. (Courtesy: Esso Petroleum Co. Ltd.)

Plate 31.2 The Automated Kettle (right). The Control Panel (left).
(Courtesy: Esso Petroleum Co. Ltd.)

operators. They are all people with manual skills. Plate 31.1 shows an instructor explaining the manipulative skills of scaffolding erection to a group of supervisors. All around are visible indication of how one part of the job proceeds from the previous part and so on, right through to completion. In this environment, an instructor at every stage can illustrate each training step. He can explain and discuss it. He can show how to do it; he can then get the trainee to do it and discuss it; then to practise and discuss it until real learning has been achieved.

Contrast this learning of a manual skill with what has been called the intuitive skill and learning required of the continuous process operator. He may have had a start-up course many years ago, lasting some months, on a particular plant. If he had this, he may since have moved from this plant to another, and perhaps to yet another, in the course of promotion or horizontal transfers. Unless he is employed on a new plant, it is unlikely that he has had a start-up course on the particular plant on which he is now employed. Coupled with that are the facts that he is a shift worker, and that he probably left an elementary school 20 or 30 years ago. The result is that he is probably in need of as much help as he can be given; to understand first the basic principles of refinery operation, and secondly the particular application of these principles to the plant on which he works.

Now, and more so in future, the main part of his work is to keep a constant watch on a great number of variables and make minor adjustments to them. This is what makes his work so different. He sees the variables on the face of instruments. Of course he also looks at the equipment itself. He goes round and checks oil bottles, checks bearings, makes an adjustment here, looks at a flange there, pinches back on a furnace burner occasionally, and when necessary he stops and starts a pump. Nevertheless his job is basically to observe changes in the environment of an automated, or a nearly automated, system. He will argue that this is not automation, since he must himself oversee the equipment and make the adjustments to the variables.

There is obviously a great difference between this kind of skill and the manual skill of the maintenance worker. The continuous process operator is never able to put his finger on the process in the way that a scaffolding instructor can put his finger on any particular application of a technique, and explain it. The process operator never sees the process which he is controlling unless he sees it on a simulator. This is one very important reason for having the three simulators, which will be described.

THE CLOSED LOOP

First of all, it is essential to understand how process automation is achieved. The concept is fairly easy to grasp. What is called a Closed Loop, or a Control Loop, is the key to the control of a continuous flow. The loop adjusts a control valve in such a way as to maintain a constant rate of flow within the pipeline. There are many kinds of controllers but they all derive from the flow controller loop illustrated by Fig. 31.1.

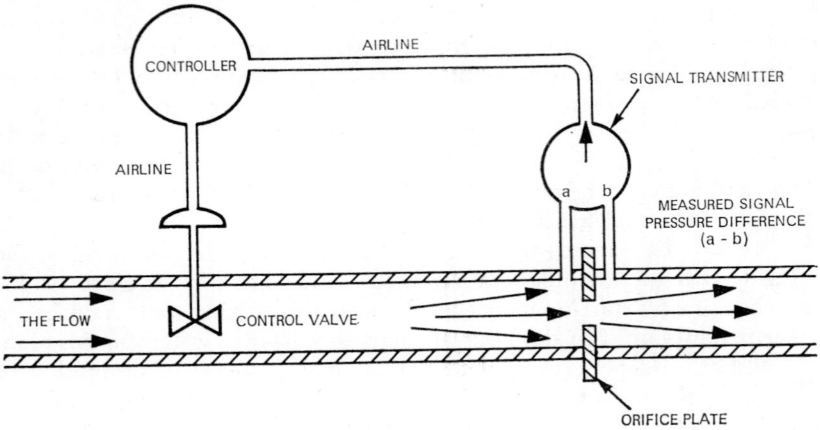

Fig. 31.1 The Control Loop

This is how it works. The orifice plate—a restricting metal plate with a hole in it—is inserted between the flanges of the pipeline. The hole in the metal plate is smaller than the inside diameter of the pipeline. The flow of liquid in the pipeline will be constricted at this point. The flow will converge and there will be an increase of pressure on that side of the plate. Conversely on the other side there will be a reduction of pressure. The measurement of this pressure difference is the key to the automatic control of the system. The measured pressure difference is conveyed to the controller which in its turn conveys a corresponding air pressure to the diaphragm of a control valve in the pipeline. The controller is calibrated so that any variation in the rate of flow in the pipeline causes the controller to send a corresponding opening or closing pressure to the valve itself. Within limits and with minor time lags, the loop thus becomes a self-adjusting mechanism which keeps a fixed amount of fluid flowing, per minute, per hour—per year!—through the orifice plate. What the closed loop cannot do, of course, is maintain a required rate of flow if the source of supply is cut or severely impeded.

THE AUTOMATED KETTLE

An application of a closed loop on a simulator will now be considered. The application has very great value because it operates real control loops from real instruments. It has a control loop controlling an incoming rate of flow of cold water. It also has temperature, level and pressure control loops. These are the four loops which, repeated hundreds of times, constitute the control system of the modern oil refinery.

In the simulator, the four loops are integrated so that a variation in one will cause corresponding adjustments in the others. It is something of a misnomer to call this device a simulator; it is in fact a 'going' process—a working model. It integrates four control loops in order to provide a specified outflow of water at a specified temperature. This process consists of cold water flowing into a five-gallon drum and then heated water flowing from it at a predetermined rate and temperature. The heating element is a steam coil. The device can be set at any desired rate of throughput, and for any desired temperature of outgoing water. It can and must be set at any appropriate desired level within the container itself. These determinants are, of course, very relevant to process refinery work. On the input side of the simulator, a control loop governs the inflow of water into the tank. The level in the tank is maintained at a desired position above the heating element by means of a float control. This sends a signal to the control valve on the output line and in this way maintains the required level of water in the drum.

Two other control systems are at work. One controls the pressure within the steam coil. If this steam pressure is reduced, say because a flow of very cold water condenses a lot of the steam within the steam coil, then another closed loop circuit operates and the steam pressure increases. There is also a temperature detector in the water itself. This is known as a thermocouple. It sends a voltage signal which is used to increase or decrease the steam pressure. The two controls, pressure and temperature, are worked in cascade which is a form of paralleling.

This is not the end of the story. The operators themselves can adjust this device as much as they like, and they will do no real damage. More importantly they see what happens and then discuss why. Their real learning occurs not so much on the device itself as on the control panel which is situated on another side of the room. All of the instruments on the control panel are actual working instruments. Each instrument has a circular indicating chart which, on normal plant operation, would move round one revolution every 24 hours. On this control panel the charts are speeded up to one revolution every half hour. This enables the operators to see exactly what is happening. They can watch the tracks made by the indicator pens on the circular charts, and they can study the model itself (see Plate 31.2, p. 314).

The Steam Kettle thus provides, in a relaxed classroom atmosphere, an opportunity to see, study, experiment with, and discuss, the functioning and malfunctioning of control systems. This is an experience which normal plant operating conditions do not permit. Perhaps the most important single benefit is in helping the operator to distinguish between a process fault and an instrument fault.

THE DISTILLATION PROCESS

The problem of simulating the distillation process itself will now be considered. Here, too, the term working model applies. To understand what happens, a little explanation is necessary. If a liquid, for example water, is mixed with salt, and it is necessary to separate the salt from the water, the liquid is boiled, evaporated and then condensed (see Fig. 31.2). Drops of distilled water form in the condensation chamber and then trickle out into the receiving vessel. When all the water in the original mixture has been boiled off, the salt will be left in the bottom of the first

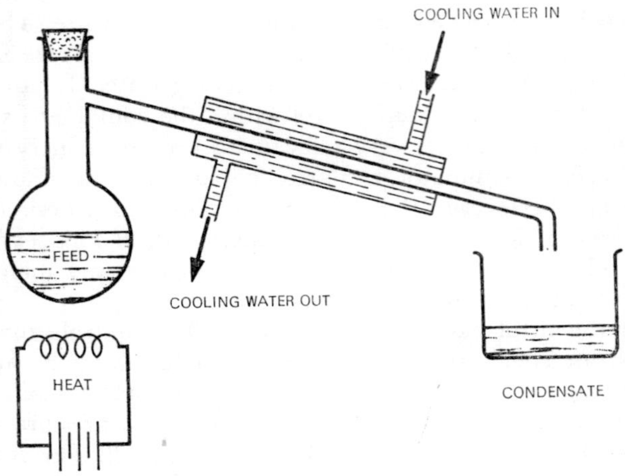

Fig. 31.2 Simple Distillation

vessel. The two components are thus completely separated. If, however, instead of mixing salt and water, whisky and water are mixed, there is then a very different result.

Suppose the mixture is 9/10 water and 1/10 whisky. It may be found so dilute that it needs strengthening. This can be done. But never again can the two components be separated completely. To enrich the mixture is, however, fairly easy. The problem lies in the extent to which one wishes it to be enriched.

If half of the mixture is evaporated and condensed there will be a richer mixture in the condensed half. This is because water boils at 100°C while alcohol boils at 85°C. They both vaporise together, but because of its lower boiling point the alcohol vaporises faster than the water. This is why the condensed mixture is richer in alcohol. If the enriched half of the mixture were now subjected to the same procedure—that is, if half of it were redistilled—then a still richer alcohol content would be produced. By continuous repetition of this procedure an ever decreasing amount of mixture richer and richer in alcohol would emerge.

However, if this type of problem is tackled in another way, a continuous process results. This is illustrated by Fig. 31.3.

The original mixture is now in flask 1 and in the process of being boiled-off. The vapours are immediately being fed into a hot and slightly richer mixture in the next flask. Each succeeding flask contains a slightly richer mixture which, though hot, is slightly cooler than the preceding stage. Consider what is happening.

The mixture in flask 1 is vaporising. The vapour is being conveyed into the liquid in flask 2, which is slightly cooler. The vapour therefore condenses. However, as it does so—and this is crucial to the argument—it gives off latent heat, and this latent heat is sufficient to make this mixture in the second flask boil. The mixture in flask 2 then repeats the process

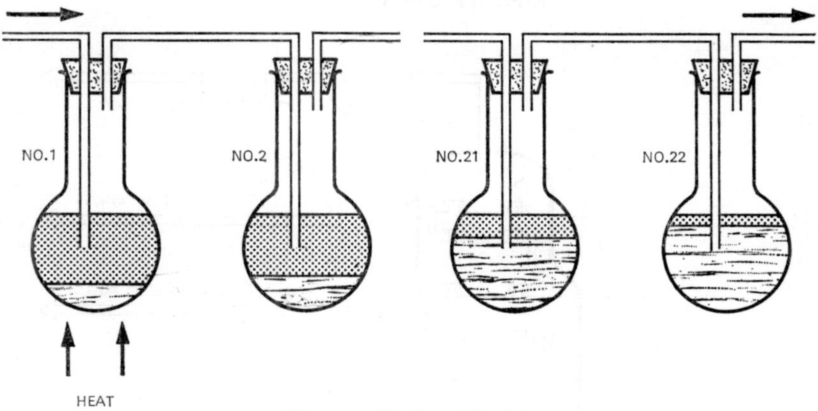

Fig. 31.3 Continuous Process

that has happened in flask 1. It boils, and its vapours are carried over to the next flask . . . and so the process could go on repeating. If instead of four there were 20 or 30 flasks connected in this way, the mixture in each would have just that much more alcohol in it than the mixture in the preceding stage.

If a very big leap is taken in the argument, it will be possible to see how an actual fractionating tower works. The crude oil, like the mixture in flask 1, is vaporised and fed-in near the bottom of the tower—any non-vaporised crude falls to the bottom of the tower as residue. The vaporised mixture rises through the first of a series of condensation trays. These trays have liquid on them. They are, however, so constructed that hot vapours may rise up and into the liquid on the trays. The device which makes this possible is called a bubble cap. The rising vapour condenses and, in so doing, gives off latent heat which causes the liquid on each tray to boil and send its vapours up to the next condensation tray where, as with the flasks, the procedure is repeated. This goes on all the way up the tower, and each distillation at each level in the tower produces an enrichment of the components. Thus, while the mixtures on the lower levels are of heavy components which comprise heavy fuel oils, those at the top of the tower are the lightest and most volatile components. They are the propanes, butanes and petrols (naphthas).

A diagrammatic view of a distillation tower, with an inset illustration to show how a bubble cap works is shown in Fig. 31.4. A circuit diagram for the second simulator which is a model fractionator is shown in Fig. 31.5.

The model fractionator, like a primary distillation tower, is a 'going' process; but, instead of its feed being crude oil, it is a clear liquid consisting of a mixture not of alcohol and water but of two colourless liquids with very different specific gravities. These are mixed, then heated and vaporised and fed as vapour into a glass tower (3 in wide by 7 ft tall). The vapours rise up the column in the way that crude oil vapors rise in an actual tower. The entire column has equally spaced trays each 3 in apart.

PRIMARY DISTILLATION

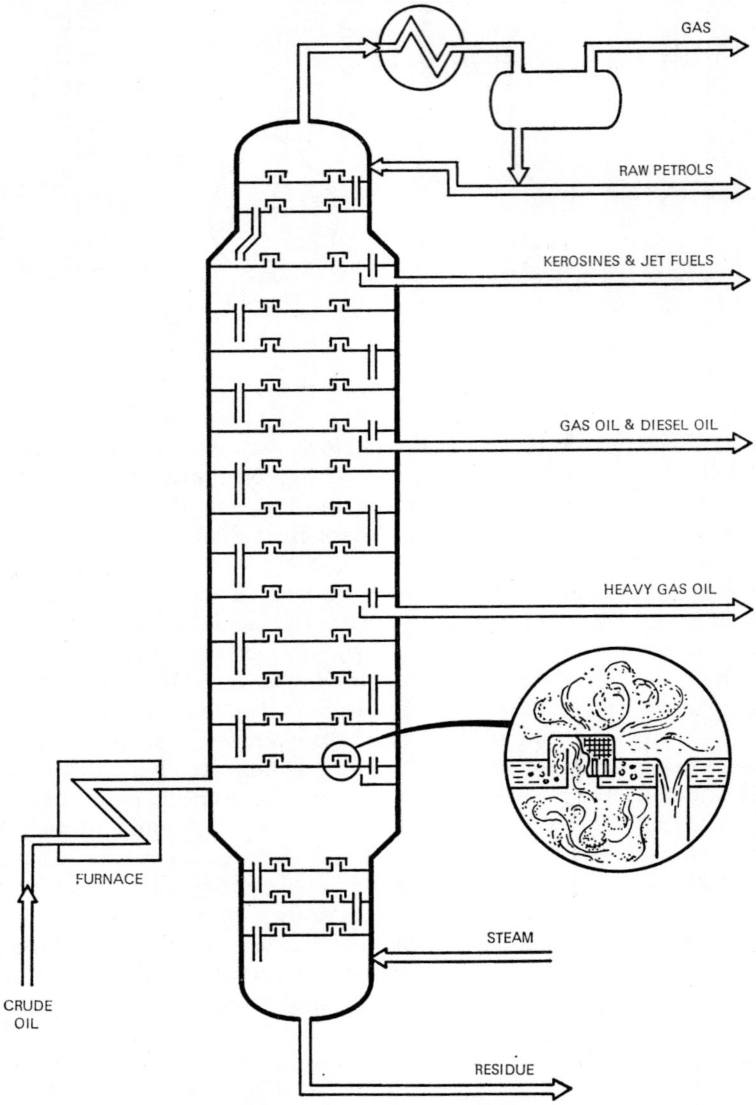

Fig. 31.4 Primary Distillation

Each tray is full of 'pin-holes' which serve instead of bubble caps. The column also has a number of controls. For example, the rate of feed is variable, the amount of heat is variable. There is a number of variable take-off points (overhead and side streams). There is also a variable feed-back (reflux) system as well as a variable 'pump-around' by means of which liquid from a lower level goes to a higher level. Most of the variables are controlled by hand-operated valves.

MODEL FRACTIONATOR

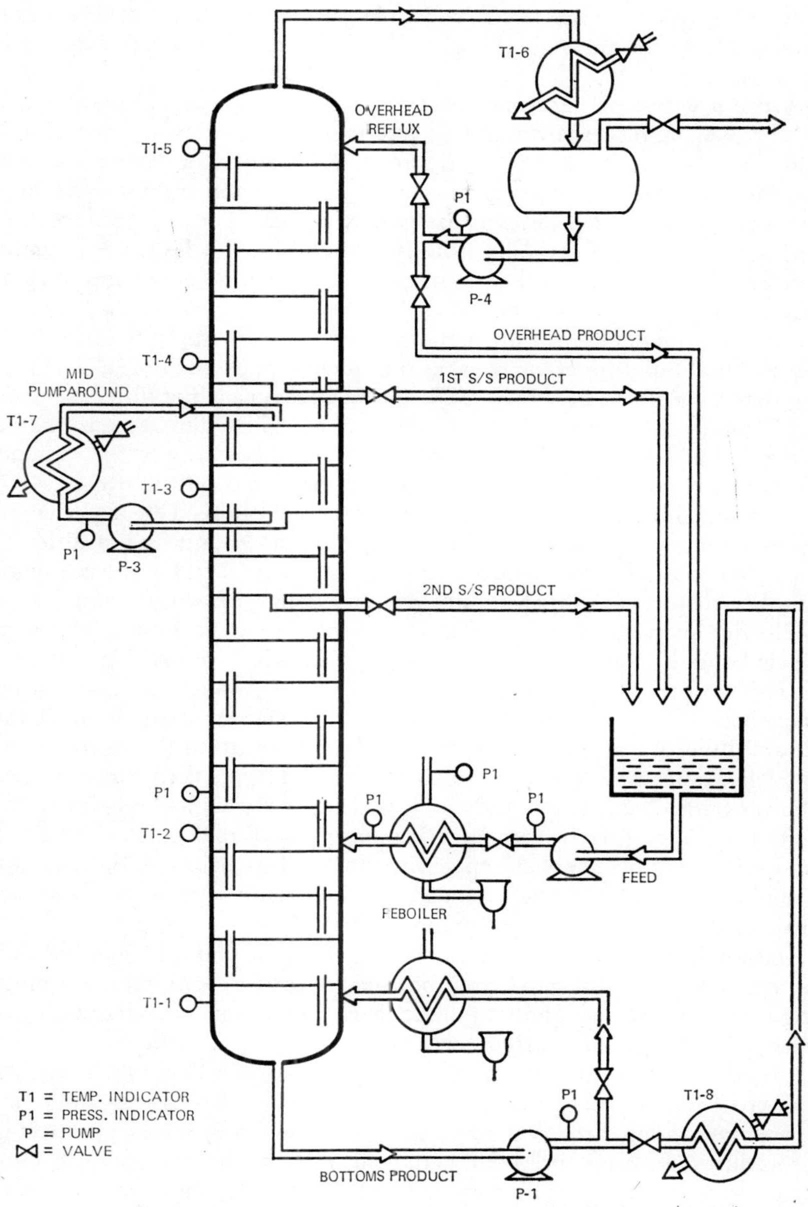

Fig. 31.5 Model Fractionator

The operator sees all these controls and manipulates them himself, at first under the guidance of the instructor. In addition, the operator can see something which he has never been able to see in years of working on an actual plant: for through the glass tower he can see the process itself occurring.

Once a year—perhaps once every 18 months—his actual plant stops. It is taken to pieces, overhauled, put together again and 'started-up'. Once it goes on stream, it is on stream incessantly for the period just mentioned unless an emergency occurs. As the plant goes on stream there are very specific procedures to be gone through. These procedures the operator learns very carefully from his supervisor. But he never actually sees crude oil going into the plant. And, of course, he never sees what is happening inside the tower.

With the model he sees the actual process. He sees the start-up; he sees the gradual build-up of vapours misting their way up the tower. Then, as condensation starts, he sees liquid begin to form on each of the tray levels, until finally the whole of the tower is simmering. This means that a balance has been established between pressure upwards of the vapour, and the weight—pressure downwards—of the liquid on the trays.

A tower thus in-balance means smooth operation. The experienced operator knows this well enough, but he also knows and fears a little the condition known as tower dumping. This happens when for one reason or another there is too much input and not enough draw-off. Liquid then builds up on particular trays and instead of vapour pressing up through the holes in the trays, the liquid falls down through them to the tray below.

This occurrence can be demonstrated on the model. It looks rather like the sudden turning on of the spray of a shower bath. Usually the liquid 'build-up' is made to occur on a tray high up in the tower of the model. There the liquid can be seen getting deeper, then suddenly this shower-bath effect occurs and the liquid drops through the holes to the next tray. This tray in turn now has too much liquid on it. There is a pause. The occurrence then repeats itself to the tray below, and then, with a slight pause between each descent, the occurrence repeats itself tray-by-tray all the way down the column.

Tower dumping in reality is a serious upset to the plant. It puts products off-grade and they have to be reprocessed. Instant recognition that dumping is taking place or is about to take place, and a quick understanding of how to put the condition right, are essential.

Another and even more serious occurrence is tower flooding. This, too, can be demonstrated on the model. Once again, quick diagnosis and quick corrective action are vital. The model not only provides practice in procedures ensuring corrective action, it also enables the operator once again to see the effect of his action. These demonstrations always fascinate the operators—particularly the more experienced operators.

Finally, the model also gives considerable scope for the study of product quality—in this case the specific gravity—of the output mixture. As in certain normal operating sequences, it can be varied by temperature or by input or outflow adjustments.

Use of this simulator undoubtedly helps to develop the skills and understanding essential for quick diagnosis and effective action by the process operator.

THE CARMODY TRAINER

The last of the three simulators is the Carmody Trainer. It consists of a console topped by circuit display panels. The device is mounted on a trolley which enables it to be pushed fairly easily from room to room. It has a total of 21 instrument faces. An instructor putting in one of these instrument faces is shown in Plate 31.3. These 'instruments' simulate the flow, level, temperature and pressure controllers and indicators of a process control room.

The essential difference is, of course, that there is no actual flow controller or flow indicator instrument mechanism behind the face. There is simply a tiny motor which is connected by wires and contacts to relays which enable the motor to be switched on in either direction. The actual switches are contained on an instructor's control box shown on the photograph. The instructor can move the main needle of any instrument either forwards or backwards. He does this by moving switches on the control box. There is a switch for every instrument face; he can, however, by a series of 'pegging and looping' operations on the contacts at the back of the instrument panel, so arrange things that when he moves one indicator a number of other instrument faces will behave as 'slave instruments' and their indicators will also move. Alternatively, he can move each instrument needle separately by means of its own switch on the control box.

With appropriate arrangements of paralleling and with frequent practice, the instructor is able to achieve a large variety and combination of instrument movements. The simulation achieved is, however, prescribed first by the limitations of the simulator itself and secondly by the instructor. It can only be as good as the instructor makes it in terms of his skill in manipulating the control box switches and the associated peg board contacts.

The basic difference between this device and the Steam Kettle simulator is that one controls in an integrated way an ongoing process—albeit a very simple one—of controlling with real instruments rate of flow and output temperature of water leaving a vessel. The Carmody controls no process. Its simulation is a limited simulation. The degree to which integration is simulated is limited, both by the scope of the equipment and by the ability of the instructor operating it.

On a normal control instrument, there is one moveable pre-set pointer and two associated moving indicators. The large (moving) pointer in Fig. 31.6 is the 'actual value' of the rate of flow. The small one above it is the pre-set pointer, 'the desired value'. Above this is a moving pointer which indicates the amount of air pressure being applied to the control valve itself. On an actual instrument, these three pointers interact. The pre-set pointer determines the position of the other two. On a Carmody instrument face, this does not happen. The only pointer which is moved

INSTRUMENT INDICATORS

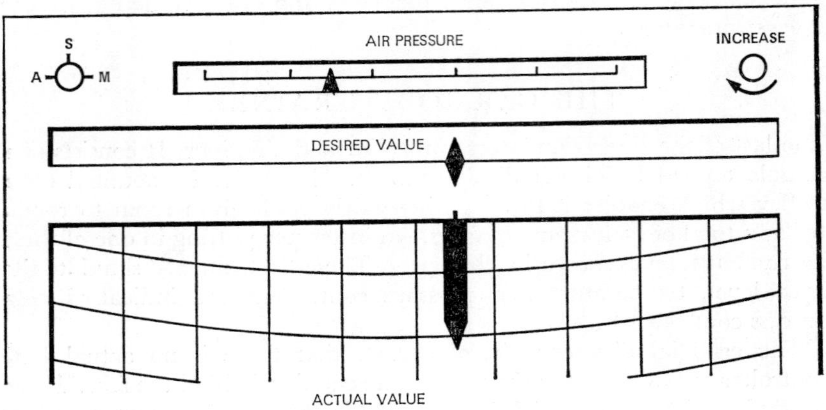

Fig. 31.6 Instrument Indicators

by switches is the 'actual value' pointer; the others are in no way integrated and, if moved at all, have to be pushed along by the finger of the instructor.

On the Steam Kettle simulator or on an actual Flow Recorder Controller on the process, the relative positions of these three pointers tell the operator certain information. For example, if a pump stopped and the process flow had ceased, the air pressure on the valve would swing to one of the two extreme positions, depending on which way the valve was set. This would tell the operator that the valve was trying to give maximum opening to the flow. Meanwhile the signal at the orifice plate would have dropped to zero and this, of course, would cause the large 'actual value' pointer to drop to zero. At the same time the desired value pointer would have remained in its original position.

No comparable integration of movements of this kind is possible on the Carmody simulator. In spite of this limitation, it is, however, a valuable aid. It is a simulator in the sense that an instructor, with a group of trainees around him, can break down any particular part of the flow circuit of a plant into simplified flow diagrams which he then displays on the magnetic board above the instrument panel. Having done this, he can then begin to explore work situations and problems in terms of the changes which he can demonstrate on the 21 instrument faces. Each instrument face he identifies with an instrument on the actual plant.

The instructor asks the members of the group to discuss what it is that they see happening and why the instruments are behaving as he has switched them to behave. Alternatively, he may have members of the group move the instruments by hand to certain positions and he himself will then begin to switch-in any appropriate movements of indicators on the instruments involved. If he wishes, he may also cause alarm buzzers to sound and red lights to come on when certain indicators reach extreme positions.

Basically, the Carmody is a discussion-provoking simulator. In this

Plate 31.3. *The Carmody Trainer.* (*Courtesy: Esso Petroleum Co. Ltd.*)

Plate 31.4. *The Carmody Trainer in use.* (*Courtesy: Esso Petroleum Co. Ltd.*)

context, it is an extremely valuable visual aid. Since no process is involved, discussion can be prolonged until all-round understanding of a problem is achieved (see Plate 31.4).

POSTSCRIPT

Each of the three described devices is very different. Each goes some way towards enabling the man in charge of an unseen process to understand a little more of what is happening. There is no easy way of training a continuous process operator. The process he controls is a complicated one. It is, therefore, exceedingly important for the operator to have his intuitive knowledge consolidated by basic observation and sound understanding of process principles. It can be seen, however, that the sometimes elusive diagnostic ability which the training officer and the process supervisor wish to inculcate in the operator can be developed by the use of simulators. Simulation exercises may in the immediate future continue to be one of the few practical ways of clarifying intuitive knowledge and of deepening the operator's basic concepts about the process and its controls.

Unless these concepts are well established, he cannot be expected to deal promptly and effectively with emergencies. In the future, increased automation will undoubtedly make his job easier. It will also make increased demands on him when the system goes wrong.

In many continuous process industries, the operator often (though not invariably) has been regarded as a semi-skilled worker. In the future, he may well need to become a technician. The transition to this level of technical competence will not be easy. Process operators today are often middle-aged men who left school at 14 and who have had no subsequent background of consistent study. For them the path of learning is difficult even where process training has been well organised. In many firms, their job knowledge has often been acquired by working alongside another operator and after that by exposure.

THE NEED TO QUALIFY

One basic problem is the absence of a training tradition; another is the unique nature of the work. Skilled craft apprenticeship may have its shortcomings but it does have a training tradition. A man who is time-served is regarded as qualified and skilled. Apprenticeship, however inadequate in the past, is the route by which the craftsman has 'qualified'.

Usually there is no equivalent route for the process operator. Many potential operators are, however, already members of a powerful 'general' workers trade union before they join a firm as process trainees. If not, they quickly become members. They join a firm as mature adults prepared to do shift work. After that their union takes care of pay rates and shift differentials. In effect this often means that once the individual is in the job he has 'made the grade'—or one of the grades and to that extent is 'qualified'. He also knows that in time he will, by his employer's standard, become fully qualified. Simply being there and working reasonably well along with other operators is often all that is necessary.

THEORETICAL TRAINING IS FEARED

When this is the background, no superimposing of an academic approach to qualification—however estimable or necessary this may be—is likely to succeed with existing operators unless big changes are made in terms of their organisation, their training and their status. Today few middle-aged operators are likely to settle down to years of theoretical study at evening classes in order to 'qualify' if there is neither urgent reason nor financial incentive for them to do so. On the contrary, the same men often show open hostility towards theoretical training.

This is because they fear it. They fear failure and humiliation. They see no point in running the risk of both if there is no inducement to do so? Not surprisingly, the majority are not (to that extent) interested in becoming 'qualified' operators. That, however, is not the whole story. They are ambivalent about training. Probably the majority would come down heavily in favour of theoretical training provided that it did not expose their inadequacies or threaten their security. They want to understand more of the complexities of their own job. At heart they dislike rule of thumb. In the absence of any tangible incentive to qualify, they are naturally reluctant to attempt an objective which they might fail to achieve.

This is only part of the dilemma which industrial training boards will in future have to face. The other tremendous fact of the situation is the accelerating pace of technological change itself. In the immediate future, new methods, such as computer controlled systems, will mean automation or near automation and will place new and heavy demands on the continuous process operator.

Almost overnight—or so it seems—he will need to lift himself even above craft level into a new status, carrying new and novel responsibilities. This leap into technical competence is the miracle which future training officers may have to bring about. It offers a terrific challenge especially as process operators usually have powerful trade union protection. The 'general' trade unions will no doubt support any progressive training measures which will enable their members to make the transition but it is equally certain that they will not acquiesce in the liquidation of those members who may be unable to make this transition.

TWO-TIER POSSIBILITY

Possibly a two-tier structure will emerge. It is even conceivable that 'general' versus craft union rivalries might be overcome and open up a second tier to the recently emerged craft instrument mechanics. This could allow them to become combined continuous process operator/instrument mechanics.

This is not such a remote possibility in view of the close association that may be expected between instrument mechanics and future computer-controlled systems for the process industries, especially as it seems likely that pneumatic instruments will remain part of the system and that the conventional instrument mechanic would be required permanently on site.

Meanwhile, computer control would probably take out of the continuous process operator's hands many of his present monitoring and instrument trimming duties. The logical next step would be that fewer operators would be required. Those remaining would then cover a wider terrain. When an emergency did arise, computers would set off alarms and the human master controller of the entire system would summon his operators to a trouble spot via portable radio communication sets carried by operators on patrol.

At that moment, it would be more necessary than ever for operators to be swift and accurate in diagnosis and equally swift in taking corrective action. With these possibilities ahead, it might be better to have two kinds of operator. One could be the process operator mechanic who would always be in a position to keep himself mentally in tune by instrument maintenance work. Because of that work he would be more likely to deal effectively with the contingency which might otherwise produce an emergency. This would not necessarily remove the need for a second or site monitoring operator.

At present, craft union barriers might prevent two such levels of process operator from being complementary. If resistance of this kind were overcome, the arrangement could probably work well. It could mean, too, that technician expertise could ultimately be acquired by some of the site monitoring operators.

FUTURE SIMULATOR TRAINING

Part of the problem could be solved by the people within the situation since a good instrument technician would be capable of learning from process simulators and of training his less skilled colleagues to do the same. Ultimately they should of course be able to deepen their knowledge of the process not only by operating simulators but by themselves designing and constructing them.

For older operators, this might be one way of achieving a clear grasp of future basic concepts which might otherwise elude them or be very superficially understood.

CONCLUSION

Training with the use of simulators may, however, only be a half measure. Intensive theoretical training and perhaps task orientated programmed learning courses may have to be added if the continuous process operator is to acquire competence and diagnostic skill in the much more automated future which lies immediately ahead.

BIBLIOGRAPHY

CROSSMAN, E. R. F. U., *Automation and Skill*, D.S.I.R.–H.M.S.O. (1960)

PART 5

ESTABLISHED ORGANISATIONS

32

British Association for Commercial and Industrial Education

P. J. C. PERRY, B.SC.(ECON.)

Director and Secretary, British Association for Commercial and Industrial Education

ORIGIN

One of the distinguishing features of British public life for the past two centuries has been the prominent place occupied by voluntary organisations in the formation and education of public opinion, and the exertion of pressures leading ultimately to Government action. This work is quite separate from the function of political parties, since it is concentrated on a narrow and clearly defined sector of public life, cuts right across party-political boundaries and unites relatively small but dedicated and expert groups of people.

The British Association for Commercial and Industrial Education (B.A.C.I.E.) is the mainstream of this tradition. It owes its existence to the realisation that, in an increasingly competitive world, the future prosperity of British industry and commerce rests in large measure on the improvement of vocational training, and its close integration with the education services at all levels.

On May 28th, 1919, the Association for the Advancement of Education in Industry and Commerce (A.E.I.C.) was established. Its 49 founder members described themselves as 'a company of people representing business houses throughout the length and breadth of Great Britain and Ireland that are at the present moment carrying out some definite education schemes for the benefit of their employees.' The President of the Board of Education, Mr. H. A. L. Fisher, in his inaugural address, welcomed the fact that the Association existed 'for the purpose of focusing and formulating ideas in regard to the best method of carrying out the Fisher Education Act from an industrial point of view.'

B.A.C.I.E.'s other parent, the British Association for Commercial

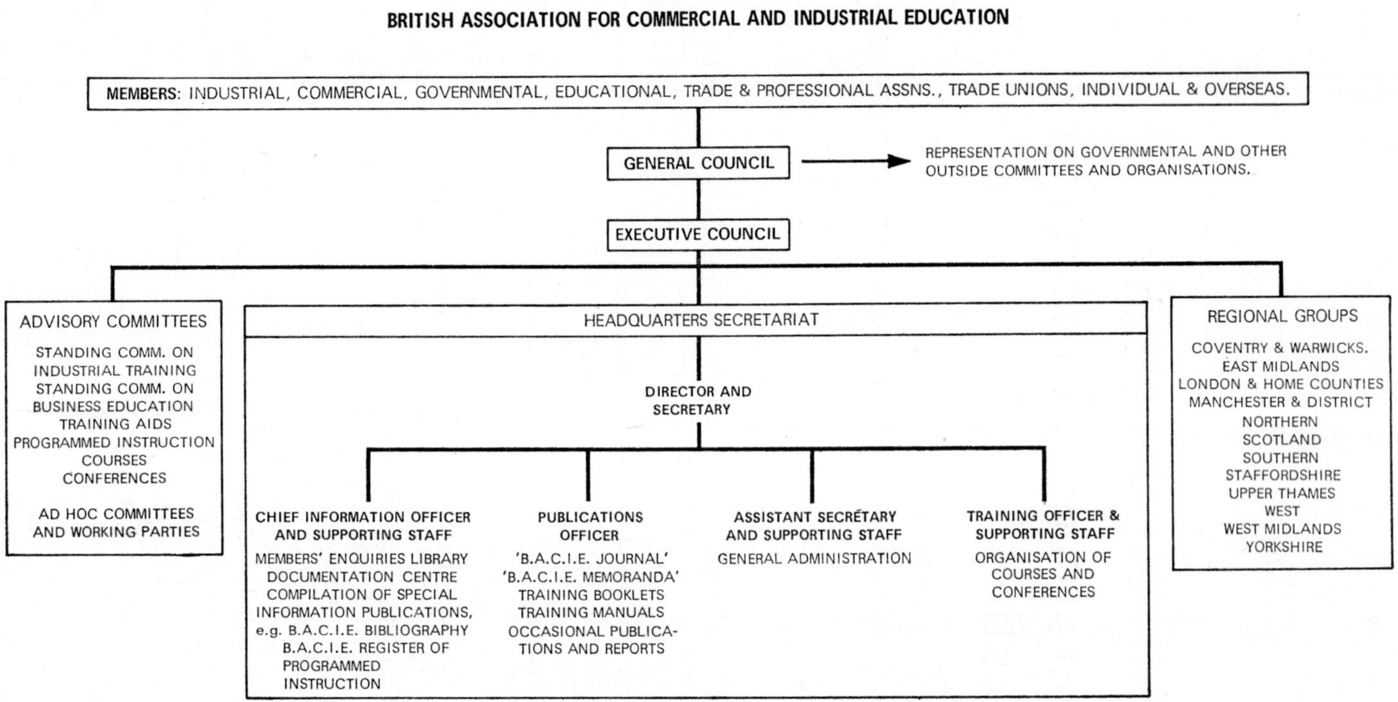

HEADQUARTERS : 16 PARK CRESCENT, REGENT'S PARK, LONDON W.1. TELEPHONE : 01-636 5351.

Fig. 32.1 B.A.C.I.E. Organisation Chart

Education, was formed on July 14th, 1931. Its biggest single achievement was the organisation of the fifth Triennial Congress for the International Society for Commercial Education in London in 1932. This was attended by 500 delegates from over 40 countries and proved an outstanding success. The new Association, however, wisely decided to combine forces with the A.E.I.C. The merger took place on April 27th, 1934, and the now familiar initials, B.A.C.I.E., made their first appearance.

The latest accession of collective strength occurred on January 1st, 1966, when the National Association for Business Education amalgamated with B.A.C.I.E. and brought some 400 teachers of commercial subjects into B.A.C.I.E.'s orbit.

From the outset, B.A.C.I.E. had powerful support. Many of the biggest names in British industry and commerce figure prominently among those prepared to pool their training experience for the common good. H.R.H. The Prince of Wales became the Association's first Patron. He was later succeeded by H.R.H. the Duke of York and, in 1952, by the present Patron, H.R.H. the Duke of Edinburgh.

At the end of the Second World War, B.A.C.I.E. began to extend the scope of its activities by bringing the nationalised industries, universities, colleges of further education, Ministries, local education authorities, professional bodies, trade associations and trade unions into membership. Overseas applicants were also admitted. B.A.C.I.E. is, indeed, the only voluntary national organisation exclusively devoted to the promotion of vocational training at all levels in the United Kingdom.

After 1946, special attention was devoted to a series of comparative studies, linked with visits to a number of foreign countries, to study their systems of vocational training. They led to the inescapable conclusion—strongly resisted by a succession of Ministers of Labour at B.A.C.I.E. conferences—that the existing British system—or lack of it—would have to give way to some form of governmental regulation. Following strong pressure, the Government set up a Committee in 1956 to study the problem. The result—the Carr Committee Report *Training for Skill* of 1958—came in for much criticism on account of its opposition to training legislation. Successive B.A.C.I.E. Conferences drew attention to the notable lack of progress, as compared with our overseas competitors.

The B.A.C.I.E. Spring Conference of 1962 had the title *Industrial Training—Whose Responsibility?* and contained proposals for training legislation. The Minister of Labour, who opened the proceedings, while expressing some sympathy, confirmed the Government's adherence to the Carr recommendations.

It was, therefore, a matter of considerable satisfaction to B.A.C.I.E. that the White Paper on *Industrial Training*, published only seven months later, contained proposals for legislation which incorporated most of the principles recommended at B.A.C.I.E. conferences.

Since the passage of the Industrial Training Act, 1964, B.A.C.I.E. has taken a significant part in assisting in its implementation, and the new industrial training boards have themselves become B.A.C.I.E. members.

ORGANISATION

B.A.C.I.E.'s membership consists in the main of organisations who pay an annual subscription and in return are entitled to nominate representatives through whom the Association's services are made available.

The Association is governed by a General Council elected by the members (see Fig. 32.1, p. 332, for details). The head office staff—at present 20—is accommodated in exceptionally fine offices, the property of the Crown Commissioners, in Regent's Park, London, W.1.

There are 12 Regional Groups which elect their own Hon. Officers and organise regional functions in addition to those provided nationally. Apart from providing facilities for functions which respond closely to local needs, the Regional Groups also help to ensure that B.A.C.I.E.'s headquarters are kept in touch with the grass roots.

Specialist committees, composed of expert members, advise the staff on a wide variety of subjects, including types and contents of courses and conferences, and developments in training aids and techniques.

SERVICES

The central feature of the Association's services is that they are treated as an integrated unit, so that courses, conferences, information and publications are complementary to each other.

COURSES

B.A.C.I.E. courses, designed in the main for 'training the trainers', have been organised for more than 20 years, The great majority are residential five-day courses. They cover a substantial and growing range of topics and are fashioned on the 'module' principle. In 1961, the Association set up a Working Party to plan a long course for training officers which might eventually be run by approved institutions and become the nationally accepted standard for industrial trainers. While the Working Party believed that a one-year course—on a par with post graduate teacher training courses—was desirable, it was recognised that the climate of industrial opinion was not yet ripe. Two experimental ten- and eight-week courses were therefore run—the second with a grant from the Ministry of Labour—and the lessons learned have had a substantial influence on the training officer courses now developed on a national scale by educational institutions, such as the Portsmouth College of Technology.

The Association, meanwhile, continues to develop a growing variety of courses. In September 1967, a training centre was opened at the Association's offices.

CONFERENCES

Four national conferences are held regularly each year: the Spring Conference (one day in London); the 'Successful Instruction' Conference and Exhibition on Training Aids (one day in a provincial centre); the

Annual Conference (two days in September, usually at a university); and the 'Trends in Training for Skill' Conference (two days at Eastbourne in December).

These functions provide unique opportunities to meet a wide cross section of the Association's members and to learn at first hand about current and future developments.

INFORMATION

The B.A.C.I.E. Information Library contains probably the most comprehensive collection of training documents in the country. This is kept up-to-date, and an especially popular service is the compilation of the *BACIE Bibliography* which is published quarterly as a detachable supplement to the *BACIE JOURNAL*. All documents listed can be borrowed by members.

The Information Officers deal with every kind of enquiry on training; and their services, as well as the reference library, are available to all members.

PUBLICATIONS

(*a*) *Training Booklets* cover a variety of basic skills in a concise and practical manner, Some 250,000 copies have been distributed during the past five years.

(*b*) *Training Manuals* are printed booklets of lecture notes used on B.A.C.I.E. courses for which requests have been received from member organisations.

(*c*) *Occasional publications* include the proceedings of conferences, reports on special topics and reprints of articles from the *BACIE JOURNAL*.

(*d*) *BACIE JOURNAL*, published quarterly, is the Association's official journal. Free to members.

(*e*) *BACIE MEMORANDA*, published seven times a year, is a six-page bulletin of news and comments. Free to members.

(*f*) *Progress Report on I.T.B.s* is a wall chart published annually containing tabulated information giving a synopsis of the main current developments and personnels of the I.T.B.s.

THE FUTURE

When B.A.C.I.E. began its work, the essential role of training as a component of industrial productivity was little understood, except by a minority of companies who saw it as more than an avoidable expense. Terms like day- and block-release, sandwich courses and analysis of training needs were largely unknown. The Industrial Training Act is beginning to be the catalyst it was intended to be.

In the era of the Industrial Training Act, B.A.C.I.E.'s functions have already begun to change. This has been due in the first place to the unprecedented increase in the Association's membership and to the fact that it consists mostly of small and medium-sized firms whose interest in training has been stimulated by the Act. Unlike their predecessors, they lack experience and seek guidance. This is being provided.

Much thought has been given to the continuing role of a voluntary organisation many of whose former activities have been taken over by statutory bodies. Experience in related fields confirms the importance and relevance of continued voluntary effort. Clearly the persuasive function has been replaced by legislative edict. But the official machinery, even when it has been assembled, will need supplementing in a multitude of ways. It will need, and indeed welcome, the informed comment and criticism of the uncommitted and altruistic expert. It cannot match the speed and flexibility to experiment which is possessed by independent bodies not relying for their existence on public funds. The importance of these functions will greatly increase in the future.

BIBLIOGRAPHY

Current B.A.C.I.E. publications include the following:

Aids to Instruction:

A Guide to the Writing of Business Letters, 5th Ed. (1968)
A Guide to the Use of the Telephone in Business, 6th Ed. (1967)
Report Writing, 2nd Ed. (1967)
PERRY, P. J. C., *Hours into Minutes* (1966)
Tips on Talking, 2nd Ed. (1967)
POWELL, L. S., *Lecturing to Large Groups* (1966)
POWELL, L. S., *A Guide to the Use of Visual Aids* (1968)
POWELL, L. S., *A Guide to the Overhead Projector* (1966)
POWELL, G. H., and POWELL, L. S., *A Guide to the 8-mm Loop Film* (1967)
Mathematics by Visual Aids (1962)
SMURTHWAITE, R., *Welcome, Stranger* (1966)
GOUGH, J. S., *Interviewing in Twenty-Six Steps* (1966)
Swedish Job Instruction Manuals, English language editions of Swedish job instruction manuals prepared by the Swedish Metal Trades Employers' Association: Cylindrical Grinding; Surface Grinding; Turning; Milling; Shaping; Bench Fitting; Drilling
A Training Officer's Guide to the Education System of Great Britain (with special reference to education for engineering trainees), B.A.C.I.E. Training Manual No. 1, 3rd Ed. (1967)
TAYLOR, E. A., *A Training Officer's Guide to Selection*, B.A.C.I.E. Training Manual No. 2 (1966)
BINGHAM, V. N., *A Training Officer's Guide to Commercial Apprenticeship*, B.A.C.I.E. Training Manual No. 3 (1968)
LUND, M., *A Guide to the Training of Office Supervisors*, B.A.C.I.E. Training Manual No. 4 (1967)
DEBENHAM, I., *A Training Officer's Guide to Discussion Leading*, B.A.C.I.E. Training Manual No. 5 (1968)
Books for Training Officers, an annotated list issued jointly by B.A.C.I.E. and the National Book League.
BACIE Bibliography, **1** (1960–62)
BACIE Bibliography, **2** (1963–67)
B.A.C.I.E. Register of Programmed Instruction, **2** (1965)

Occasional Papers and Reports:

STEERS, C. E. B., *Clarifying Objectives in Supervisory Training* (1966)
HAHN, K., *Education and Changes in Our Social Structure* (1960)

Policy in Perspective, 1964, Report of the B.A.C.I.E. Annual Conference (1964)
Continental Comparisons, Papers presented at the B.A.C.I.E. Annual Conference (1961)
Cmnd. 1892: *The Next Step*, Report of the B.A.C.I.E. Special Conference on the White Paper, Industrial Training; Government Proposals (Jan. 1963)
A Standard Method of Costing the Training of Apprentices
European Apprenticeship, C.I.R.F. Monograph, 1, No. 2
Industrial Training Boards: Progress Report No. 2 (Mar. 1967)
The Industrial Training Act in Retrospect, Spring Conference (1967)
Industrial Training Boards: Progress Report No. 3 (1968)

33

British Institute of Management

R. M. McKINNON

Formerly Head of Press and Publications Division, British Institute of Management

ORIGIN AND OBJECTIVES

The basic objective behind the work of the British Institute of Management (B.I.M.) has been described many times, but nowhere more cogently than by Mr. John Bolton, then Chairman of the Institute, who said in the Annual Report for 1966:

> 'The difficult and prolonged struggle facing Britain today understandably places a heavy strain on managements and managers in industry and commerce; and the task of achieving our national economic goals is more than ever dependent on the extent to which our nation's managers can bring to bear professional skills and enlightened dedicated effort in creating the necessary expansion in the country's real wealth. The achievement of these ever higher standards of managerial performance is the Institute's main task and the opportunity to serve the nation has never presented a more exciting challenge to us all.'

The B.I.M. is not of course the only body engaged in the task of raising managerial standards in this country—and indeed a number of its activities are run in collaboration with other bodies—but over the years it has become a national clearing house for information on management policies, practices and techniques. That apart, its role in the drive to expand management education facilities becomes more and more important with each year that passes.

THE CURRENT POSITION

Today, then, the British Institute of Management is the largest independent body of its kind outside the U.S.A. Founded in 1947 on the recommenda-

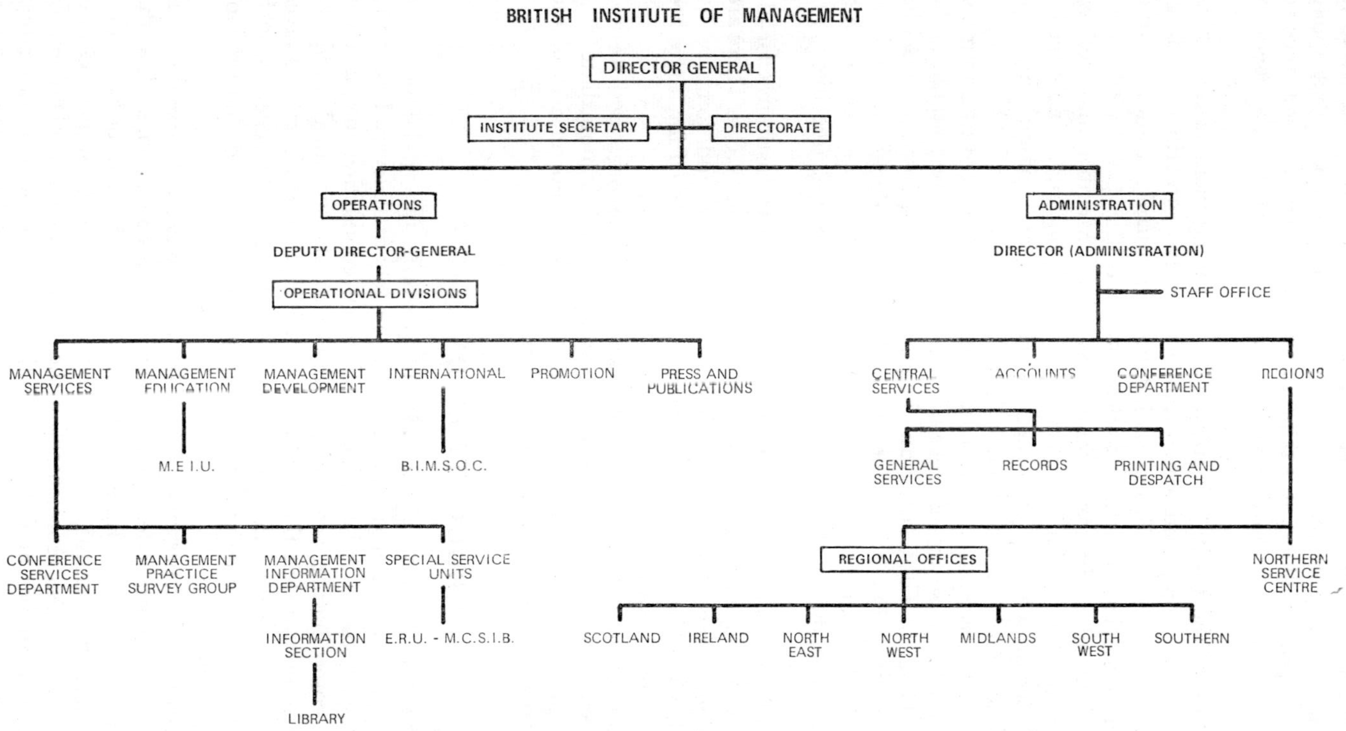

HEADQUARTERS: MANAGEMENT HOUSE, PARKER STREET, W.C.2. TELEPHONE: 01-405 3456

Fig. 33.1 B.I.M. Organisation Chart

tion of the Baillieu Committee set up by the Board of Trade, the Institute in its early years received Government financial support, but today though non-profit making, it is self-supporting. It does, however, receive a Government grant to further activities aimed at helping management in general.

A feature of the Institute is its rapidly expanding membership which in March 1968 comprised some 10,000 collective subscribers (a term used to cover private firms, nationalised industries, trades unions, employers' associations, trade associations and the like) and 20,000 individual members in three grades—Associate, Member and Fellow. It is no platitude that individual members are being increasingly recognised as people who are interested in achieving and maintaining professional standards in the practice of management.

B.I.M. policy is controlled by a National Council assisted by a number of committees. There are also seven Regional Councils, responsible to the National Council. These advise and help in developing Institute activities in the regions. Policy is implemented by the Director-General, Mr John Marsh, C.B.E., and the senior executive and specialist staff who form a substantial proportion of the 250 people on the payroll (see Fig. 33.1, p. 339).

Most of the senior and executive staff work in the Institute's Headquarters at Management House, Parker Street, London, W.C.2. There are also regional offices covering the whole country. In addition, and to ensure that members' needs and interests are fully catered for, there are 50 B.I.M. Branches operating over the same area.

At Management House, there are several divisions organised to deal with specific aspects of management, and to provide immediate and comprehensive service to members. The operations and services of the Education Division will be described first.

EDUCATION DIVISION

The Education Division of the Institute performs the vital function of promoting education and training for management at all levels. It operates as a bridge between industry and institutions conducting management courses, co-ordinating academic studies with the practical needs of industry to ensure that their interests coincide. This Division also gives advice to companies on internal management development schemes, such as management succession, appraisal, job rotation and 'on-the-job' training.

The Institute is represented by its honorary officers and senior staff on national committees dealing with management education, such as the Confederation of British Industry (C.B.I.), the Department of Education and Science, the Central Training Council and the University Grants Committee. It is also represented on the Committee on the Diploma of Management Studies and the Foundation for Management Education (F.M.E.) and others. In 1963/4, the Institute played a central part in the events leading to the establishment of the two British Business Schools in London and Manchester. It provided the secretariat for the Franks and Normanbrook Reports which it published, and then together with the Federation of British Industries (F.B.I.) and the F.M.E. launched the successful appeal to industry and commerce for £5,000,000 to promote management studies at university level.

INFORMATION SERVICE

For over six years now, the B.I.M. has been providing its members with an information service on management courses as they have become available in the U.K., Europe and the U.S.A. Recognising the importance of B.I.M.'s work in this field, the Government has made a grant to the Institute's Management Education Information Unit with a view to extending its facilities to a larger number of companies. Additional advisory and information staff have been recruited, and up-to-date information retrieval equipment installed. To date, the unit has details of about 600 institutions (including four company training centres) and 140 overseas organisations running management courses, together with information on over 3,000 actual courses.

To assist in its work, the Institute has set up an Advisory Committee consisting of representatives of the Department of Education and Science, the Ministry of Labour, the Board of Trade, the C.B.I. and the F.M.E., to which it is hoped to add academic and industrial training board representatives.

One of the most popular features of B.I.M.'s education programme over the last 12 years has been its Management Development Division's Modern Management Course at Oxford; well over 3,000 younger executives from all branches of industry have attended this course, which combines advanced study of management skills and principles with practical case examples. Central to B.I.M.'s future education policy will be the aim of consolidating a real working partnership between industry and the universities and colleges.

WORLD LINKS

The Institute through its International Management Division also has formal and informal links with management associations, institutes and training centres throughout the world. It has reciprocal relationships with sister institutions in many industrialised countries, such as the U.S.A., France, Germany, Holland, Austria, South Africa and Japan, as well as links with management centres in a number of developing countries. In the international field, this work has four main aims:

1. To help British companies to examine special management problems encountered overseas, and to keep abreast of developments and trends.
2. To keep them informed of general conditions prevailing in overseas countries where they may wish to operate.

3. To help executives newly-appointed overseas to adapt themselves to new environments.
4. To assist overseas-based companies with information on current British management practices and conditions.

Conferences and symposia are held regularly in conjunction with management centres abroad to enable British businessmen to exchange views and experiences with their opposite numbers from foreign countries. Close collaboration with America's National Industrial Conference Board (N.I.C.B.) and the American Management Association has yielded mutually profitable information on various aspects of business management.

Liaison between overseas organisations and British management institutions is the province of the B.I.M. Secretariat for Overseas Countries (B.I.M.S.O.C.). Since 1963, the Institute has been responsible for managing this unit on behalf of the Ministry for Overseas Development, for advising the Ministry's Committee on Technical Education and Training in Overseas Countries (T.E.T.O.C.), and for making recommendations to the British Council when necessary. B.I.M.S.O.C. deals every year with a great many visitors and enquiries from developing countries.

MANAGEMENT SERVICES

In connection with B.I.M.'s priorities in Great Britain, Mr J. Bolton spoke for all when he stressed that in tough and competitive times management's task was becoming increasingly more complex and exacting. 'This task,' he said, 'calls for informed decision-making and enlightened risk-taking.' The Management Services Division of the Institute aims to lighten the busy manager's burden. It provides the majority of B.I.M.'s services to company members, operating very much like the workings of a computer unit. The library at Management House is the storage unit for the largest collection of management literature in Europe; it contains a great deal of unpublished or company material generally unobtainable in this country, multiple copies of standard works, and most British and American publications on management. In addition, it receives over 350 periodicals from all over the world from which abstracts are made and published, and is a unique source of reference material on management subjects. Specialised information and research on current developments in management is made available to company members through a comprehensive range of information notes and summaries.

Two units have been set up in response to the great demand for detailed information on salary structures and management consultancy: the Executive Remuneration Unit is developing the salary surveys carried out by B.I.M. since 1955, and the Joint B.I.M./C.B.I. Management Consulting Services Information Bureau is keeping track of the rapidly increasing number of management consultancy services available to management.

Companies in numerous industries and trades have so far benefited from the services of the Centre for Interfirm Comparison (C.I.F.C.). Set up by B.I.M. in 1959 in association with the British Productivity Council, the C.I.F.C. prepares confidential reports which enable the managing director to see how overall success in his company compares with that of other industrial firms.

COURSES AND CONFERENCES

Through the facilities of the Management Services Division, B.I.M. 'programmes' and interprets current data according to members' requirements. A major facet of the Institute's work is analysing present industrial systems, and forecasting future needs; this is done mainly through conferences, courses, seminars and lectures, which are held in London and other main industrial centres. These offer:

1. A forum for the examination of current and future problems.
2. A broader context in which the manager may place his own function.
3. An opportunity for making personal contacts and exchanging views and experiences in an informal atmosphere.
4. A working knowledge and appreciation of new management techniques.

Among speakers and seminar leaders are prominent industrialists, educationists, management consultants and other specialists. Highly qualified experts from Europe and the U.S.A. are also invited to participate. Conferences may range over several themes, or be devoted to discussing a single subject. There are courses of one or two days for directors, senior managers and specialists, and one-week residential courses for the middle and junior executive levels. A special service offered to those at the highest policy-making levels in large-scale industry is a series of meetings held in the board rooms of host companies; similar discussion meetings for directors of substantial local companies are organised in each of B.I.M.'s regions. The popularity of B.I.M.'s courses may be demonstrated by the attendance at seminars in 1967: over 550 were staged and were supported by over 6,000 directors and senior executives.

PUBLICATIONS

Amongst other services provided by the Institute is the publication of a wide range of books and booklets on management subjects, as well as regular information bulletins for individual and collective members. B.I.M.'s activities are publicised in the press, and covered by a monthly report in *Management Today*.

CONCLUSION

As mentioned at the beginning, numerous B.I.M. activities and projects are carried out in collaboration with Government agencies and other bodies in the management field. It has been shown, for instance, how the C.I.F.C. is a joint effort of the Institute and the British Productivity

Council. Other examples include joint manpower planning conferences with the Ministry of Labour and various action conferences with the British National Export Council and other bodies.

No organisation can be an island if it is to work effectively in the field of management.

BIBLIOGRAPHY

Current B.I.M. publications include the following:

General Management:

DRUCKER, P., *The Effective Business: the Effective Executive* (1964)
URWICK, L., *A Short Survey of Industrial Management* (1962)
A Bibliography on Management Subjects
MENZIES, SIR ROBERT, *The Interdependence of Political and Industrial Leadership in the Modern State* (1964)
ATHERTON, H. S., *Running the Smaller Firm* (1965)
DIEBOLD, J., *Focus on Automation* (1965)
BRECH, E. F. L., BRECH, R., ATKINSON, F. J., and ROGERS, F. J. H., *Higher Management Lectures* (1965)
HUMBLE, J. W., *Improving Management Performance* (1965)
KIPPING, SIR NORMAN, *Management and the Public Interest* (1965)
Leading Questions, a management symposium of speeches and articles from 13 chairmen and managing directors (1965)
MILLWARD, J. G., *Municipal Work Study*

Management Education:

KOONTZ, H., *Requirements for Basic and Professional Education for Scientific Management* (1964)
MISSELBROOK, B. D., BARRY, W. S., WEDDLE, W. A., and ROBERTS, T. J. (with a foreword by D. J. Mann), *Making Management Development Effective* (1964)
EDWARDS, SIR RONALD, *Universities and the World of Business* (The First Urwick Lecture) (1965)
The Conspectus of Management Courses (1968)
Management Courses in the U.K. used by Senior and Middle Managers from the Developing Countries (1965)
How to Lead Case Studies (1963)
Management Training Techniques (1962)
MARSH, J., PARSONS, J., GRATWICK, J., HOPKINS, R. R., and ROSE, P., *Five talks for sixth forms* (A career in management) (1965)
HALL, SIR NOEL, *Academic Freedom in Management Education* (The Second Urwick Lecture) (1966)

Personnel Administration:

Absence from Work: Incidence, Cost and Control (B.I.M. survey) (1961)
Absence from Work: Recording and Analysis (B.I.M. survey) (1955)
The Cost of Labour Turnover (1954)
CLEGG, H., and NARASIMHAN, P. S., *Implications for Management of Longer Standard Holidays* (1963)
CLEGG, H., *Implications of the Shorter Working Week for Management* (1962)
Job Evaluation: A Practical Guide, Revised (1961)

Marketing:

The Field Sales Supervisor (A Study Group Report) (1959)
CORBIN, A., *Implementing the Marketing Concept* (1966)
MCCLELLAND, W. G., et al. *Management Training and Development in Retailing* (1963)
O'SHAUGHNESSY, J., *Work Study Applied to a Sales Force* (1965)

Production:

CORDER, G. G., *Organising Maintenance* (1963)
Production Control in the Small Factory (1962)
Application of Production Control: A practical manual of general use to batch production managers (1963)
CURRIE, R. M., *Simplified P.M.T.S.* (1963)
BATTERSBY, A., *A Guide to Stock Control*, Pitman (1962)
CURRIE, R. M., *Work Study*, Pitman, 2nd Ed. (1963)
CURRIE, R. M., *The Measurement of Work* (1965)
SMITH, K. M., *A Practical Guide to Network Planning* (1965)
BRECH, E., DE PAULA, C., and WHITE, N., *Management of Research and Development* (1964)

34

City and Guilds of London Institute

D. E. WHEATLEY, M.A.(CANTAB.), B.SC.(LOND.)

Director (Further Education and Training), City and Guilds of London Institute

INTRODUCTION

This chapter outlines the origin, aims and work of the City and Guilds of London Institute as the national (U.K.) body which provides schemes for part-time technical college courses and related examinations taken by operatives, craftsmen and technicians under training in industry.

ORIGINS

The Institute* was set up in 1878 'for the advancement of technical education' by the Guilds of the City of London and the City Corporation. The Guilds had since medieval times had responsibility for education and training of apprentices in their respective trades. Their intention in founding the Institute was to provide a mechanism for collective action on a national basis and to form a 'ginger group' for technical education at a time when neither central nor local government showed any interest in the subject, or indeed any awareness of its importance to the nation. The effectiveness of the 'ginger' is indicated by the fact that today, in round terms, about half-a-million people under training in industry attend technical colleges for part-time classes based on 'City and Guilds' syllabuses. The Institute is still an independent body—it was granted a Royal Charter in 1900—but has established many direct links with the public education service as this has developed, and now works in very close association with it. An organisational chart is shown in Fig. 34.1.

*The Institute founded, built and staffed the City and Guilds College, which was incorporated as the Engineering Department of the Imperial College of Science and Technology in 1908. The College is now the largest university engineering faculty in the Commonwealth. Other foundations are the Finsbury Technical College (closed in 1926) which was a pioneer and prototype of technical colleges in this country, and the City and Guilds of London Art School, Kennington, London S.E.11.

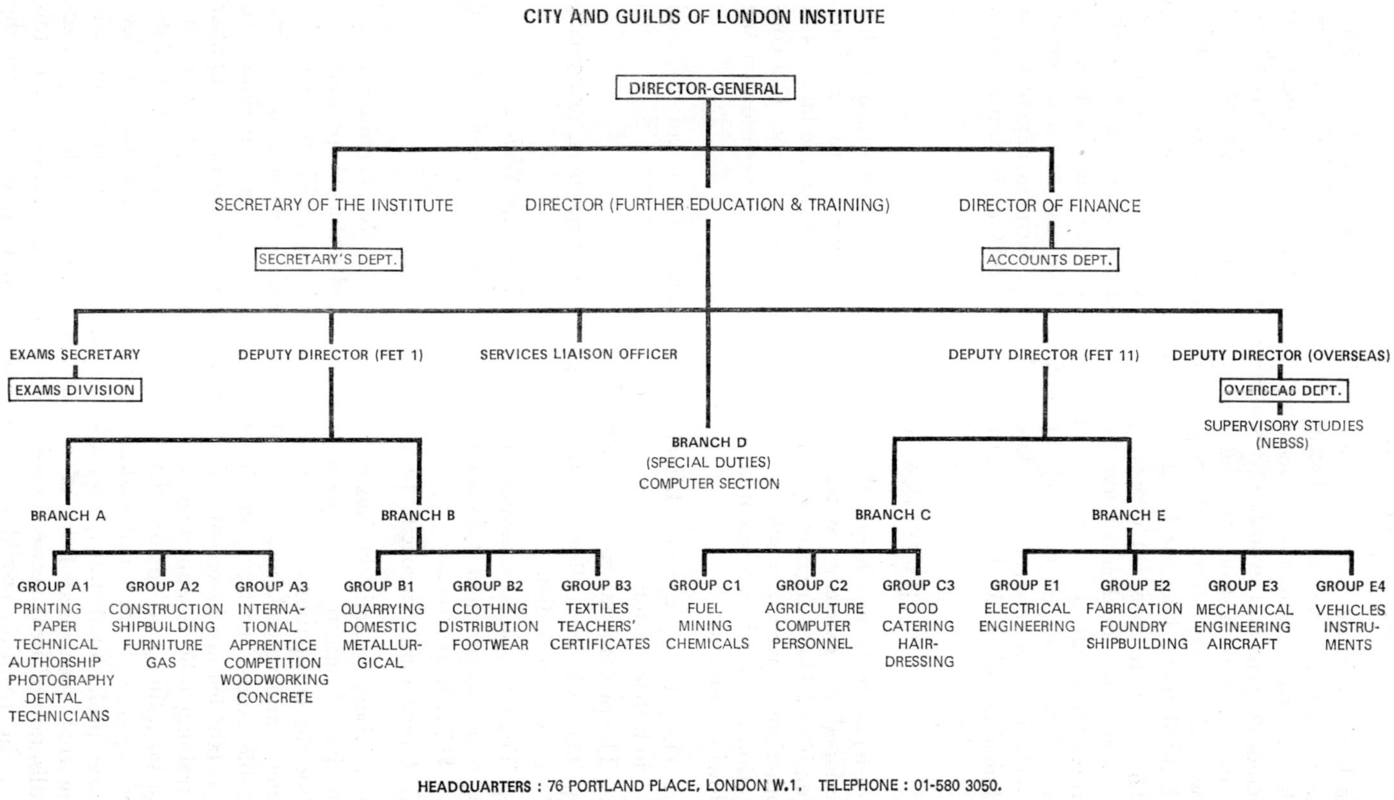

Fig. 34.1 City and Guilds of London Institute Organisation Chart

TRAINING SCHEMES

The Institute is not (nowadays) a technical teaching body. It publishes schemes for part-time courses of technical education to be held in technical colleges, holds examinations based on the regulations and syllabuses the schemes contain and awards certificates to successful candidates. There are over 250 such schemes—the number is constantly growing—and between them they cover all the major productive and extractive industries and distribution. They are listed in the Institute's 'Form 4'.

The Institute does not concern itself with education for commerce, nor with professional education. Its schemes are designed in the main to meet the needs of operatives, craftsmen and technicians, although of course these terms are not easily applied in every case, as for example, in such fields as retail distribution, work study, farm management, or technical teaching, in all of which there are City and Guilds examinations.

ADVISORY COMMITTEES

In order to devise its schemes and keep them up-to-date, the Institute has developed an extensive system of nationally representative advisory committees. There are over 150 of them and each acts as 'the father and mother' of a particular scheme or group of related schemes. The constitution of the advisory committee invariably includes provision for the following elements:

1. The industry concerned, including its technical association(s) and industrial training boards.
2. The professional body(ies) concerned.
3. The technical colleges, national association of teachers, principals, governing bodies.
4. The six regional examining bodies in England and Wales.
5. The Government Education Department(s)—by appointment of H.M. Inspectors.
6. Other nominees specific to the particular field.

The advisory committees thus provide opportunity for accredited representatives of industry (employers, employees and technical specialists) to define the occupational characteristics and training requirements of the students and to collaborate with the educationist members in devising suitable course patterns, curricula, syllabuses and examinations. An important feature in recent years has been the increase in the number of training officers appointed as advisory committee members, by industrial training boards or other industrial bodies. The servicing of the advisory committees and the administration of the examinations employs a large professional staff of educationists who develop expertise in their somewhat specialised field of activity, not least in effecting cross-fertilisation of ideas between committees and between industries. In this way, the Institute provides a considerable and unique forum for the exchange of information and ideas between industry and the education service on a national basis—an essential requirement if courses are to keep pace with technological change.

THE EXAMINATIONS BOARD

All the new and revised schemes prepared by the advisory committees are submitted for approval to the Institute's Committee for Technical Education. The Committee is possibly the most widely representative body concerned with technical education in the U.K. and includes representatives of the education departments for England and Wales, Scotland and Northern Ireland, the national teacher associations, the associations of local education authorities and the regional advisory councils and regional examining bodies in England and Wales. The Institute's relationships with these organisations thus ensures that its schemes are approved for implementation in their colleges. The courses and/or examinations are held at some 1,200 local centres, mostly technical colleges, under the direction of local education authorities, but they also include approved industrial establishments, centres for the Services throughout the world, ships at sea and centres in Commonwealth countries.

EXPLORATORY COMMITTEES

The Institute constantly receives from industry requests that new schemes for courses and examinations should be introduced in order to promote the satisfactory establishment of courses of study for particular categories of industrial personnel. If such a request is approved as conforming generally to Institute policy, an 'exploratory' committee is set up with a constitution similar to that indicated above for an advisory committee.

The exploratory committee draws up the scheme with particular reference to:

1. The precise identification of the type of industrial worker for whom provision is to be made.
2. The characteristics of the work and the technical background it requires.
3. The numbers of potential students.
4. Their general educational standard.
5. Their probable future employment and the possibility of promotion.
6. The relevant industrial apprenticeship and training schemes.
7. The resources of the technical colleges and the possible effects on them of the proposals of the committee.
8. The most appropriate method of examination (e.g., whether practical examinations are desirable or feasible).
9. The extent to which teachers may appreciate guidance on the treatment of the subject, particularly if the proposed course has novel features.

The report of the exploratory committee is submitted to the Committee for Technical Education and if the Committee is satisfied that the proposed scheme is educationally and industrially sound, it is formally adopted by the Institute and details are published.

The exploratory committee is then reconstituted as the standing advisory committee for the subject. In particular, the advisory committee is responsible for keeping the structure of the scheme and the detailed syllabuses under constant review, ensuring that the examinations remain relevant to current industrial and educational practice, maintaining the closest possible contacts with colleges providing courses and the industry supplying students, recommending examiners for appointment by the Institute, moderating (through a confidential sub-committee) the question papers drafted year by year, and generally advising the Institute on all matters connected with the scheme.

The Institute regards it as a fundamental requirement to maintain the closest possible contacts with the colleges and industry and to foster the reciprocal exchange of information and ideas. It is equally important that the extent of the representation on the Institute's committees should be realised and that the representatives should be regarded as a 'channel of communication'. Training officers have of course a particularly valuable contribution to make in this connection and the Institute is always pleased to have their views, either direct or via an industrial representative on a committee.

'TAILOR-MADE COURSES'

The Institute's method of working, as described above, ensures that its schemes for courses and examinations are 'tailor-made' to meet the needs of particular categories of industrial employee. Consequently, the schemes do not readily fall into a small number of tidy categories. Their differences may apply not only to the industrial status of the student (e.g., operative, craftsman, technician), but also features such as:

1. The length of the course.
2. The number of grades of examination.
3. The standard expected on entry to the course.
4. The content of related science and mathematics.
5. The inclusion of practical and project work.
6. The completion of a course as a condition of entry to the examination.
7. Whether courses need to have prior approval by the Institute.
8. The pattern of the examination.
9. Whether course work in laboratory or workshop is to be included as part of the examination.

(The list is not exhaustive.)

It is almost impossible to classify the Institute's schemes. They form a continuous range, meeting the needs of operative, craft, technician, higher

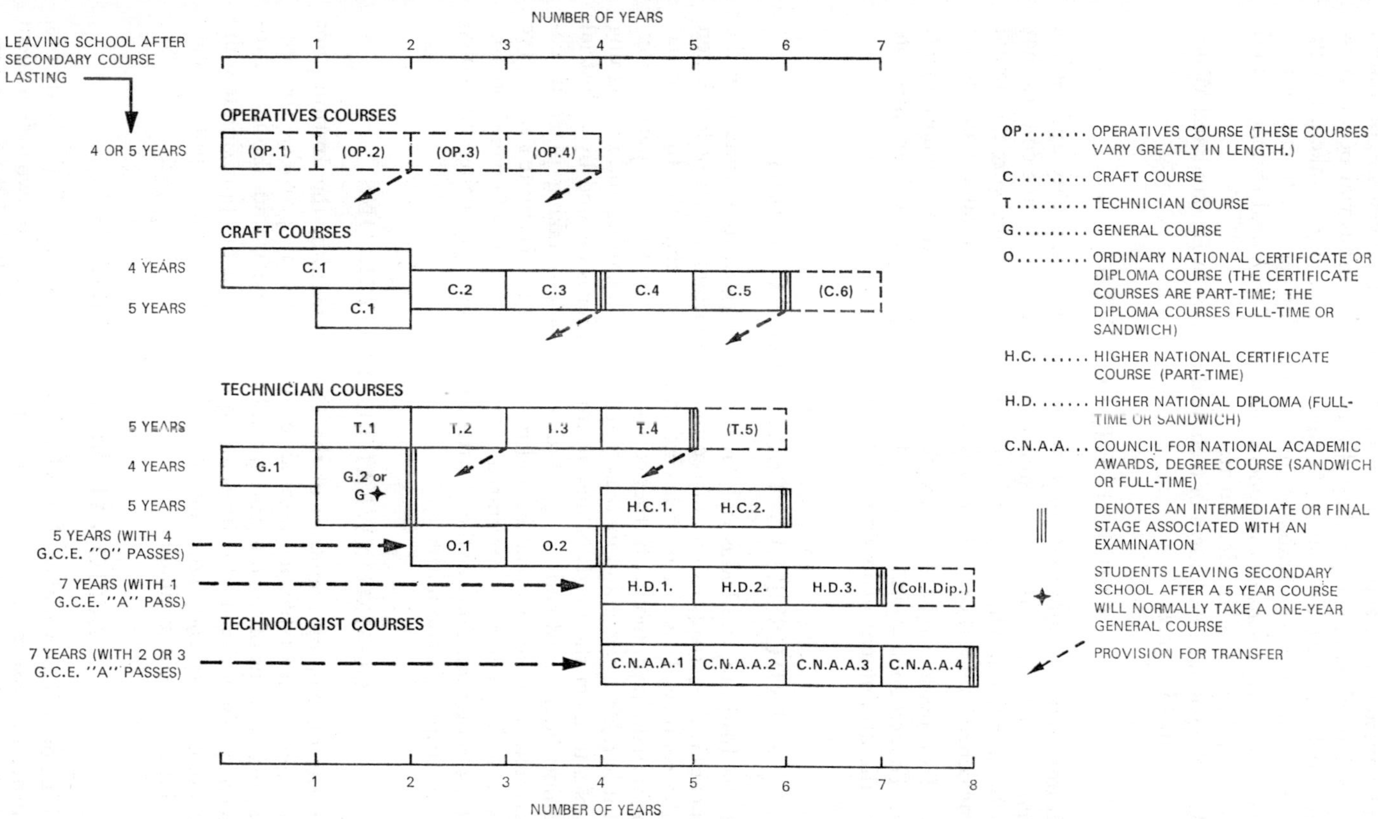

Fig. 34.2 Outline of the new pattern of technical courses (England and Wales) (after D. E. Wheatley, (ed.), 'Industry and Careers', *Iliffe Books Ltd., 776 (1961)) (modified)*

technician, and some supervisory occupations. Their diversity is, of course, a reflection of the corresponding diversity of the industrial occupations with which they are associated. At the same time, care is taken, having regard to the resources available in the education service, to ensure that every scheme conforms to the general pattern of technical education and puts no unjustifiable burden on it.

A useful general breakdown of technical courses in England and Wales is shown in Fig. 34.2.

COURSE OBJECTIVES

It is important to note that the *primary* purpose of any City and Guilds scheme is to promote the establishment of the most effective courses that can be envisaged. The design of the examination follows *after* the design of the course and it is carefully contrived to satisfy two main requirements:

1. To provide a reliable and valid appraisal of students' achievement.
2. To have such an influence on the course of study as will best promote the attainment of its educational objectives.

IMPACT OF THE INDUSTRIAL TRAINING ACT, 1964

For the Institute, the Industrial Training Act, 1964, must rank with the Education Act, 1944, as one of the most important pieces of legislation of the century. It is not only a vindication of all that the Institute has sought to do in promoting further education and training as a partnership, but it heralds a new era of increased activity in designing and redesigning courses to meet the needs of new training patterns devised by industrial training boards. Already a great deal of work is taking place in co-operation with various boards, and members of the Institute's staff serve in various capacities on committees or working parties of several boards, just as representation of the boards are increasingly being appointed to the Institute's committees.

COMMONWEALTH PROVISIONS

For the developing countries of the Commonwealth, the Institute not only makes available its normal examinations, it provides specially adapted versions of these to suit special requirements and also has introduced specially designed courses and qualifications at approximately the level of ordinary and higher national diplomas for use in countries where the latter are not available.

NATIONAL SECRETARIATS

Arising out of its normal work, the Institute has a number of other activities. It provides the secretariat and administrative facilities for two independent examination boards—the National Examinations Board for Supervisory

Studies and the Examinations Board for the National Certificate in Agriculture. It also is responsible for the organisation of British participation in the International Apprentice Competition.

THE CITY AND GUILDS BROADSHEETS

The Institute is aware that training officers and others in industry, and equally the members of staff of technical colleges, must find it difficult to keep constantly up-to-date with the rapid changes taking place in the revision and introduction of new City and Guilds schemes. It therefore makes special arrangements for them to be kept advised of such changes by publishing a *Broadsheet* five times a year. The *Broadsheet* contains summary accounts of new developments and information on how more detailed information may be obtained. In addition, the summer issue each year is accompanied by new editions of the publications *Form 4—List of Subjects* and *Form 2—New and Revised Schemes* and *Institute Publications.* The Institute will, on request, put training officers on the distribution list for the *Broadsheet.*

BIBLIOGRAPHY

DEPARTMENT OF EDUCATION AND SCIENCE, *The Educational System of England and Wales*, H.M.S.O., Rev. Ed. (1966)

DEPARTMENT OF EDUCATION AND SCIENCE, *Further Education for School-leavers*, H.M.S.O. (1964)

Arrangements and Conditions for the Award of National Certificates and Diplomas, a series of 20 pamphlets each covering one subject. Some of these are available free from the Department of Education and Science, others are available on sale from H.M.S.O.

Further Education for Craftsmen, City and Guilds of London Institute (1965)

Further Education for Technicians, City and Guilds of London Institute (1965)

The Technical Colleges and their Government, Confederation of British Industry (1960)

Further Education for those in Industry and Commerce, Confederation of British Industry (1961)

CENTRAL OFFICE OF INFORMATION, *Technical Education in Britain*, H.M.S.O. (1962)

WHEATLEY, D. E. (ed.), *Industry and Careers*, Iliffe Books Ltd. (1961)

DEPARTMENT OF EDUCATION AND SCIENCE, 15–18 *Report of the Central Advisory Council for Education (England)*, 'The Crowther Report', H.M.S.O. (1959)

DEPARTMENT OF EDUCATION AND SCIENCE, *Better Opportunities in Technical Education: Cmnd.* 1254, H.M.S.O. (1961)

DEPARTMENT OF EDUCATION AND SCIENCE, *A Plan for Polytechnics and Other Colleges: Cmnd.* 3006, H.M.S.O. (1966)

DEPARTMENT OF EDUCATION AND SCIENCE, *Day Release*, 'The Henniker-Heaton Report', H.M.S.O. (1964)

DEPARTMENT OF EDUCATION AND SCIENCE, *General Studies in Technical Colleges*, H.M.S.O. (1962)

DEPARTMENT OF EDUCATION AND SCIENCE, *Statistics of Education*, H.M.S.O. (Annually)

SCOTTISH EDUCATION DEPARTMENT, *Technical Education in Scotland—The Pattern for the Future: Cmnd.* 1245, H.M.S.O. (1961)

The Training Officer's Guide to the Education System in England and Wales, British Association for Commercial and Industrial Education, July 1966

35

The Industrial Society

JOHN GARNETT, M.A.

Director, The Industrial Society

AIMS AND OBJECTIVES

More than 7,000 industrial and commercial companies, nationalised industries, central and local government departments, employers' organisations and trade unions subscribe to The Industrial Society. The Society is an independent and non-profit-making body founded in 1918. Its objective is the improvement of relations in all places of employment so that people may give of their best to their work.

More than anything else, the future of this country depends upon the more effective use of its manpower. Thus, the Society's objective has never been more important and the contribution the Society can make is large; as through its members the Society is involved in the working life of ten million people, nearly half the country's total labour force. The Society has a full-time staff of 100, and the director and all members of the advisory and training staff have had considerable practical experience in industry and commerce (see Fig. 35.1 for organisational details).

The Society specialises in the action which needs to be taken in order to achieve the following:

(*a*) *Effective leadership*—increasing the efficiency with which managers, including supervisors, manage people—by training, by improving management structure, through management by objective, and management and supervisory development.

(*b*) *Positive management–union relationships*—improving industrial relations by ensuring that the proper roles of management and trade unions are recognised and accepted, particularly those of foremen and shop stewards: building up responsible trade union membership: establishing regular management–union consultation, grievance and negotiating procedures.

(*c*) *Effective terms of employment and conditions of work*—making sure that terms of employment, particularly wage systems and plans for security of employment, encourage the maximum flexibility and co-operation

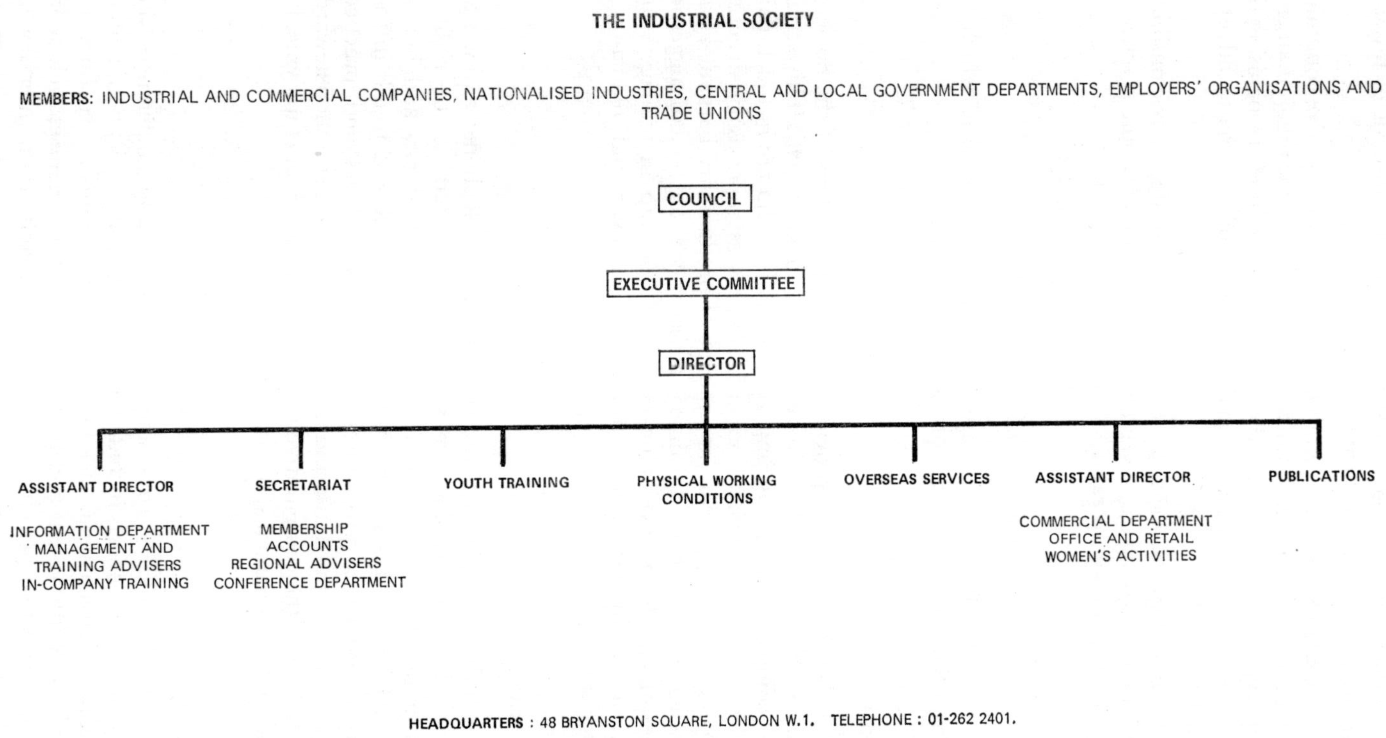

Fig. 35.1 Industrial Society Organisation Chart

among employees: ensuring that there are proper standards of physical conditions of work and effective food services.

(*d*) *Adequate communication*—establishing the need for effective communication: emphasising the importance of communicating through managers and supervisors and setting up the necessary procedures: recognising the role of elected representatives and the value of systematic advisory machinery.

(*e*) *Young employee development*—encouraging attitudes of responsibility and reliability in young employees: encouraging young people at school of high potential to take up careers of management in industry.

SERVICES PROVIDED

Under the following headings are set out the main services which the Industrial Society offers its members:

ADVISORY

A comprehensive service is provided by the 33 members of the Society's advisory staff, many of them known experts in their field. This service enables companies to consult impartial specialists who are in touch daily with developments elsewhere on such subjects as supervisory structure in offices and works; gaining acceptance of productivity bargains; payment schemes; shop steward training; advisory committees and suggestion schemes; methods of developing effective leadership in managers and supervisors, including target setting; staff turnover and reduction of absenteeism.

INFORMATION

Every month, this department answers some 1,000 telephoned and written enquiries. It gives guidance on such diverse matters as the right to search employees, wage structures, pensions, redundancy and holiday practice. It also makes surveys into current practice and trends in such matters as grievance procedures, promotional policies and practices, sickness payments, the duties and training of shop stewards, shop stewards' credentials, and the comparative terms and conditions of employment of manual and non-manual workers.

TRAINING

Each year, the Society runs some 400 courses and conferences. For management, the subjects include target setting, communication, managing people, management meetings and organisational structure. On conditions of employment and personnel procedures, courses include job evaluation, effective personnel record systems, shift work, interviewing, industrial and factory law, industrial health and pensions. There are over 70 courses run for supervisors, mainly on basic supervision. Apprentices and young trainees are catered for with four residential conferences held at Keble College, Oxford, and one held at the University of Sussex. Office workers are covered by courses on communication, office management, personal secretaries, staff turnover, and consultation and organisation amongst supervisory, clerical and technical staff.

Canteen organisation and administration, catering costs and modern catering management courses are run by the Society's physical working conditions department which also administers the courses on employee amenities, i.e., washrooms and lavatories and the methods and costs of various cleaning systems. The publications department runs two basic courses on the production of house journals and one national conference for house journal editors each year.

The needs of shops stewards, both men and women, and full-time trade union officials are also catered for. And there is an annual retail conference and courses on management of sales representatives and the recruitment, selection and training of retail staff.

IN-COMPANY TRAINING

The Society has expert trainers who are able to go into industrial or commercial organisations and design individually tailored training schemes which they then conduct with the assistance of the company's senior staff.

For example, the Society's in-company training advisers performed this service for a large mail order and manufacturing organisation. The company asked the Society to make recommendations on the likely effects of the Industrial Training Act. Two advisers carried out a survey visiting every division of the group and interviewing senior managers. In their report, they recommended that a member of the main board be given responsibility for training; that a central training manager be appointed; that divisional managers meet regularly to co-ordinate training plans; that high potential managers be given experience in more than one division to provide a pool of talent for future board members; that divisional management succession charts be drawn up; that all supervisors should go on certain training courses soon after their appointment; and that a systematic induction procedure for new employees be set up. These recommendations were accepted, and assistance is still being given to the newly appointed group training co-ordinator.

FILMSTRIPS

The Society has produced more than 100 sound filmstrips for use on management and supervisory training courses. They are available to members and the public on sale or hire. They are divided into the following five sections:

(*a*) *Human problems at work.* Case studies taken from incidents which have actually happened and have raised some difficulty, either in dealing with a human relations problem or in some detail of administration such as bad timekeeping, absenteeism, induction methods, transfers between departments and general discipline.

(*b*) *People and production*. Talking points on issues such as executive overwork, the introduction of new equipment and efficiency in administration.

(*c*) *Training*. Filmstrips on the techniques of induction, interviewing, secretarial practice, the use of the telephone, and so on, including several titles illustrating programmes in the Ministry of Labour Training Within Industry for Supervisors scheme.

(*d*) *Sales training*. Filmstrips on maintaining good customer relations; for example, *Over the Counter*, the story of two new sales assistants in a retail store, and why one succeeds and the other fails.

(*e*) *School to work*. Filmstrips on preparing for an interview, fitting into the new working environment, etc., designed for school-leavers and those in their first job. For example, *Joining a Union*, an older employee answers the questions posed by a new boy who has been approached by the shop steward and invited to join the union.

All the Society's filmstrips are designed to stimulate discussion within a training programme. At the end of each filmstrip with the aid of a series of questions which are included with the filmstrip, the group leader encourages each member of the group to participate in the training scheme. The Society does not provide set answers to the problems posed. The groups are expected to discover their own solutions within the policy of their company.

PUBLICATIONS

In addition to the running of courses and the administration of all the Society's publicity, the publications department produces a monthly magazine *Industrial Society* which provides up-to-date information on all matters relating to employment. The Society also produces numerous books and booklets which are written by specialists in factory and office management. These range from *Advice to a Girl on Starting Work*, a simple, easy-to-read pamphlet for the school-leaver and the new junior entrant explaining some of the basic facts about the unfamiliar world of work, to *Lavatories and Washrooms in Factories and Offices* which sets out legislation and good general practice as represented by British Standards and Industrial Society recommendations. Also the Society publishes a series of *Notes for Managers* which covers the manager's responsibility for communication, various aspects of training, target setting, industrial relations, effective joint consultation and job evaluation.

OVERSEAS DEPARTMENT

The Society has a special overseas department. Its main responsibility is to supplement the management training schemes of companies expanding their industrial and commercial interests in developing countries overseas, and indigenous enterprises in those countries.

The services of this department, therefore, consist of carefully planned tutorial courses in the personnel management and industrial relations fields, designed primarily for men of some experience who are considered likely to benefit from broader advancement training.

In addition, the department offers short vacation courses on management and supervision to overseas students and trainees completing their studies in technical subjects at British educational institutions.

CONCLUSION

The procedures which the Society recommends at its courses and conferences, in its publications and in its recommendations to individual organisations are not theoretical. The Society promulgates good practice which has been decided and used to practical advantage by its member organisations.

Further details concerning the activities of the services offered by the Society may be obtained from The Industrial Society, 48 Bryanston Square, London, W.1 (Tel. 01-262 2401).

BIBLIOGRAPHY

Current Industrial Society publications include the following:

Selected Information Surveys and Reports:

No. 146—*Service Recognition and Gifts to Employees* (1968)
No. 145—*Pension Funds* (1967)
No. 144—*Conditions of Employment for Full-time Sale Assistants in the Retail Distributive Trades* (1967)
No. 143—*Staff Overtime* (1967)
No. 141—*Productivity Agreements, the Process of Negotiating and Communicating* (1967)
No. 140—*Shift Work* (1967)
No. 139—*Model Redundancy Agreement* (1966)
No. 136—*Duties and Training of Shop Stewards* (1966)
No. 129—*Some Notes on the Operation of the Industrial Training Act* 1964 (1968)
No. 127—*Promotion Policies and Practices* (1965)
No. 125—*The Computer—A Threat or a Challenge* (1965)
No. 119—*Adjustment to Retirement* (1964)
No. 113—*Foreign Language Training—Companies' Arrangements* (1964)
No. 100—*Better Housekeeping at Work* (1963)
No. 99—*A Case Study of the Promotion and Training Scheme in a Group of Self-Service and Service Stores* (1963)
No. 98—*The Transition from School to Work* (1962)
No. 58—*Catalogue of Industrial Films* (16 *mm*) *and Filmstrips* (35 *mm*) *in Individual Sections under Subject Headings* (1967)

Notes for Managers Series:

GARNETT, J., *Training your Supervisors* (1963)
GARNETT, J., *The Manager's Responsibility for Communication* (1968)
FRANK, E., *Methods of Training your Staff* (1964)
FRANK, E., *Using Staff Training Techniques* (1964)
HENDERSON, J., *Being a Supervisor in a Factory* (1968)

HENDERSON, J., *Being a Supervisor in Retail Distribution* (1964)
HENDERSON, J., *Being a Supervisor in an Office* (1964)
A Guide to Employment Practices (1965)
HENDERSON, J., *Effective Joint Consultation* (1965)
FORREST, A., *The Manager's Guide to Setting Targets* (1968)
CLARKE, M., *Industrial Relations* (1966)
HENDERSON, J., and CORNFORD, B., *The Manager's Responsibility for Safety* (1966)
MERRIE, A. H., *Job Evaluation* (1968)

Periodicals:

Industrial Society—The Society's Journal (Monthly)

Other Books and Pamphlets:

DAVIES, D. LL., *Advice to a Boy on Starting Work* (1966)
FRASER, P., *Advice to a Girl on Starting Work* (1968)
SKELLEY, O. K., *Conference Planning and Administration* (1966)
MCCLURG ANDERSON, T., *Manual Lifting and Handling* (1967)
DOWNIE, J., *Training of Young People in Industry* (1965)
PEPPERELL, E. M., *Using Secretarial Services* (1964)
Using Sound Filmstrips (Designed for training officers) (1962)
Works Committee Member's Handbook (1966)
Legal Problems of Employment (1966)
RUTTER, E. G., and OTTAWAY, K., *Sick Pay* (1967)
Successful Suggestion Schemes (1966)
Status and Benefits in Industry (1966)

Two Useful Handbooks:

Supervisor's Handbook (1968)
Apprentice's Handbook (1968)

36

The Institute of Personnel Management

E. TONKINSON, B.SC. (ECON.), M.COM.

Director, The Institute of Personnel Management

INTRODUCTION

This chapter outlines the work of the Institute of Personnel Management (I.P.M.) as a professional management association and, in particular, its contribution to training and the services it offers training specialists.

PERSONNEL MANAGEMENT AND THE TRAINING FUNCTION

The Institute defines personnel management as 'a continuing responsibility of all those who manage people as well as a description of the work of those who are employed as specialists. It is an integral yet distinctive part of management concerned with people at work and their relationships within the enterprise'. More particularly, it is concerned with the development of policies and their application in such fields as organisational and manpower planning; recruitment, selection and termination of employment; training and career development; terms of employment; wage and salary policy; labour relations; working conditions and employee services.

The key resource of a business is its manpower; in this lies its experience and potential. General management must, therefore, place high on its list of priorities the husbanding and development of a company's manpower resources. The final responsibility for such personnel policy rests with the board but its importance to the well being and growth of a business has led to the development of functional specialists in this field as advisers to general management and as practitioners of specific skills such as selection, training, industrial relations, etc. It has become general practice in recent years, especially among progressive companies, for these functions to be co-ordinated by a member of the company's top management team who often bears the title of Personnel Director or Group Personnel Manager.

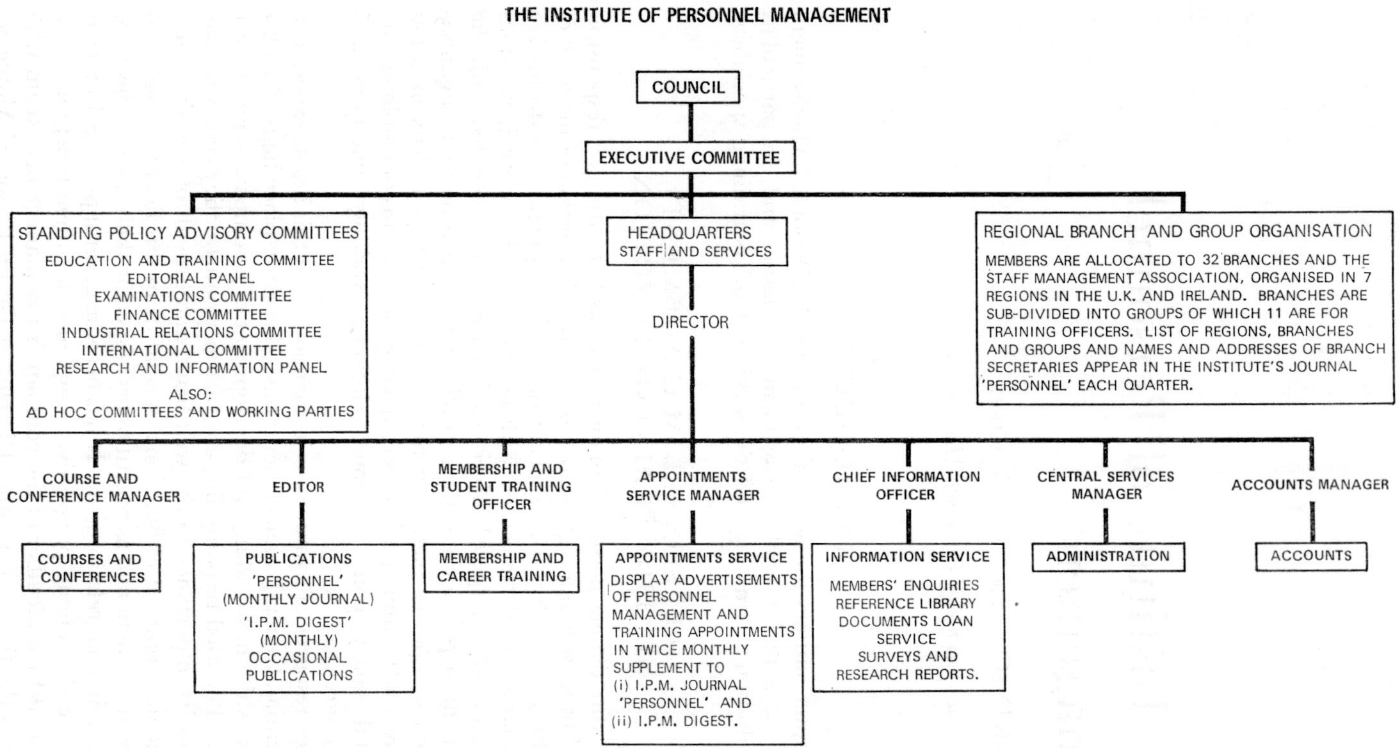

N.B. IN THE DIAGRAM ONLY HEADS OF SECTIONS ARE SHEWN. EACH SECTION HAS ITS COMPLEMENT OF SUPPORTING STAFF - TOTAL STAFF 44

HEADQUARTERS : 5 WINSLEY STREET, OXFORD CIRCUS, LONDON W.1. **TELEPHONE** : 01-580-3271.

Fig. 36.1 I.P.M. Organisation Chart

This has led to a problem of semantics since the word 'personnel' is used to describe not only the overall function or its control at top management level but also a functional department or role which is usually concerned with one or more of the specific aspects already mentioned. The Institute holds the view that all 'human resources specialists' are concerned with personnel management.

The Institute recognises that as general management becomes increasingly sophisticated and professional, it will demand more expertise from its advisory specialists in the personnel field. In fact, over the past 50 years its aim has been to encourage through training and experience high standards of qualification and performance from its members. A recent I.P.M. publication describes the training and experience necessary for the fully trained and competent specialist training officer and states that 'the relationship between training and other personnel activities is so close that it would be unfortunate if the training specialist were to be located away from the department responsible for these activities'. (*Selecting and Training the Training Officer*, by Nancy Taylor, p. 20.)

Obviously, however, great variety will be found within companies in the way specialist departments are organised. Many personnel managers combine responsibility for education and training with other responsibilities. Similarly, many education and training officers find themselves in various relationships with personnel departments. They may combine both training and other personnel functions or occupy a specialist position under a senior personnel executive. In other companies, training may be separated organisationally from other aspects of personnel work even at senior levels, although at the policy making level of the board, there must be co-ordination. Again, there are companies without a personnel department and in the present situation an education and training officer may be the first personnel specialist to be appointed. Whatever the organisational pattern may be, a training officer will soon be concerned with other aspects of personnel management such as selection, induction, manpower planning and development. There will be a common need to use and develop the variety of specialist knowledge within the personnel department and a close relationship between training and other personnel activities should be encouraged to ensure the full effectiveness of both.

In considering the career development of all personnel specialists, including training officers, experience in related personnel fields is clearly essential. Moving promotable people from one personnel specialism to another must continue and, in particular, those destined for senior appointments must have knowledge and experience of other personnel functions which they will be controlling. With increasing specialisation in personnel management, there is a clear danger of specialists working in a vacuum with their efforts unrelated to the work of others. This threat is particularly acute in the training field because of its recent rapid development as a specialist activity. The separation of training from the mainstream of personnel management, in the Institute's view, would be against the long-term interests of both companies and training officers themselves.

THE INSTITUTE'S SERVICES FOR TRAINING

The Institute is a voluntary professional management association with nearly 10,000 members. Since its foundation in 1913, it has played a significant role in the development of training in this country. It welcomes the rapid increase in this development throughout industry and commerce in recent years, and is geared to play its full part in furthering progress both by co-operating with the appropriate Government departments, the Central Training Council and the training boards; and by establishing through training, investigation and research high standards of qualification and performance among its members.

ORGANISATION

The Institute is financed by the subscriptions of its members and governed by a Council and Executive Committee made up of representatives of its regions and branches throughout the country. Its Honorary Officers, elected annually by vote of corporate members, consist of the President, the Immediate Past President, the Treasurer and a number of Vice-Presidents (at present six), who exercise a functional role over various aspects of the Institute's work. A National Committee on Education and Training, chaired by a Vice-President, advises Council on the Institute's policy on industrial training. The Committee represents at national level the views of practising specialists in this field; it assists in the promotion of activities for such specialists and the expansion of the Institute's services aimed at the particular needs of those responsible for training in their organisations. An organisation chart for the Institute is shown on p. 362 in Fig. 36.1.

BRANCHES

The Institute's 34 branches and a special section, the Staff Management Association (S.M.A.), are responsible for organising various activities such as study groups, conferences and meetings, a number of which are devoted to some aspect of training. Within the branches and the S.M.A., there are 46 groups organised on a geographical or special interest basis. Twelve such groups for education and training specialists have been set up and are attracting considerable interest and attention both from members and non-members alike.

MEMBERSHIP

Membership of the Institute is on an individual basis. There are five grades, two of which carry corporate status and allow the member to participate fully in governing the Institute. These Corporate grades of membership, Member and Associate Member, are available to training officers on equal terms with other personnel specialists. Entry to these two grades requires certain qualifications and/or experience specified in the

Institute's Articles of Association. Other membership grades include Graduate membership, entry to which is through the successful completion of a course of training (including the I.P.M. examination) approved by the Institute. After suitable experience, Graduate members are eligible for Associate Membership and later on for Membership. Student membership is appropriate for those taking a course of training recognised by the Institute. The Affiliate grade is appropriate for practising personnel specialists, including training officers, who do not fulfil the Institute's requirements for other grades, or for those in jobs where a knowledge of personnel management is desirable.

TRAINING FOR A CAREER

One of the main aims of the Institute is to encourage the development of formal training for specialists so that, through knowledge and experience, they are fitted to make an increasingly effective contribution to the formulation and execution of policy within their companies. Since 1946, the Institute has co-operated with a number of universities and technical colleges in the provision of training and recognises a number of courses for Graduate membership of the Institute. Some of these are full-time postgraduate courses leading to a diploma, while others are part-time non-diploma courses. Full-time courses normally last one year; part-time courses last two years and most incorporate periods of supervised practical training.

In 1955, to supplement opportunities for career training, the Institute introduced its examination, the standard of which is approximately that of a pass degree and which is held twice a year. The syllabus for the examination contains a paper on education and training. Students who pass the Examination are admitted to Graduate membership and are eligible to apply for Associate membership, provided they hold a personnel or recognised specialist appointment (which includes training) and have done so for at least two years. More and more companies require that candidates for personnel appointments should have completed a course of training successfully or have passed the Institute's examination.

SERVICES FOR MEMBERS

The Institute offers various services covering the whole personnel function, many of which can be of direct assistance to those with education and training responsibilities. During the past few years, it has developed more specific services within the training field; these services are discussed in detail below.

PUBLICATIONS

The Institute's monthly journal, *Personnel*, contains authoritative articles and notes on specific training matters of paramount importance to training officers. The Journal is sent to all members and its circulation exceeds 10,000. A monthly *IPM Digest* keeps members informed of current trends and carries regular features on training. The Institute normally produces at least six new publications each year, which are on sale to the public and may be purchased at reduced prices by members. From the Institute's current list of over 50 occasional publications, a selection of titles of special interest to training officers will be found in the Bibliography at the end of this chapter. A full list can be obtained from the Publication Sales Department.

INFORMATION SERVICE

The Information Section answers over 3,500 enquiries a year from members and undertakes surveys of current practice and developments. The reference library at I.P.M. headquarters includes publications, surveys, reports and other extensive material on all aspects of personnel management which can be consulted by members. The library is expanding rapidly, particularly in the training section, to meet increasing demand from members. A full scale investigation on training within companies, supervised by Miss Joan Woodward of the Imperial College of Science and Technology, is at present being carried out among members.

COURSES AND CONFERENCES

In addition to the opportunities provided for formal careers training, a comprehensive programme of short courses and seminars, mainly non-residential, is arranged in London. In the current programme (1967–1968) of over 50 short courses, the special needs of the training officer are catered for in such courses as Operator Training; Supervisory and Management Training; Developing Management Potential; Training in the Hotel and Catering Industry; Job Evaluation; Job Analysis; Staff Appraisal; Manpower Planning; Selecting the Right Candidate; Selection Testing; Interviewing Techniques Laboratory. Many of these courses are repeated during the year and are likely to feature in future programmes. Full details of these courses, some of which are recognised by the training boards for grant purposes, can be obtained from the Course and Conference Manager. The programme is reviewed and revised annually to meet the needs of members.

The Institute holds an annual National Conference, usually in Harrogate, which is normally attended by 1,000 delegates; it is one of the largest management conferences held in this country, and a substantial part of the programme each year is devoted to some aspects of training. This Conference is organised by the Institute's Headquarters Staff, which also organises occasional conferences, sometimes in association with other bodies such as the Ministry of Labour, British Institute of Management, Industrial Society and the Organisation of Employers' Federations and Employers in Developing Countries. In addition, the six Regional Committees of the Institute also organise annual regional conferences and some of these have training themes.

APPOINTMENTS SERVICE

This service is available to members and employers. Display advertisements of personnel management vacancies, including education and training appointments, appear each month in both *Personnel*, the Institute's journal, and the *IPM Digest*, each of which are sent to all members. The service is widely used by companies who see it as a valuable means of recruitment. Approximately 900 jobs are advertised each year and it is the Institute's experience that employers increasingly require candidates to be members of the Institute as a condition for appointment.

CONCLUSION

The Institute exists as an association for all who are qualified in the specialised areas to do with people at work and aims to cater for the needs of its members employed as specialists in personnel management. Training is an integral part of personnel management and the I.P.M. provides for specialists in training the full services of a professional institute, as well as the opportunity for development side by side with other personnel specialists.

The administration of the Institute and its central services is carried out by its Director and 38 permanent staff at its Headquarters at 5 Winsley Street, Oxford Circus, London, W.1 (Tel.: 01-580 3271) from which address further information can be obtained.

BIBLIOGRAPHY

Current I.P.M. publications include the following:

For Training Officers:

Induction, Introducing Industry to the Individual
HERZBERG, E., *Some Principles of Training* (1964)
DAVIES, A. T., *Industrial Training* (1963)
SEYMOUR, W. D., *Operator Training in Industry* (1964)
ARMSTRONG, J. R., *Supervisory Training* (1967)
STERLAND, E. G., and CRAWFORD, S., *Apprentice Training* (1963)
MACKENZIE DAVEY, D., and MCDOWELL, P., *Programmed Instruction* (1965)
ROBERTS, A., *Retraining Older Men in the Art of Writing* (1966)
GRAY, D. H., *Manpower Planning* (1966)
TAYLOR, N., *Selecting and Training the Training Officer* (1966)
ROBERTS, T. J., *Developing Effective Managers* (1967)
EDINBURGH GROUP REPORT, *Perspectives in Manpower Planning* (1967)

37

The Institute of Supervisory Management

E. L. G. THORNE, B.SC.(ECON.), B.SC., C.ENG., F.I.T.O., A.M.I.PROD.E., A.M.B.I.M., A.M.I.W.M.

Director of Education, The Institute of Supervisory Management

ORIGINS AND OBJECTIVES

The Institute of Supervisory Management (I.S.M.) is the professional institute for Supervisors. It is a national body whose aim is to promote the appreciation of supervisory management as a profession through education, association and research. The Institute is concerned to establish and maintain the highest standards of qualification and performance. Membership services include local activities, publications, advice and information, appointments, courses and conferences. The Institute is a self-supporting, voluntary association governed by a Council elected from its members. It is non-political and does not act as a negotiating body on salaries and wages.

Until March 1966, the I.S.M. was called the Institute of Industrial Supervisors, and known as 'the I.I.S.' It was formed in 1947, but originated in 1944 from two discussion groups which met at Birmingham and Smethwick after T.W.I. Courses. The Institute now has over 9,000 individual members, including 400 members of its new Teachers branch—The Association of Teachers in Supervisory Studies.

Like most professional bodies, the Institute's membership is organised into sections and regions. There are over 70 sections in 11 regions. All of them carry out a full programme of lectures, discussions and social activities of different kinds and often hold joint meetings with the local sections of B.I.M., I.P.M., and the Institution of Works Managers (I.W.M.). An organisation chart for the Institute is shown opposite in Fig. 37.1.

THE CHANGED NAME

Until the middle 1960s, a high proportion of the Institute's membership was derived from the engineering industry. This was inevitable since it

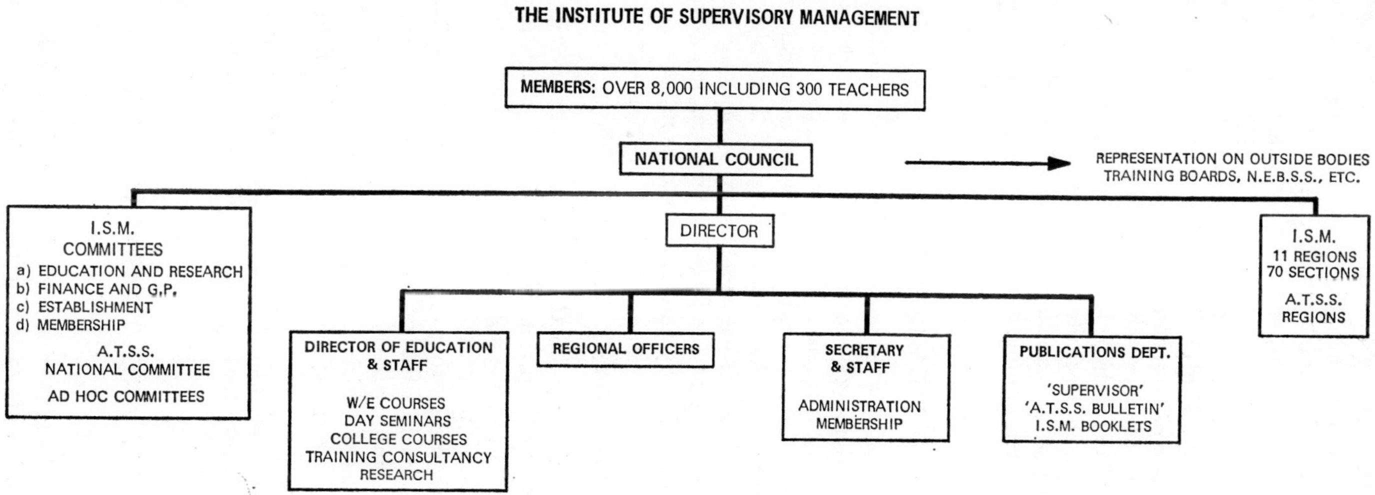

HEADQUARTERS: KING EDWARD HOUSE, NEW STREET, BIRMINGHAM 2. TELEPHONE : 021 643 7845

Fig. 37.1 I.S.M. Organisation Chart

was in the engineering industry that the supervisor had emerged as an entity in his own right. More recently, however, a growing number of applications for membership has been received from people employed in a much wider range of occupations, and from varying levels of supervision. Thus, people from local government, atomic energy and other public authorities, at supervisor, accountant and even director level have entered into membership. This gives the Institute a much broader base, and it was in recognition of this change that the name was changed to the Institute of Supervisory Management. This change, which coincided with a great deal of discussion on the role of the supervisor and the nature of his job, demonstrates the Institute's unequivocal conviction that the Supervisor is both a part of management and a professional in his own right, entitled to a corresponding status and a progressive career in his chosen profession.

MEMBERSHIP

Membership of the Institute is awarded in various grades—Student, Graduate, Associate Member, Member and Fellow, depending upon education, training and experience. The Institute recognises clearly that academic achievement must be closely allied with practical experience. Supervision is essentially a 'do-ing' as well as a 'knowing' job. The I.S.M. will, therefore, always leave membership open to the man who is experienced and competent at his job, but who has not been able to follow sustained courses of study and sit examinations.

THE PAST TRADITION

So far, it can be seen that the I.S.M. closely resembles most professional institutions. However, the Institute has always been much more closely involved in job directed training and education than other similar bodies. Whilst most professional institutions insist on academic achievement for membership, through the passing of examinations, the I.S.M. has only recently become involved in this process. On the other hand, it has for many years been organising a very wide range of week-end and mid-week residential courses, which have been generally recognised in all parts of the country as the most relevant and realistic in a field peculiarly open to exploitation.

Today, the I.S.M. remains the only body specialising in the education and training of supervisory management at all levels, backed by a large and active membership.

EDUCATION AND TRAINING FUNCTION

RESIDENTIAL COURSES

In recent years, particularly since the introduction of the Industrial Training Act, the I.S.M. has experienced an increase in the demand for training and has expanded its activities accordingly.

Thus its well established range of residential courses has been greatly extended in terms of their number and of subjects covered. Details of these courses are given at the end of this chapter. Recent additions include a new range of courses dealing with specialist functions, e.g., 'Maintenance', 'Network Analysis', 'Value Analysis' etc. It can be seen from this that the I.S.M. is also providing a training service for the functional specialist in the technological field, such as the development engineer, project leader, etc., who also has a supervisory function to perform. The Institute has detected a need for courses designed to meet the changes taking place in the job of the ordinary supervisor. Typical of these are the courses on 'Accident Prevention', 'Power Press Regulations' and particularly the 'Avoidance of Disputes' course, which is closely linked to the York Memoranda of the engineering industry.

The I.S.M. has thus avoided the pitfall of providing 'general' courses purporting to cover the jobs of all supervisors in all industries. It is accepted that the job of the supervisor can be analysed and that his job will vary from time to time. Training officers can, therefore, pick from the range of courses a programme designed to provide a specially devised continuum of training for specific supervisors in specific jobs at specific times.

SUPERVISORY TRAINING CONSULTANCY SERVICE

For some years now, the I.S.M. has been giving professional advice to individual firms on the organisation, content and style of supervisory training programmes. It has thus built up an unrivalled store of knowledge and experience in this highly specialised field. Now that the training boards (particularly the Engineering Industrial Training Board) have laid down that Grant schemes will be based upon the analytical approach to the job, from which training arrangements should be derived, the Institute has developed a consultancy service to undertake both job analysis and the training function. From its experience, it has developed a technique designed to make the very complicated procedures advocated for larger organisations meaningful to the medium-sized and smaller firm. From an analysis of the job of the individual supervisor, and after discussion with higher management, the Institute's staff prepare training programmes utilising all known sources, including colleges, consultants, etc., to give a realistic training programme individually tailored to the needs of the supervisor in the firm concerned.

IN-PLANT COURSES

Inevitably, as a result of this consultancy service, the Institute has become involved in a great deal of 'In-Plant' supervisory training, and is rapidly establishing itself as the leading organisation in this field.

Additionally, it runs a series of seminars designed to give training officers and others instruction on analytical methods. These seminars are run with the co-operation of the training boards, and have proved to be of real value to those firms which do not have the sophisticated facilities of the larger organisation.

SUPERVISORY TRAINING AIDS

In common with other aspects of training, the use of new training techniques in the supervisory training world is developing rapidly. The Institute has developed a range of case studies, exercises and visual aids which help to enhance the realism and effectiveness of training programmes, and take into account the fact that the students are mature and responsible people.

NEW FIELDS

In addition to the work described above, the I.S.M. Education and Training Department is currently engaged in discussion with industrial training boards, manufacturers' associations and others concerned to develop new ranges of courses to suit the following:

1. Special industries (e.g., Iron and Steel).
2. Supervisors from specific groups of firms (e.g., the small business).
3. Special groups of supervisors, such as Office Supervisors (a long neglected member of the profession).
4. The young potential supervisor, usually an ex-apprentice, who will have high technical skill but little supervisory experience.

ASSOCIATION OF TEACHERS OF SUPERVISORY STUDIES (A.T.S.S.)

Perhaps the most interesting and significant development in the work of the Institute recently has been the creation of the A.T.S.S. This body was formed at the suggestion of a group of teachers in colleges, and now has over 300 members (about equal division between teachers and training officers) with flourishing regional branches in the main conurbations. Members of this Association are also full members of the I.S.M., at the appropriate grade. The Association has the following aims and objectives:

1. To raise the standards of teaching in supervisory studies and to promote research into related matters, but particularly into teaching techniques and their effectiveness.
2. To hold meetings, conferences, schools and visits, locally, nationally and internationally, for the interchange of ideas and experience among teachers and training officers engaged in supervisory studies.
3. To prepare publications containing case study and other material, to carry on an advisory service.
4. To ensure that the problems of supervisory studies are properly presented to bodies conducting examinations in this field.
5. To form branches in any part of the world.

The A.T.S.S. is essentially a practical body. While it is concerned with theory and research, its prime aim is to raise quickly the standard of supervisory teaching throughout the country, and not to be an isolated academic body developing abstract theory. Like its parent body, it aims to be a 'do-ing' organisation.

THE I.S.M. AND THE NATIONAL EXAMINATION BOARD IN SUPERVISORY STUDIES (N.E.B.S.S.) SCHEME

For many years, the bulk of the educational work undertaken by the Institute was in the specialised short course field, and whilst advice was given to technical colleges and other institutions on examination courses, the Institute did not become greatly involved in this aspect of supervisory training (although many local I.S.M. Sections had a very close working relationship with their local colleges). However, with the demise of the B.I.M. Certificate in Foremanship and Supervision, the I.S.M. was invited to serve on the newly established National Examination Board in Supervisory Studies, together with representatives from other interested bodies. While N.E.B.S.S. was getting under way, the I.S.M. instituted a very highly successful Certificate Course which was adopted by a large number of technical colleges in the Midlands.

This scheme has now been suspended, following the launching of the N.E.B.S.S. scheme. The Institute now invests a great deal of effort through its Education Department and Regional Officers in ensuring that the N.E.B.S.S. scheme is understood, adopted and run in a meaningful way by local colleges and firms throughout the country. Additionally, the Institute has now established an expanded I.S.M. post-Certificate and Diploma Course at a number of colleges with specially wide experience in supervisory training. This is an extension of the N.E.B.S.S. Course in which there is increased student participation. This scheme, together with 'pre-N.E.B.S.S.' courses, now provides for a continuum of supervisory training in technical colleges, as shown in the chart below.

PRE-N.E.B.S.S. SHORT COURSES
(e.g., Communication, Principles of Supervision, etc.)

|

N.E.B.S.S. CERTIFICATE COURSES

|

N.E.B.S.S. ENDORSEMENT CERTIFICATES

Production Control.	Costing.	Work Study.	Industrial Relations, etc.

|

I.S.M. DIPLOMA COURSE

Entrance to the Diploma Course is by examination and interview. The award of the Diploma (by examination and interview) gives admission to the appropriate grade of Membership of the Institute.

OTHER SERVICES

In addition to the activities described, the Institute publishes a range of high standard booklets related directly to the job of the supervisor. They are used widely by firms and colleges for teaching purposes.

The I.S.M. monthly journal, *The Supervisor*, provides both a link with membership and a vehicle for the publication of articles allied to the supervisory function, and has established a high reputation amongst senior managers in all fields of industry.

CONCLUSION

The I.S.M., with its history of close involvement with supervisors at the practising level, coupled with its expanding activity in the training and education field, has an unparalleled store of knowledge and experience in this particular area of training. The area demands an approach which is both practical and visionary. This the I.S.M. aims to continue to promote in order to meet the challenging and expanding supervisory training activities which lie ahead.

BIBLIOGRAPHY

Current I.S.M. publications include the following:

WALKER MORRIS, J., *Job Evaluation*, Ed. Rev. (1968)
SMITH, J. E., and SMITH, J. F., *A Matter of Costs* (1966)
SMITH, J. E., *A Look at Budgetary Control* (1967)
HILLS, B. A. C., *Jones Learns about Balance Sheets* (1960)
WHINCUP, M. H., *Safety and the Law* (1964)
WYLIE, T., *A Concise Guide to Industrial Relations*, Ed. Rev. (1966)
WYLIE, T., *Group Exercises in Industrial Relations* (1964)
Report Writing, 2nd Ed. (1967)
COMPTON, H., *Effective Speech* (1967)
Social Attitudes to Work, 2nd Ed. (1968)

The Supervisor, monthly journal of the Institute, is available to non-members.

Weekend Residential Courses:

Series 1

- The Practice of Supervisory Management
- Human Relations
- Cost Control, Cost Reduction
- Industrial Relations, Industrial Law
- Work Study
- Effective Speaking
- Effective Writing
- Principles of Accident Prevention
- Supervision in the Office (Basic)

Series 2 (Advanced Series)

Job Evaluation and Incentives
Organising for Production
Weighing-up People and Jobs
Increasing Cost Consciousness
Men, Money and Trade
The Supervisor and Decision Making
Supervision in the Office (Advanced)

Series 3 (Special Series)

Operator Training and the Supervisor
Critical Path Analysis
The Maintenance Supervisor (Mid-week)
Avoidance of Disputes
Safe Working of Power Presses
Value Analysis
Operations Research

One-Day Courses:

There is also a wide range of one-day courses for training officers including day seminars for analysis of training needs.

38

Industrial Training Service

J. P. de COURCY MEADE, O.B.E., M.A.(OXON.)

Director and Secretary, Industrial Training Service

INTRODUCTION

The Industrial Training Service (I.T.S.) is the specialised consultant and advisory service sponsored by the Central Training Council and the Ministry of Labour. Its main purpose is to provide expert and practical help with the development of modern systematic training schemes to industrial training boards, employers' organisations or to individual companies which will not only fulfil the requirements of the Industrial Training Act, but which will produce the maximum economic benefit to be obtained through effective and appropriate industrial training.

ORIGINS

The I.T.S. was formerly known as the Training Advisory Service of the Industrial Training Council (I.T.C.) and was established by the I.T.C. in 1960. The I.T.S. is a non-profit making company limited by guarantee, financed mainly from fees charged for consultancy work. The Chairman and Directors of the Service are appointed by the Minister of Labour on the advice of the Central Training Council. It receives a grant from the Ministry of Health and Social Services in Northern Ireland.

STAFF

The I.T.S. has a staff of 32 experienced training development officers available to advise and give help on a consultative basis with the practical problems of developing effective training or retraining for operatives, apprentices, technical staff, supervisors or managers. These officers work in six regional teams based in London, Birmingham, Wetherby, Glasgow, Cardiff and Belfast. Addresses are given at the end of this chapter. All the officers of the Service are ready to help each other, whether in the same region or not, with the specialised knowledge or experience he or she may have. Projects can be undertaken for an industry or for a company having several establishments involving simultaneous work in several different areas. In each case, special groups of officers are set up with one officer acting as the leader and contact for the group.

ACTIVITIES

The main activities with which the I.T.S. is specially able to help are as follows:

ASSESSMENT OF TRAINING NEEDS

An assessment of training needs of industries or companies can be made, on the basis of which the broad lines of training *policy* can be formulated and priorities established. This usually includes a preliminary survey and discussions. It is followed by the development of training schemes.

DEVELOPMENT OF TRAINING SCHEMES FOR OPERATIVES

This involves helping managements with:

1. Analysis of repetitive or process jobs in terms of skills and knowledge or procedures required for competent performance leading to a syllabus.
2. Planning of training programmes in detail including preparation of instructional plans, basic exercises, special devices leading to time-tables.
3. Selection and training of instructors, selection and testing of trainees, records.
4. Target setting, making programmes work and ultimately recording all the material in the form of a manual.

DEVELOPMENT OF TRAINING SCHEMES FOR SUPERVISORS

Assistance may be given to companies with the introduction of the analytical approach described in the I.T.S. booklet *Supervisory Training—A New Approach for Management* by J. P. de C. Meade and F. W. Greig, H.M.S.O. (1966). This may involve helping managers with:

1. Analysis of supervisory jobs with the co-operation and participation of managers and supervisors concerned.
2. Appraisal of individuals.
3. Planning of individual training programmes including formal courses, discussions, projects, tutorial guidance.

Alternatively, the I.T.S. may assist senior training officers or managers to carry out these procedures themselves.

DEVELOPMENT OF TRAINING SCHEMES FOR MANAGERS

Assistance is given with the introduction of the processes of analysis of managerial jobs, assessment of company management training needs, appraisal of performance of individuals, preparation of programmes of management training or education, and development (including objective setting where appropriate).

Training may be given in techniques which managers at all levels may need to fulfil their essential jobs within companies, such as discussion work, setting up and controlling projects, interviewing, objective setting, etc.

DEVELOPMENT OF TRAINING SCHEMES FOR APPRENTICES AND TECHNICAL STAFF

The Service can assist with the identification of patterns of skills and knowledge required, preparation of a detailed syllabus of basic training, design of exercises, setting of targets, planning of a training centre or special bay, arrangements for planning and monitoring experience, choice of appropriate courses of further education, selection testing and recruitment procedures, selection and training of instructors, design of records and reporting systems (work log books), etc.

TRAINING OF TRAINING OFFICERS

The Service is able to provide integrated training for training officers including formal course instruction in the techniques and methods of modern systematic industrial training, on a sandwich basis, interspersed with planned project work under tutorial guidance carried out in the trainee training officer's own company.

Training may be arranged for an individual training officer in his own company if the I.T.S. is working there on the development of training schemes for the company.

Alternatively, formal course instruction may be arranged 'off-the-job' for a group of individuals from different companies while project work is carried out in their own companies under tutorial guidance from I.T.S. officers.

GROUP TRAINING SCHEME PROMOTION

The I.T.S. can give practical help to bring together a nucleus of companies in the same or different industries to form a Group Training Scheme, by carrying out an initial survey of needs, helping to recruit and train a training officer, design of initial training schemes, etc.

CONFERENCES

One-day working conferences are arranged for limited numbers of managing directors and senior executives of companies dealing with the practical problems of such subjects as 'planning systematic training for repetitive operative jobs', 'Planning systematic training for process operative jobs', 'Planning systematic training for supervisors'. About 30 conferences a year are held in main industrial centres.

HOW TO USE THE SERVICE

Employers may wish to use the Service for one or more of the types of work described above. Enquiries may be directed to the Head of the Service or to any of the Regional Officers (see addresses at the end of this chapter).

Initial discussions and, where necessary, short preliminary surveys, as well as proposals of work and estimates of cost, are made without charge or obligation.

WORK DONE

Since the Service began, it has made contact, through conferences or advisory visits, with some 3,500 companies in 75 industry groupings.

Consultancy work in depth has been carried out in over 150 companies in the following industries:

Agriculture		I.T.B.
Brewing	*	
Carpet Manufacture	*	I.T.B.
Chemicals	*	
Clothing	*	
Confectionery		
Construction	*	I.T.B.
Engineering	*	I.T.B.
Food		
Footwear	*	
Furniture and Timber	*	I.T.B.
Ceramics, Glass & Mineral Products	*	I.T.B.
Iron & Steel	*	
Jewellery		
Jute	*	
Knitting, Hosiery & Lace	*	I.T.B.
Leather	*	
Metal		
Paper & Board	*	
Printing	*	
Rubber		
Soap		
Water Supply	*	I.T.B.
Wool & Allied Textiles		I.T.B.
Woven Wire		

* = Work for Employers' Organisation or Joint Council.
I.T.B. = Work for or sponsored by an Industrial Training Board.

USEFUL ADDRESSES

Addresses of the Service's Head Office and the six Regional Offices are given below.

Office	*Address*	*Telephone*
Head	53 Victoria Street, London, S.W.1.	01-222 5421
N. Ireland	35 Royal Avenue, Belfast, 1.	0232 28794
Midlands	Third Floor, Scala House, Holloway Circus, Birmingham, 1.	021-643 5839
Wales & S.W.	49 Churchill Way, Cardiff.	0222 43441
	13 St. Thomas Street, Bristol, 1.	0272 20218
Scotland	105 St. Vincent Street, Glasgow, C.2.	041-221 4067
London & S.E.	39 Victoria Street, London, S.W.1.	01-799 3882
North	43 Market Place, Wetherby.	0937 3531
	Central Administration Building, Kingsway, Team Valley, Gateshead-on-Tyne, 11.	0632 878089

39

The Institution of Training Officers

V. T. STEVENSON, F.I.T.O., F.R.S.A., F.R.G.S., F.R.ECON.S., M.I.S.M., A.M.B.I.M., A.R.AE.S.

A Vice-President of the Institution of Training Officers

HISTORY OF THE INSTITUTION

During the past quarter of a century, the initiation, organisation and implementation of education and training in industry and all allied spheres of activity have become so highly developed that training practitioners have found the need for an institution devoted primarily to their interests. This need became increasingly apparent as the Industrial Training Act approached the Statute Book. In the spring of 1964, the year the Act received Royal Assent, a number of training officers and others concerned with training in industry met to consider the situation. The outcome of this meeting was the founding of the (British) Institution of Training Officers in August 1964, whose title was subsequently modified to simply the Institution of Training Officers.

The Council of the Institution organised its first public meeting on training matters on May 19th, 1965. Support for this event was so enthusiastic that it was decided to make a one-day conference on training matters an annual event thereafter. The success of the 1966 Conference, held also in London, on March 2nd, more than justified this decision.

During 1965, applications for membership of the Institution increased to the point where the appointment of a full-time Secretary became necessary and he established the Institution's office at 55 Station Road, Beaconsfield, Bucks (Tel. 0494 6 3994). Also during that year the Institution's official journal, *The Training Officer*, was first published. Appearing quarterly in 1966, it started bi-monthly publication from the beginning of 1967 and monthly publication from January 1968.

During 1966, Standing Committees of the Council were set up which consolidated and extended earlier efforts, including machinery for the establishment of area branches of the Institution. Meanwhile membership continued to increase.

From 1967 onwards a valuable contribution to the advancement of training philosophy and techniques has been made through the presentation, from time to time, of papers on training subjects by experienced persons, and the holding of public conferences including the annual conference. An organisational chart for the Institution is shown below in Fig. 39.1.

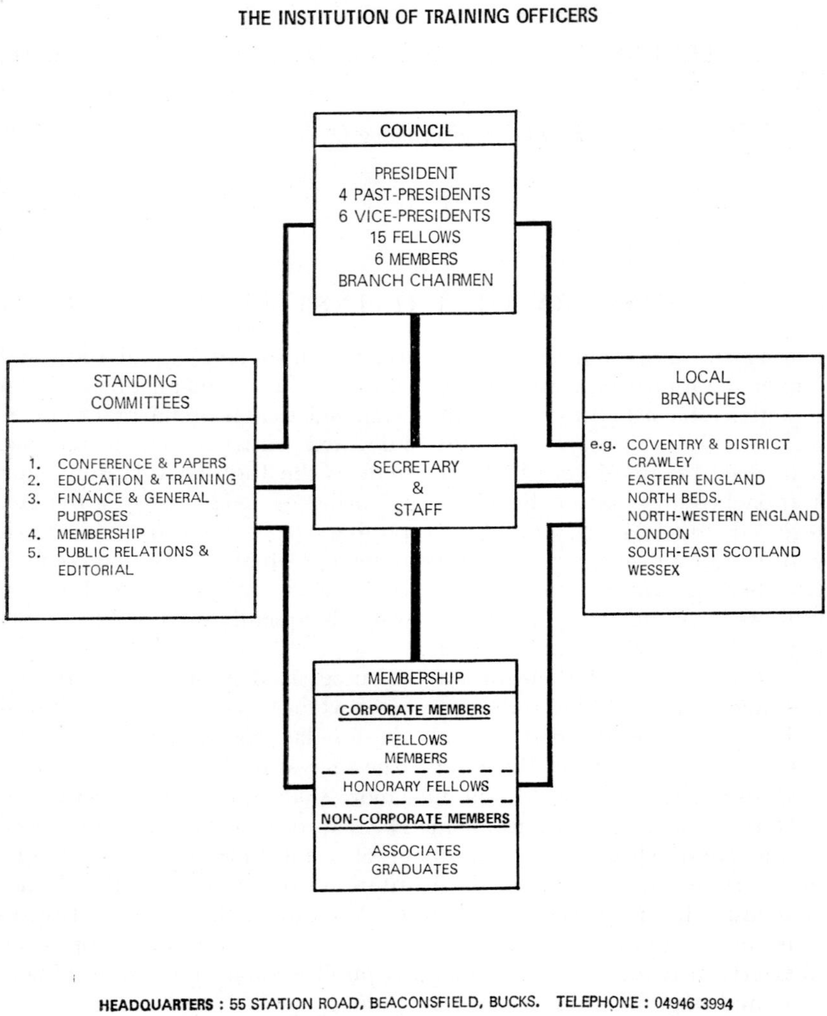

Fig. 39.1 I.T.O. Organisation Chart

OBJECTIVES OF THE INSTITUTION

The Institution is intended to promote and encourage the development and diffusion of knowledge about the practice of training in Industry, Commerce, the Public and Social Services, and the Armed Forces. As stated in its Constitution, the Institution has been founded:

1. To establish and maintain a high standard of qualification and performance among training officers.
2. To promote and encourage the setting and maintaining of adequate standards of training.
3. To provide a means of communication between training officers and other individuals and bodies having education and training responsibilities.
4. To encourage investigation and research into training methods and other matters connected with training and to assist in such research.
5. To observe and compare methods of training in countries throughout the world.
6. To represent the professional training officers' opinions to other bodies concerned with education and training and to co-operate with such bodies for the furtherance of industrial education and training.
7. To hold meetings for the reading and discussion of papers bearing upon training and related subjects and the application thereof and to print, publish and distribute proceedings or reports of the Institution or any papers, communications, works or treatises on training, its application or subjects connected therewith.

MEMBERSHIP AND ITS REQUIREMENTS

The almost limitless variety of interests and occupations encountered in the working environment of this country leads to a correspondingly great variety of interests, occupations, and functions which concern members of the Institution. At one extreme, there is the large corporation with its thousands of employees performing a wide range of tasks, and at the other, there is the small firm employing a mere handful of workers. There is on one hand the organisation doing extremely varied and complicated work and on the other is the firm which manufactures products so simple that a child could make them. There are the research groups, the designers, the makers, the factors, the sellers, the service-givers. All of these and also every organisation, whether large or small, sophisticated or simple, have a need for education and training. This need usually calls for the part-time or full-time services of trained persons. Their requirements are met by the Institution of Training Officers through one of its different membership gradings:

Honorary Fellows—usually Non-Corporate Members

Fellows } Corporate Members
Members }

Associates } Non-Corporate Members
Graduates }

Academic attainment, industrial experience and technical expertise undoubtedly are important, but the most critical factor which distinguishes one training practitioner from another is the degree of responsibility he or she exercises in the performance of their duties. This issue carries much weight when membership gradings are being considered and established. Hence:

Fellows are normally persons with considerable responsibility for education and training, such as training directors, group training officers, training managers, senior training consultants and senior officials of training boards.

Members of the Institution will have been for at least two years *practising* training officers, training consultants, training advisers and the like with more limited training responsibilities and experience than those of Fellows.

Associates are persons of 26 years of age and above whose qualifications satisfy the Institution's requirements for Corporate Membership, except that they have not been training officers for the necessary two years. Alternatively, they are persons engaged in or associated with industrial training who are unlikely to qualify for Corporate Membership. Such persons are drawn from many parts of the industrial training field, from training schools, from universities, from technical colleges, from the Civil or Social Services, or from the Armed Forces. Normally, they implement industrial training schemes or their service equivalent, or help to provide education or training complementary to such schemes.

Graduates of the Institution are normally new entrants to the training field, aged 24 and upwards, who are expected to qualify for Corporate Membership after spending at least two years doing work of the responsibility expected of a Member.

Honorary Fellowship of the Institution is awarded to those who have made an outstanding contribution to the practice, organisation or administration of training or who have given distinguished service to the Institution or to the achievement of its objects. It is the Institution's highest honour and unless a recipient was previously a Corporate Member normally involves Non-Corporate status.

Because of the stress laid on the training function, it is in the interest of a candidate seeking admission to membership of the Institution to state as fully as possible the precise nature of his own function.

Applications for membership of the Institution should be addressed to its Secretary, who will supply an application form and Notes for the Guidance of Candidates. The Membership Committee normally meets monthly. Hence, it is usually possible to advise applicants about their membership within eight weeks of the Secretary receiving an application.

ORGANISATION OF THE INSTITUTION

As the diagram on p. 382 shows, the Institution of Training Officers is governed by a Council. This consists of a President, four Past-Presidents, six Vice-Presidents, 21 Corporate Members and the Chairmen of Local

Branches. Five Committees advise the Council on particular functions of the Institution and are composed of members of Council and members specially co-opted for their knowledge and experience. They are:

1. The Conference and Papers Committee.
2. The Education and Training Committee.
3. The Finance and General Purposes Committee.
4. The Membership Committee.
5. The Public Relations and Editorial Committee.

THE LOCAL BRANCHES

The work of the headquarters of any professional institution is made much more effective when supported actively by members and well-wishers operating in their home areas. The Institution therefore encourages the formation and operation of Local Branches.

The Local Branches are presided over by Corporate Members of the Institution who occupy the offices of Chairman, Vice-Chairman, Secretary and Treasurer. Each Branch seeks to meet local needs by organising regular meetings between training officers and others concerned with industrial education and training.

Typical of the approach used is that of the Wessex Area Branch, which draws its members from West Sussex, Hampshire, the Isle of Wight, Dorset and East Somerset. During the 1965–1966 session, the theme of its meetings was 'The Implementation of the Industrial Training Act'. In order that members might learn how other countries are dealing with industrial training, the Wessex Branch's theme for the 1966–1967 session was 'Industrial Training Abroad', and Branch members were addressed by experts from countries as far apart as Sweden and Japan. During the 1967–1968 session contemporary issues in the U.K. formed the theme.

The reader is recommended to write to the Secretary of the Insitution for information concerning a Branch in his area.

SERVICES OFFERED BY THE INSTITUTION

OFFICIAL JOURNAL

The Institution's monthly journal, *The Training Officer*, deals exclusively with industrial education and training matters, particularly as they apply to the training practitioner.

PUBLIC CONFERENCES

From 1967 onwards, at least one public conference will be held annually.

PRESENTATION OF PAPERS

February 1967 saw the start of the regular presentation of papers on industrial education and training subjects by persons of eminence.

Usually read in the Assembly Hall of the Royal Commonwealth Society, 16–20 Craven Street, London W.1, these papers are expected to provide significant contributions to the literature of training.

INFORMATION SERVICE

Enquiries concerning education and training matters are invited from members of the Institution and from the public and should be addressed to the Secretary of the Institution.

SPECIALIST GROUPS

As the Industrial Training Act is increasingly implemented, it is expected that a need for specialist groups will arise within the Institution to deal with specific problems of sectionalised interests within the training field. The Council of the Institution is very alive to this need and will undoubtedly sponsor such groups when the demand for them is made known.

CONCLUSION

The list above is brief. It is inevitable that, in the relatively short time since its foundation, the Institution should not yet have become all-comprehensive in its field. No organisation could be that in approximately three years. What is significant, however, is that, already, marked and steady growth in membership has justified the aspirations of its founders. The Institution may, therefore, be expected to play an increasing part in the crucial years ahead in helping to fulfil the objectives of the Industrial Training Act in giving industrial training and its practitioners the status they deserve in our modern society.

Industrial Broadcasting and the Training Officer

MARGARET MARSH, B.A.

Further Education Organiser (Radio), British Broadcasting Corporation

INTRODUCTION

Radio and television have for many years been accepted as part of the normal range of educational aids in the schools. Further Education broadcasting on its present scale is a much newer service and many training officers may not know of the possibilities it opens up for their work. The purpose of this chapter is to explain the Further Education service with particular reference to industrial training.

Further Education broadcasting covers an enormous range of subjects, and at first sight an autumn term's programme of 36 series of broadcasts can look confusing and irrelevant to industry, ranging as it does from Fanny Cradock on Cooking to Chinese for Beginners. Here, however, is a selection of programmes relevant to industrial training.

SOME SELECTED SERIES OF BROADCASTS

The number of programmes in each series is shown in parentheses. [R] indicates Radio.

Technology series:

Clean Cut (5) —A detailed examination of new ideas in metal cutting.

Carry on Building (6) —A study of building techniques and planning.

The New Electronics (5) —A study of a range of solid state devices, how they are made, how they function and how they can be used.

Low Cost Automation (5)

Dairy Farming (10) —A systematic refresher course for farmers and stockmen.

Engineering Science —For technical colleges, to support the first year of engineering craft training. (A two-term service.)

Production Engineering and Building

Management series:

This is Marketing* (6) —Theory and practice for students, and for people working in market research, advertising, sales and distribution.

[R] Case Studies in Industry
[R] Human Relations in Industry } Problems of industrial relations in modern industry.

[R] Industrial Training in Action—Some practical aspects of the application of the Act.

The Managers (5)
The Supervisors (5) —Work study, training, safety, communication for foremen, junior managers and supervisors.

Management Research (5) —Including incentives, working groups and industrial organisation.

Personnel Management (6)
Small Business (6) —To clarify management problems of small firms.

Industrial Affairs (8) —Industrial relations and the Industrial Training Act.

*Won the Shell Award 1967 for the best television programme on industry and trade.

Some background information about the philosophy and arrangement of these programmes is essential in order to give perspective to the contribution that Further Education broadcasting can make in this field. Whilst many general programmes are of value to the serious student of the industrial scene, they do not set out to achieve more than a quickening of interest or an enlargement of information. With Further Education broadcasting, there is a formal definition laid down by the Postmaster General for the authorisation of extended hours of broadcasting. The core of this is that the broadcasts 'should be planned in series and be designed to help the viewer and listener to achieve a progressive mastery of some skill or body of knowledge'. The result is that these series are always planned as a result of careful pre-broadcast study of the needs of specialised sections of the community.

An example was the study carried out by the Further Education staff of the B.B.C. into the needs arising out of the development of the Industrial Training Act. These studies have built-in possibilities for further study, which are developed by accompanying booklets, linked correspondence courses or support material for discussion groups.

PROGRAMME TIMES

(*a*) *Evening and weekends.* Most of the programmes are broadcast at times suitable to people at home rather than to an organised student audience. Radio 'Study Session' goes out each evening from 6.30 to 7.30 p.m. on Radio 3 (Third Network) with Saturday morning repeats on Radio 4. On B.B.C.-1 Television, most of Sunday morning is taken up with Further Education broadcasts with repeats on Saturday mornings or late weekday evenings. On B.B.C.-2 Television, however, there are programmes at 7.00 to 7.30 each evening under the general title 'Outlook', many of which are suitable for organised discussion groups.

(*b*) *Special needs.* Special times are provided for special needs. On Radio 4 (Home Service), there is a series for married women thinking of returning to professional life. In 1967 a television series on Pig Farming was broadcast at 12.30 p.m. for the working farmer's mid-day break, and a series on 'The Supervisors' was repeated in 1967 at 4.00 p.m.

(*c*) *Technical colleges.* Finally, a service for technical colleges is broadcast in the daytime at various times fitting in with the pattern of schools television. For example, a series on Engineering Science for first-year apprentices related to the syllabus recommendations of the Engineering Industrial Training Board is being broadcast in 1968/9.

Diverse demands are made on this fairly limited band of time. There are two broad categories of programme: personal needs—learning a language, a practical 'hobby' skill, appreciation of the arts: and national needs—political understanding, technology, professional up-dating of teachers and doctors. It is in this second category that the increased concentration on the needs of industry falls. This is a development arising out of advice from the formal consultation machinery which the B.B.C. has for this service. This is the Further Education Advisory Council, which includes representatives of the whole range of adult, further and higher education. It includes the National Advisory Council for Education in Industry and Commerce, the T.U.C. and the A.T.T.I., and has one of its special committees giving advice on programmes designed to contribute to technical, commercial and vocational education.

OBJECTIVES

'Contribute to' is the keynote of this service. Broadcasting cannot, even if it had unlimited air space, staff and money, do the same job as the lecturer, tutor, instructor or training officer, any more than it can replace the teacher in a school. It does in many situations complement work done on training courses but its particular and unique value is that it can reach people in their homes. This means that, for example, the 1967/8 Management series on Sunday and Saturday mornings could be seen by people who have received no management training, or, who are most unlikely to be free to attend courses because they work for one of many thousands of small firms which make up three-quarters of British industry. An audience of this kind presents a challenge compared with a captive audience in a college or on a training course. The programme has to compel the viewers' interest, and must not be so packed with information that the whole thread is lost if children burst in or if the telephone rings. This influences the general approach, sometimes criticised by professional educators, expecting perhaps the distillation of the best lecturer or tutor in management studies. If the programme is successful, it will be doing something quite different; the foreman, the shop steward, the personnel officer can identify himself with, for example, a situation portrayed in a case study or the environment shown on film. He can be stimulated to relate what he has seen to his own situation, with a heightened and sharper impact than that of a lecturer or a book.

This of course is where the limitation comes in. The radio or television set cannot be asked to repeat something, explain more fully or start a discussion with you. This means that something more needs to be done if the programme is to make the fullest possible impact.

ORGANISED USE OF PROGRAMMES

There are three possibilities. Firstly, people can watch or listen at home and meet later to discuss it; this has been done by a small number of firms, with only limited success. It is far from ideal but does encourage more critical viewing or listening, and incidentally is a useful way of organising feedback for the broadcasters.

Secondly, when a series is shown at a suitable time, groups can either be organised specially or the broadcasts can be used as part of an existing course. An afternoon Industrial Affairs series, for instance, was used by about 500 groups, including some existing N.E.B.S.S. courses, shop stewards or management courses in technical colleges, and other courses specially convened within firms or in co-operation with University Extra-Mural or Workers' Educational Association lecturers. Series of this kind are supported by tutors' notes which give full details of the content of each programme, so that the tutor or leader can prepare his own material and plan the discussion. This group use seems so far to be a useful exercise; it seems that a broadcast followed by such discussion provides a stimulus of a kind that a programme watched in isolation, or a lecturer on his own cannot offer. Perhaps, apart from the obvious advantages of visual presentation, particularly for students not used to formal education, the lecturer is on the 'same side' as his group and temporarily has a different relationship with them. At the very least, the students can come back at the programme with more freedom than they can at a lecturer, and they can all unite in criticising it. This group activity, new as it is, was successful enough for the B.B.C. to repeat 'The Supervisors' in afternoon periods during the autumn of 1967. But, of course, no one time can suit everybody, nor can every programme in a series be suitable for every kind of level or occupation.

The third possibility is the purchase of films or tapes. Neither the B.B.C. nor Independent Television has the resources to set up and administer a loan service, but the B.B.C. has two organisations, Radio Enterprises and Television Enterprises, to whom enquiries about purchase of tapes and film should be directed (the addresses are given at the end of this chapter). The cost of 16 mm film prints varies according to the scale of demand and contractual obligations, but it is hoped it will remain at £45 per half-hour film. It is now permissible to tape further education radio programmes for use in a recognised further education establishment, by payment of a 35s. od. recording fee payable to the Mechanical Copyright Protection Society. For contractual reasons, it is not yet permissible to video-tape television programmes off transmission. The advantages of film and tape for training purposes are obvious, and many potential users express frustration at the complications of taping and buying. The situation is changing as the service of industrial programmes develops and as demands from the world of education and training become more clearly established. There is thus a growing awareness of the possibilities of audio-visual techniques in education and training. It should be remembered, however, that the B.B.C. is a *broadcasting* organisation, and must always consider the extent to which licence revenue can be used for specialist secondary use, apart from the very real anxieties that professional organisations have about the contractual problems.

There is an inherent value in the fact that the programmes are *broadcast*, and for that reason further education programmes directed at industrial audiences, even clearly defined minority target audiences, are not at the moment designed as *training* programmes in the narrow sense of the word. The very fact of a series being broadcast has been found to have a stimulus in itself. When a series on Retirement was broadcast in 1965, some companies organised groups to follow it to help them to formulate company policy on preparation for retirement. Firms who had previously done no shop steward training used the 1966 Industrial Affairs series for 'on-the-job' shop steward training.

Necessarily, a great deal of this is tentative, and may seem nebulous to practical people. It must be emphasised again that further education broadcasting for the carefully defined vocationally motivated target audience is fairly new, and can still be said to be in a formative stage. This means that it depends on the advice and views of the consumers for its effective development. There are two kinds of consumers, the organisers and the receivers of industrial education and training, whose reactions may not be identical.

The providers of training are also of two kinds, those in the world of organised education and those based on industry—their demands and provision are not necessarily based on mutual agreement. But just as effective provision can only develop out of continued discussion between industry and further education, so the role, target and contribution of broadcasting, can only become defined by continual exchange of views.

INFORMATION

To give advice about a service, it is necessary to be informed about it. The B.B.C. provides three stages of information. In March of each year, the Further Education Liaison Office publishes a complete outline of all programmes for the following September to June which is widely circulated, through industrial mailing lists as well as organised educational establishments. Three months before a new series designed for a vocationally motivated audience is due to be broadcast, more detailed information becomes available, with programme headings and synopses. For those who have written asking for them (and for certain series suitable for group or institutional use), there are tutors' notes with much more detail available shortly before the programme is broadcast. Twice a year in October and March there is a pull-out supplement to the *Radio Times* giving again all programmes under general subject groupings.

Again, the communications network, like all informational networks, is not all inclusive, but the indications are that there is a growing awareness that a service to industry exists and is developing. It is in a development stage, and it is not possible to indicate more than possibilities of its development. But looking at the experience of school broadcasting and development in Further Education broadcasting in other areas, it is possible to indicate some idea of the way in which things might develop. The link with correspondence courses (as in the Radio 'O' and 'A' level English programmes linked with the National Extension College), the sale of television film loops in the case of the television Middle School Physics series and Engineering Craft and Science series, the preparation of film strips to accompany a radio series as with a new art series, the preparation of folders of material to accompany a radio series on industrial case studies, indicate some of the possible lines of development.

But perhaps more fundamental is the gradual recognition of radio and television as 'accepted' methods of education, and a realisation that they can provide useful source material in different situations for the training officer. Broadcasting is not a complete service any more than sending a man on a three-week training course as a training officer can make him a training officer. But it does provide a useful stimulus and enrichment to systematic learning, and its potential contribution to direct training is considerable. Its eventual contribution will depend very much on how much people from industry are prepared to use, evaluate and influence it.

USEFUL ADDRESSES

Further Education Liaison Officer, B.B.C., London W.1 (for enquiries about Further Education programmes for industry).

Further Education Officers: Greater London, South East and South West:

B.B.C., London W.1.

Midlands and Wales:

B.B.C. 52 Carpenter Road, Edgbaston, Birmingham 15.

North, Scotland and Northern Ireland:

B.B.C., Woodhouse Lane, Leeds 2.

General Manager, Radio Enterprises, B.B.C., London W.1 (for enquiries about sale of radio programmes).

General Manager, Television Enterprises (Education and Training Films): B.B.C., Villiers House, Haven Green, London W.5 (for enquiries about films of B.B.C. Education programmes).

BIBLIOGRAPHY

ROBINSON, J., and BARNES, N. (eds.), *New Media and Methods in Industrial Training*, B.B.C. Publications (1968)

Case Studies in Industry, a folder of sixty facsimile documents illustrating six industrial case studies, B.B.C. Publications (1968)

Index

The Vocational Aspect of Secondary and Further Education
A journal published on behalf of the Editors and of the Principles of the Colleges of Education at Bolton, London (Garnett College), Huddersfield and Wolverhampton. This journal delves behind the educational scene and explains what educational facilities are available for your trainees; it lays bare the successes and failures of courses run in conjunction with local industry. Price £1 10s. 0d. per annum. 3 issues each year.

Value Engineering
The title is self explanatory and this is the only journal in Britain or America to deal solely with this subject.
Do you employ value analysis techniques?—if you do, make sure you don't miss out on this one. Bi-monthly. Price £3 10s. 0d. p.a.

Programmed texts

We have several linear texts of special interest to the Training Officer—these are:
The Manager and Programmed Learning—price 10s. 0d.
The Industrial Training Act—How it affects you—price 10s. 0d.
Discounted Cash Flow—price 15s. 0d.
Break Even Charts—price 15s. 0d.
Effective Communication—price 11s. 0d.
Critical Path Method—price 15s. 0d.
Programmed Book Keeping and Accounts—30s. 0d.
Workshop Mathematics—23 texts

Case histories in programmed instruction

This is a series of twelve case histories published monthly at an individual price of 10s. or a subscription price of £5 5s. 0d. Some of the 1968 titles are listed below:

1 *Improved Quality Control at J. R. Freeman* by J. H. Newark
2 *Programmed Learning for Clerical Work in the GPO* by A. A. Oates and C. F. Robertson
3 *From Craftsman to Technician—A PERA experiment* by B. W. Blunder
4 *The Use of Programmed Instruction* by I.B.M.

and

Take home book-productivity progress series

Take Home Books—a fortnightly series of booklets on general knowledge subjects, for your employees—such titles as 'Keeping off the Sick List', 'The Arithmetic of Buying a House.' Productivity Progress is strictly for the training officer—such titles as 'Shop Steward Training', 'Measuring Productivity'.
4 subscription options so phone for details.

1966

Journal orders post free: books post free cash with order.
SEND NOW FOR YOUR ORDER TO:
The Training and Technical Publications Division,
Pergamon Press Ltd., Headington Hill Hall, Oxford.
Telephone your enquiries to Oxford 64881 and ask for
Mr. J. D. Warsany or Mr. K. Cordell.